A SOCIAL HISTORY OF AMERICAN TECHNOLOGY

Ruth Schwartz Cowan

New York Oxford
OXFORD UNIVERSITY PRESS
1997

OXFORD UNIVERSITY PRESS

Oxford New York
Athens Auckland Bangkok
Bogota Bombay Buenos Aires Calcutta
Cape Town Dar es Salaam Delhi
Florence Hong Kong Istanbul Karachi
Kuala Lumpur Madras Madrid Melbourne
Mexico City Nairobi Paris Singapore
Taipai Tokyo Toronto

and associated companies in

Berlin Ibadan

Library of Congress Cataloging-in-Publication Data

Cowan, Ruth Schwartz, 1941–
A social history of American technology / Ruth S. Cowan.
p.   cm.
Includes bibliographical references and index.
ISBN 0-19-504606-4. — ISBN 0-19-504605-6 (pbk.)
1. Technology—Social aspects—United States—History.
I. Title.
T14.5.C69 1996
303.48'3'0973—dc20   96-5505 CIP

3 5 7 9 8 6 4 2

Printed in the United States of America

on acid-free paper

For
Neil M. Cowan
LC in LGA

. . .

In the labor of engines and trades and the labor of fields I find the
    developments
And find the eternal meanings.
. . .

Strange and hard that paradox true I give,
Objects gross and the unseen soul are one.
. . .

The hourly routine of your own or any man's life, the shop yard, store
    or factory,
These shows all near you by day and night—workmen! whoever you are,
    your daily life!
. . .

In them the realities for you and me, in them poems for you and me, . . .
In them the development good—in them all themes, hints possibilities.

Walt Whitman, *Leaves of Grass* [Book XV]
"A Song for Occupations"

Contents

Acknowledgments

THIS BOOK IS INTENDED TO INTRODUCE READERS to a fascinating scholarly pursuit—the history of American technology. I could not have conceived it, let alone written it, without the devoted labor of my predecessors and colleagues, several hundred passionate historians: collectors, organizers, arrangers, analysts, and students of American material culture. My thanks to all of them—but most particularly to those who paved the way by creating and sustaining the two wonderful institutions that created and still sustain our field: the Society for the History of Technology and its lively journal, *Technology and Culture*.

Carroll Pursell and Mel Kranzberg encouraged my first efforts to explore American technology; I dearly wish Mel had lived long enough to read what resulted from his inspiration. Conversations, stretching over years, with Judy McGaw and John Staudenmaier have often been crucial and always been enlightening. Because of Dan Kevles and the Sherman Fairchild, Jr. Foundation I was able to spend a year at the California Institute of Technology—in which supportive atmosphere more than half of this book was written. My Stony Brook colleague Ned Landsman and my Caltech colleague Doug Flamming graciously read and commented on two chapters about which I was initially quite worried. Robert Friedel, patient man, read and commented on the whole of the very long first draft; his authorial grace and encyclopedic knowledge saved me from more gaffes than I care to recall. My ideas and my prose have also been tested and honed by the reactions, pro and con, bored and engaged, of the students who sometimes struggled and sometimes soared through my courses in the history of technology.

My last thanks, springing from the fullest heart, are reserved for the members of my family. Sarah Kiva Cowan, May Deborah Cowan, Jennifer Rose Cowan, and Neil Michael Cowan: helpful fact and footnote checkers, patient finders of misplaced (never lost!) note cards, expert surfers of the net, penetrating questioners, sprightly conversationalists, wise in the ways of word processing programs, ever-ready schleppers of manuscript boxes, superskilled (and very funny) organizers of cluttered offices, studies, bookcases, desktops, floor piles, and filing cabinets. You are a perfect scholarly support system, providing calm when I panic and inspiration when I flag—for which I am unendingly grateful.

Glen Cove, New York   R.S.C
June 1996

I

IN THE BEGINNING

IN THE MIDDLE OF THE EIGHTEENTH CENTURY IN EUROPE, some taxonomists—the people who classify and name the different species of animals and plants—had an argument. Having decided that human beings, men and women, ought to be classified with the animals, they could not agree on a name for our species: *Homo politicus*, the primates who create governments? *Homo sexualis*, the primates who are perpetually in heat? *Homo sapiens*, the primates who think? Or *Homo faber*, the primates who make things? Eventually they settled on thinking as the crucial characteristic that sets human beings apart from the orangutans and the chimpanzees, which seem anatomically to be our closest relatives.

This book focuses on one of the human characteristics that the eighteenth-century taxonomists rejected, making things. Our opposable thumbs, our hands, and the things that we make and manipulate with our hands are as much a part of our humanity as our brains and the thoughts we have had, our governments and the constraints they have put on our behavior, even our sex hormones and the ways in which they have influenced our lives.

Technology has been a fact of human life as long as there have been human lives. From the time that human beings emerged as a separate species on this earth, they have been trying to control, to manipulate, to exploit, and sometimes even to subdue the earth with tools. Technological change has occurred so rapidly in the twentieth century that we sometimes think of ourselves as living in a characteristically technological age, surrounded as we are by automobiles and superhighways, skyscrapers and plastics. A

1

moment's thought should convince us, however, that in reality we are no more (or less) technological than any of our ancestors. In all times and all places, human beings have attempted in some fashion to use tools to control the natural environment in which they were living—and this is as true of the human beings who first learned how to rub two stones together to make a spark as it is of those who subsequently created the atomic bomb. We like to think that in times past people lived more "natural" lives than we do today, but in point of fact, log cabins, tepees, and grass huts are as "artificial" as hydraulic cement, atomic bombs, antibiotics, and computers. They are all equally products of human hands, of human artifice, of making things, of *homo faber*.

Which is why we should be suspicious whenever we see the labels "handmade" or "natural" on a product, no matter what that product may happen to be, no matter how "old-fashioned" it may happen to look or feel or taste. *All* products are both handmade and natural because all human beings are naturally equipped to make things with their hands. Is a piece of cloth more handmade because it has been woven on a wooden loom rather than a metal one? Is bread that has been baked in a wood-fired brick oven more natural than bread baked in a metal oven fired by "natural" gas? And when we actually do encounter something that is more handmade than something else (for example, bread dough that has been kneaded by someone's hand rather than a machine), why do we assume instantly that the former is better than the latter? Studying the history of technology may help us to understand why so many people seem to believe that being handmade and natural somehow means simultaneously both "better" and "traditional."

A Social History of American Technology

We use the word technology to denote those things that people have created so that they can exploit or manipulate the natural environment in which they are living. Technology is a more general word than tool. Tools are used to produce things, but both the things that are produced (like bridges, houses, gears, woolen cloth, and clean laundry) and the things that are used to do the job (like wrenches, hammers, drill presses, looms, and washing machines) are included in the term technology. Domesticated animals and plants are technologies that people created in order to secure food supplies; medications are technologies that people created in order to improve their health. Even languages and the things that contain languages (such as books, letters, computer software, and student essays) are technologies: they are things that people have created so as to better control and manipulate the social environment.

Technological systems are arrays of technologies. Once primitive human beings had passed beyond the use of digging sticks they had passed out of the realm of technology into the realm of technological systems. A single tool, even the most primitive of them, is usually not sufficient to get the job done. A hammer, for example, must be applied to something (a nail, a board), which is itself also a tool, in order to function as a hammer; a can needs a can opener; even the original digging stick may have needed to be sharpened with a stone. All technological systems, of necessity, have people embedded in them. A hammer is not really a hammer until someone picks it up and uses it. Technological systems can, and sometimes do, become quite large and quite complex: the paradigm would be a personal computer which requires an entire electrical network plus software, diskettes, printers, and modems—as well as programmers, inputters, trainers, and manufacturers—in order to function properly.

If the things we make and manipulate with our hands, technologies and technological systems, are as much a part of our humanity as the ideas we think and the governments we create, then they also ought to be part of our history. The history of technology is an effort to recount the history of all those things, those artifacts, that we have produced over the years. The *social* history of technology goes one step further, integrating the history of technology with the rest of human history. It assumes that objects have affected the ways in which people work, govern, cook, transport, communicate: the ways in which they live. It also assumes that the ways in which people live have affected the objects that they invent, manufacture, and use. A social history of technology, in short, assumes a mutual relationship between society and technology; it also assumes that changes in one can, and have, induced changes in the other.

American history has been recounted from many different perspectives over the years—biographical, economic, intellectual, to name just a few—but rarely from the perspective of technology. This is odd because for over 250 years American technology has been regarded—by Americans and by foreigners—as the hallmark of our culture, one of the things setting us apart from our European brethren and one of the important factors contributing to our extraordinary prosperity. We who have been born and raised in the twentieth century should be particularly interested in understanding our history from this perspective because we are in a unique position to comprehend how profoundly technological change—think of birth control pills and computers—has affected our way of life. We are also in a unique position—think of our country's size and our superhighways, or our role in the cold war and our space program—to understand how profoundly our way of life has affected our technology.

The first section of this book examines the technological character of part of North America during the very long historical period before the industrial revolution began and the United States was born.

In the first chapter, I try to describe, albeit very briefly, the material culture of the many peoples who were indigenous to the continent before the Europeans arrived—and the profound differences between the technologies of those two groups of people, natives and settlers. By way of introduction, I have also used that chapter to say something about the differences between the North American and European environments, differences that ultimately put a uniquely American stamp on some of the technologies that subsequently developed here.

The second chapter is devoted to colonial agriculture since that was the enterprise in which the vast majority of European Americans were engaged in the years between settlement and independent nationhood. In part because the United States was born as an agricultural, rural culture—and its unique governmental system derived from that culture—colonial agriculture has been enveloped in the mists of patriotic nostalgia. I contend in this second chapter that by examining the way in which the colonists actually wrested their livings from the ground, we can develop a much more realistic understanding of our national past.

The third chapter, finally, focuses on artisans: the colonists who made the various tools with which the others tried to control their environments. Although they were only a tiny fragment of the population, colonial artisans are a historically crucial fragment, for they provided—to use an odd but suitable metaphor—the technological womb from which industrialization was born. In the second section of this book, readers will discover that the American industrial revolution was very different from the British, French, and German industrial revolutions. We cannot begin to understand those differences—so crucial to the subsequent history of the West—unless we understand the unique circumstances in which colonial artisans worked.

1

The Land, the Natives, and the Settlers

THE UNIQUE CHARACTERISTICS OF AMERICAN TECHNOLOGY, those things that set it apart from English, Japanese, Canadian, Mexican, or any other technology, derive, at least in part, from the unique characteristics of North American geography. That particular part of the North America now called the mainland United States is a natural environment that people have been manipulating and to which they have been adapting for thousands of years. It is also a vast stretch of land, reaching from the Pacific Ocean to the Atlantic, across three climate zones (semitropical, temperate, and semiarctic), encompassing at least a dozen very different ecological regions.

Virtually every observer who has ever recorded an impression of this geographic area regarded it as very specially blessed: blessed because the climate, in most places and at most times, is reasonably temperate; blessed because much of the land is fertile and many of the woods and streams, bountiful; blessed because there are rich lodes of minerals and fuels on the surface of the land and far below it; blessed, finally, because in its varied terrain, the land is awesomely beautiful. From time immemorial this land has sustained human existence; most of the continent was inhabited by 10,000 B.C.—about the same time that cultures were beginning to form in Egypt and Mesopotamia.

The earliest settlers made their homes in the area that is now Alaska but as the years wore on, as the centuries and the millennia passed, the descendants of these early settlers migrated southward and eastward, eventually filling the whole of North America. People often say that necessity is the mother of invention, but in fact, opportunity can be just as powerful a

5

spur. As the native Americans migrated to different geographical regions they met with opportunities to feed and clothe themselves in new ways, and they responded to these opportunities by modifying the technologies they had brought with them, creating both new technological systems and new cultures. Some remained nomads, wedded to the traditions of their ancestors, but others learned to domesticate plants, to farm, to form settled communities—even small cities—and to give up their nomadic heritage. In this they were very similar to their long-separated cousins across the seas, who were, during those very same millennia, passing from what is called the Paleolithic—nomadic—stage of culture to the Neolithic—agricultural—stage.

Yet across the seas other changes were occurring as well; the Neolithic peoples of Europe and Asia were creating various technological systems (specialists in prehistory like to say that they were passing from the Bronze Age to the Iron Age) and these changes were not echoed in North America, creating profound differences not only between the peoples of the two continents, but also between the environments in which they were living. Thus, when the descendants of the Neolithic agricultural peoples of Europe arrived in North America in the seventeenth century they carried a set of tools and a set of cultural assumptions that were utterly foreign to the descendants of the Neolithic agricultural peoples who were native inhabitants of the continent.

The Land and the Native Inhabitants

The various geographic regions of the United States are today very different from each other, and they were probably even more different from each other in the past. The Pacific coast (the one first inhabited by the earliest settlers) is very rugged terrain; there are few beaches and many mountains dropping off suddenly into the sea. The southern section of this coast has a dry climate, but in the north it is quite wet; from north to south there are many fertile mountain valleys (some of them immense), and from east to west several large rivers flow down from the mountains into the sea.

Farther east, separated from the coastal ranges in some cases by deserts and in others by arid lowlands, there is another range of mountains, considerably more menacing, now called the Rockies. East of the Rockies there are plains and prairies, the former virtually treeless, the latter somewhat more wooded, but both dominated by grasses and interrupted by some very large rivers, the Mississippi and its tributaries. The prairies end in the heavily wooded Appalachian highlands, much less forbidding mountains than the Rockies, and these highlands are themselves bounded, yet farther east, by the Atlantic coastal plain. This plain is very narrow in the north, but in the south, in the modern day Carolinas, it is almost 500 miles wide. Unlike the Pacific, the Atlantic coast is full of beaches, protected coves, bays, and inlets.

As human beings migrated into these very different geographic regions, the cultures and the technologies that they developed differed from each other markedly. New environments create opportunities for people to modify their old tools so as to sustain and maintain themselves in new ways. Although we now make the mistake of lumping them together in one generic term—"Indian"—each native American tribe was in fact unique. They differed in their languages, their religions, their social systems, and also their technologies. Indeed had these technological changes not occurred, anthropologists and archeologists would today find it difficult to distinguish some native American cultures from others since it is from their tools that we have come to know them.

When the first human settlers arrived in North America, they carried tools and technological skills with them. By our standards those tools were primitive, but they were nonetheless tools—and very large quantities of them (principally chipped stone spearheads) still survive. From the form and the distribution of those surviving tools, archeologists guess that those early settlers were nomadic (no house foundations have been found), land based, big game hunters who knew how to make fires. They probably also knew how to fashion wood (into digging roots or the shafts on which spearheads were mounted), but these wooden instruments no longer survive.

In the frozen wastes of the Arctic regions, the first native Americans developed technological systems focused not on land but on the sea and on sea mammals. They used harpoons (a spear with a detachable head attached to a flexible line) to catch whales and sea lions; the boats with which they prowled the icy waters searching for prey were made of the hides of animals previously caught. They learned how to fashion shelters out of blocks of ice, how to build sleds, and how to liquefy the body fat of their prey in order to provide fuel with which to cook.

Centuries, perhaps millennia later, the native Americans who had migrated to the arid southwest possessed technological skills that were largely unknown to their brethren farther north. Some of these people built single dwellings—and others, whole villages—out of clay, called adobe. Instead of animal skins, they wore woven cotton for clothing. They had learned not only how to domesticate the cotton plant but also how to cultivate the crop with hoes and how to straighten the cotton fibers with combs. They had invented spindle whorls (wooden shafts over which perforated disks of wood, stone, or clay have been slipped) for twisting and stretching the fibers (to make thread) and looms for weaving it. These southwestern tribes were also skilled potters; they knew how to fashion complex vessels out of clay and then to decorate and fire the vessels so that they would be both beautiful and watertight.

Along the middle of the Pacific coastline (in the areas which are today the states of Oregon and Washington and the province of British Columbia), very different cultures and technologies emerged. These people subsisted on fish, but unlike the natives of the north, they were not nomadic. They built houses out of wood planks, which means that they had learned not only to fell trees but also to cut planks, using various stone-based woodworking tools such as axes and cudgels. The tribes of the Pacific Northwest used nets (which they fashioned out of vines) and weirs (underwater traps made of stone or wood) to catch fish. They did not practice agriculture, but they ate various wild roots and berries, and the techniques that they used to preserve these plant products by drying served also to preserve the vitamin and mineral content of their food. As a result, the members of these tribes were exceptionally well fed; there are no records of famine ever having beset them.

On the plains and the prairies, there were yet other tribes who lived a nomadic life, hunting the buffalo. These people were skilled leather workers. Their dwellings, their clothing, and their footwear all derived from the skins of the animals they hunted. On the plains and the prairies, native Americans had learned how to stretch the hide of an animal by pulling it between stakes driven into the ground; they removed the hair and flesh by passing a sharp tool over the hide, then dressed it to make it soft and easy to manipulate by applying suitable "chemicals"—sometimes the brains of the animal that had provided the skin, sometimes human urine.

Some of the prairie tribes were only partially nomadic and farmed for part of the year. They had developed the skills needed to clear land, to plant crops, and to harvest. They also possessed the tools for grinding grain (principally corn) so that it could be eaten, as well as the utensils needed for cooking it.

The tribes of the eastern coastal plain and of the Appalachian highlands had adapted themselves to woodland living. They practiced settled forms of agriculture, but they did so without destroying the forests in which they dwelled. When a section of forest was chosen for farming, its underbrush was cleared by burning and its trees were killed (not uprooted) by girdling (cutting a strip of bark and surface tissue out of the circumference of the trunk). The soil thus created was exceptionally rich (composed of generations of leaf loam combined with vegetable ash); the dead trees remained standing but did not come into leaf. Under these circumstances no preparation of the soil was really needed; holes could be dug with a simple stick and seeds planted. These Eastern tribes also knew how to interplant crops (corn and beans together, for example) so that one species could provide support for the other and weeds could be kept to a minimum.

The native Americans of the eastern coastal plain were exceptionally adept at utilizing wood and other tree products. They built their houses from wooden poles, sheathed in overlapping layers of bark, stitched together as shingles. They sweetened their food with the sap of some trees and made their houses and their boats watertight with the sap of others. Their boats, their utensils, and their weapons were all fashioned either from wood or from bark, while their baskets were made from the stems of plants that grew between the trees. Like the tribes of other regions, they had adapted their technological systems to the unique ecological characteristics of the environment that they inhabited.

From Thomas Hariot, *A briefe and true report of the new found land of Virginia* (1590). "The manner of makinge their boates in Virginia is verye wonderfull," Hariot wrote. Fire was used to fell a tree (*rear right*), remove its branches (*rear left*) and burn a trench down its center (*front*). Then shells were used to scrape and smooth the trench. "For whereas they want Instruments of yron," Hariot continued, "yet they knowe howe to make them [boats] as handsomelye . . . as ours." (Courtesy Ward Melville Memorial Library, State University of New York at Stony Brook.)

Thus over the course of many centuries, as the native Americans migrated into the various regions of North America, technological as well as cultural change had occurred. Some of the native Americans were nomads, while others lived in permanent villages. Some ate small fish and game, while others hunted large mammals both on land and at sea. Some knew how to clothe themselves as protection from the frigid winds of the Arctic, and others lived, barely needing clothes, in tropical zones. Some dug up wild roots and collected wild berries, while others planted the seeds of domesticated plants such as corn, beans, and pumpkins. The primitive spears of the first settlers had been made more versatile as harpoons and then as arrows fitted to bows. People who had once lived in tribal nomadic bands had settled down in some places and become farmers; in those places they developed the skills of spinning, weaving, potting, and woodworking—skills their ancestors had not had.

Yet the pace of change had always been slow, so slow that the natural environment had had time to recover from whatever ecological damage each technological change had produced. One of the technological systems that had developed in Europe and Asia (but not in North America) in those prehistoric centuries was the use of mineral ores—copper, tin, lead, iron—to create metal tools—knives, axes, plowshares, swords, guns. Lacking bronze and brass and iron, the native Americans had been unable to devastate forests or uproot sod or alter the contours of the land or decimate animal populations. Lacking large numbers—the population density of native Americans was always low—native Americans had been able to sustain themselves without damaging, or even profoundly altering, their natural environment.

Thus, over the course of several millennia, because their populations were small and their rate of change profoundly slow, the native Americans had managed to live, as nearly as it was possible for humans to do so, in harmony with nature. The technological systems that they had perfected were not capable of profoundly or permanently disturbing the environments in which they were living. Those sixteenth- and seventeenth-century European explorers who believed that the wealth of the North American continent was largely untapped were absolutely right—despite the fact that people had been living on that continent for a very long time. The native tribes had manipulated the environment in order to feed, clothe, and house themselves, but they had not extensively exploited the land, had not taken out of it all, or even part, of what it was capable of yielding.

The European Settlers

From a technological point of view, the sixteenth- and seventeenth-century European explorers and settlers must have looked like visitors from another planet to the natives of North America who first greeted them. Although apparently members of the same species—having two arms and two legs, opposable thumbs, and erect posture—these newcomers were, in virtually all other respects, frighteningly weird. They did not eat the same foods, wear the same clothes, speak the same languages, or utilize the same implements for even the simplest of human endeavors, like cooking or eating.

Consider, just for a moment, not the people but some of the things that must have come over to North America on the *Mayflower* in 1620. The Pilgrims brought with them several books (Bibles)—an object no native American had ever seen before, containing pieces of a technological system (written language) in which no native American had ever been embedded. They also brought axes, hatchets, knives, saws, hammers, augers, and a jackscrew—all implements made of a material, iron, that the natives could neither produce nor manipulate. At least one wooden plow was on board, an implement intended for turning over the soil, a job that the natives of the eastern seaboard had never needed to accomplish.

Either the *Mayflower* itself or a subsequent supply ship carried pigs and cows and oxen, so that the Pilgrims might be supplied with meat, milk, and draft labor. Domesticated animals were yet another technological system that must have startled the native Americans, who, with their intimate knowledge of deer, wolves, and beavers, had probably never dreamed that there were animals that would sit still long enough to be milked or to have a yoke thrown over their shoulders. The Pilgrims protected their bodies with clothing made of woven cloth—a material the natives, clad in leather and furs, had never seen and did not know how to make. In addition, the Pilgrims possessed implements—guns—the sight, sound, and effects of which must have terrified the people who had suddenly become their neighbors.

If the objects were strange, so too was the culture—the whole system of symbols and behaviors that had been built over around and through these objects. The culture of the Europeans must have conflicted with, or at least seemed incomprehensible to, the culture of the native Americans. The Europeans, whether they were French, Dutch, Spanish, or English, were accustomed to communicating in written languages, something that their forebears, descendants of the Neolithic peoples of the Middle East and the Mediterranean, had been doing for at least two millennia. When Europeans wanted to signify that some agreement had been made between two parties, they wrote it down. When they wanted to exchange pieces of real estate, they composed a deed. When they wanted to exchange valuable goods without too much inconvenience, they composed bills of exchange. When they wanted to establish some new way of behaving, they developed instructions and wrote them down. When they were feeling introspective or momentous, they created diaries. When they wanted to communicate with people at a distance, they wrote letters. The most prominent symbol of their entire culture, the object that meant more to them than any other, the focus of their communal and individual spiritual lives was—a book.

The native Americans also needed frequently to communicate with each other, or on occasion with their new neighbors. They too exchanged valuable commodities or pieces of real estate; they too had need to learn and to teach new behaviors; they too felt introspective and momentous; they too had religious ceremonies and spiritual longings. But none of these was accomplished through, or symbolized by, writing, an activity as meaningless to the natives as it was meaningful to the settlers. The native Americans and the Europeans were not just speaking two different languages, they were embedded in two different language systems, one that was entirely oral and the other that was partly oral and partly written. Between those two systems communication was difficult indeed—perhaps, ultimately, impossible.

But poor communication was not the only source of difficulty. The Europeans came from societies that had been practicing settled agriculture for many, many centuries; the traditional practices of European agriculture were ancient, going back, in some cases, to the traditions of the peoples who had been farmers and shepherds in Babylonia, Egypt, Greece, Rome, and Palestine. Thus when they wanted to farm, the Europeans cleared the land of trees: chopped them down, uprooted them, cleared them out all together. At the beginning of each growing season, they broke through the winter-crusted earth and turned it over so that seeds would root and plants would grow. When they planted vegetable seed, Europeans planted it neatly in rows, the better to be able to cultivate during the growing season. Each field that they planted had a regular time to be left fallow, after which it would be planted again. They used their fields over and over again, renewing the soil, when it was depleted, sometimes by letting it lie fallow, sometimes by plowing under decayed plant material, and sometimes by collecting and distributing animal manure. Since their animals were valuable commodities (for food, labor, and manure), they built shelters as protection from predators. Europeans also created fences to keep their animals out of the fields in which their crops were growing.

All of these agricultural practices were alien to the native Americans. They did not build fences and did not understand what straight lines of wood or stone were intended to signify. They did not own animals and so considered a pig or cow or ox found wandering in the woods (which the colonists frequently allowed their animals to do) to be fair game. The natives, especially those in the vicinity of Plymouth and Massachusetts Bay, believed that the English were insanely overworked, laboring from dawn till dusk to dig rocks and tree roots out of fields. To the natives, the English seemed profoundly stupid since they were very poor hunters and did not know how to regulate their diets in winter so as to go for several days without eating. The English, in turn, regarded the natives as irresponsible since they did not labor especially hard to store large quantities of food for the winter.

European settlers tried to build permanent homes for themselves and, having struggled so hard to clear the land, tried to keep their farms fertile indefinitely. The natives didn't bother; when yields proved scanty on one piece of land, they gave it up, let it revert to its natural state, and moved somewhere else. Accumulation of goods and attachment to objects such as houses, fences, and plows seemed to them to be the height of foolishness; "They love not to bee encumbered with many utensils," as one seventeenth-century observer put it.[1]

Nothing, however, was quite so strange, or created quite so much hostility, as the habit Europeans had of giving parcels of land in perpetuity to individuals and their offspring—of owning land. From the European, most especially the English, point of view, a person could become the owner of a piece of land in one of three ways: by developing it for use, by buying it, or by receiving it as a gift. Once that person owned the land, he or she had control of it absolutely: to sell it, give it as a gift to someone else, rent it, alter it. From the native point of view, people owned the produce of the land that they had worked, but not the land itself. Native families did not expect to live on or work any given piece of ground for any great length of

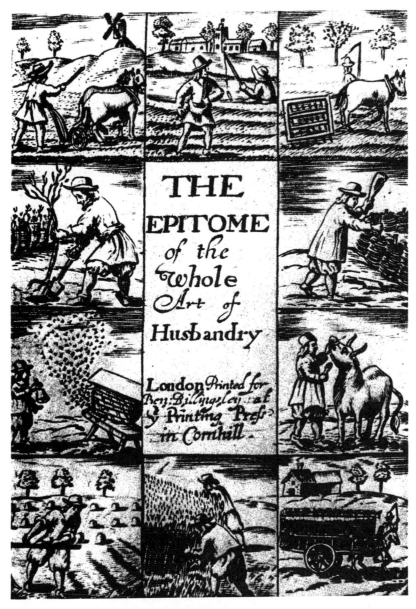

Some of English tools and domesticated animals that might have astonished the native inhabitants of eastern North America. This title page from an English book by Thomas Blagreve, *The Epitome of the Whole Art of Husbandry* (1675), shows (*clockwise from upper left*) a plow, straight furrows, a harrow, fencing, an ox, a cart, a sickle, a scythe, a bee hive, a shovel. (Courtesy New York Public Library.)

time; hence they had no reason to think about buying or selling, donating, or setting restrictions on any piece of territory.

We will probably never know what Chief Powhatan thought he was doing when he "sold" Manhattan Island to the Dutch for a dozen or so pack-horses loaded with "Indian goods," but we can be fairly certain Peter Stuyvesant thought something different. The governors of Plymouth and Massachusetts Bay, citing Genesis ("And God blessed them: and God said unto them: 'Be fruitful, and multiply, and replenish the earth, and subdue it'"), believed that the natives only had ownership rights over the very few plots that they were actively working at the time the settlers arrived. The land, as John Winthrop put it, "lay open to any that could or would improve it [since] the Natives in New England inclose noe Land, neither have any settled habytation, nor any tame Cattle to improve the Land by."[2] Everything that was not being "subdued"—clam banks, fishing ponds, berry-picking areas, hunting grounds, uncleared forest—could be, and was, allocated to settlers, justifiably (from the English point of view) and without guilt. Unfortunately for both parties, the natives were incapable of seeing the justice in that arrangement.

Subduing the land was something that Europeans had been doing for centuries before the first of them tried to accomplish it in North America, and over the course of those centuries, the landscape of Europe had been changed profoundly. Where there had once been vast, dense forests, there were now in Europe mile after mile of cultivated fields. When the forests disappeared, so too did the game animals—deer and pheasant, ducks and squirrels—that had once inhabited the forests; so too did many of the nuts and berries that had once been a standard part of human diets. By the seventeenth century, Europeans could only feed themselves through agricultural practices; they had lost the ability (which the native Americans still possessed) to live directly off the land. To Europeans, the forests were dangerous, terrifying places that had to be subdued before civilization, or even survival, could be accomplished.

Over the centuries, Europeans had also become accustomed to and dependent on a social structure characterized by a high degree of differentiation: in Europe, some people specialized in farming, while other people pursued the crafts or were merchants, soldiers, priests, or physicians. This differentiation was possible at least in part because European agricultural techniques were efficient enough to produce surpluses of food; some people could be spared from food production to pursue other endeavors.

Over the course of the centuries, a social system had developed in Europe in which farmers—the people who produce the commodity that everyone needs to survive—had become dependent on numerous products and services provided by nonfarmers. In England, France, Spain, and Holland, there were people who specialized in butchering, baking, and candlestick making—as well as building ships, weaving sailcloth, shoeing horses, spinning flax, weaving wool, grinding grain, and carving stone.

The technological systems of seventeenth-century Europe, and the differentiated social system with which they were intertwined, may seem simple to our jaded twentieth-century eyes, but they were very complex when compared with native American systems—and the Europeans who crossed the Atlantic were very dependent on them. Most Europeans could not live as the Indians could, "unencumbered by utensils." (Those few Europeans who enjoyed living in the native style were usually explorers and trappers—loners—not founders of permanent settlements.) In order to farm, a European husbandman needed a plow; and in order to maintain the plow, he needed a blacksmith. In order for the blacksmith to be supplied with wrought iron, he had to buy some from the owner of an iron furnace, who couldn't make wrought iron unless he also employed skilled craftsmen. Neither the husbandmen nor the furnace owner could have gone about his business without depending on merchants (since iron ore is not evenly distributed in every locale in which people need it), who were themselves dependent on carpenters (to make carts) and teamsters (who loaded and drove carts). There was no level of European life into which such networks of social and technical dependency did not spread.

Europeans could neither feed nor clothe themselves without the help of skilled craftsmen and complex machines. The hard outer shell of most dried grain seeds must be mechanically broken or human stomachs cannot digest it. Native American women accomplished this task by grinding their corn (after it had been removed from the cob and dried) between two stones, or in a stone or wood device similar to a mortar and pestle. By the seventeenth-century, Europeans had long forgotten how to do this themselves; in the early Middle Ages, some enterprising Europeans had learned how to build water- or horse-powered gristmills for the purpose. These mills represented large investments both of money and labor because they were (for their time) large structures of wood or stone (then very expensive materials) containing millstones, which had to be quarried, transported, cut, and attached to gearing systems, as well as waterwheels, which only skilled craftsmen could build and maintain.

So it was also with clothing. If a native American boy needed a new set of trousers, he could kill a deer, his uncles and brothers would help his father skin it, his aunts and sisters would help his mother stretch it and treat it, and his mother would then sew it together. The tools involved were the bows and arrows needed to fell the deer, the sharpened stones needed to skin it, and the awls (made from bone) needed to pierce the skin so that thongs might be threaded through. Most native American men and women possessed the skills that were needed both to fashion those tools and to process those hides.

If an English girl needed a new dress, the process of providing it for her was very different, and no member of her immediate family could have owned all the tools that were necessary, let alone known how to make or repair them. Cloth was purchased from a merchant, who obtained it from a dyer, who purchased it from a fuller (who owned a mill in which cloth was pounded and treated with chemicals to shrink it). The fuller received the cloth from a weaver (who owned and operated a loom), who had received the thread from spinsters (who used spinning wheels that had to be built by carpenters), who purchased raw wool from sheep that had been raised and shorn by yet other skilled craftspeople. Only then might the girl's mother be ready to do the sewing that the dress required, although even then she needed to own a needle (purchased from a peddler, who got it from a needle maker) and some thread (which she might have spun herself).

When Europeans were transplanted across the Atlantic Ocean, they were cut off from the network of craftspeople on whom their lives up until then had depended. Just imagine what it might have been like to break a needle or run out of thread when you are three thousand miles—and several months—away from the nearest spinning wheel or needle maker. Imagine trying to sharpen an ax when you have no access to a grinding wheel and you need the ax to chop down some trees. Worse yet, imagine needing to chop down those trees in order to build something that will shelter you from the elements when you are not a carpenter and haven't the faintest idea how to tether the pieces of wood together so the shelter will remain standing when the winds begin to blow.

The Europeans needed to learn basic survival skills from the native Americans, and they needed to learn those skills very soon after arriving. Few if any craftspeople were tempted to make the treacherous trip across the ocean. Had they been tempted to come, many of their crafts could not have been pursued in the New World either because the equipment they needed was too difficult to transport or because the essential raw materials were initially too difficult to obtain.

Had the European settlers not adopted some of technological skills of the native Americans, they never could have survived. They needed to learn how to build traps to catch wild game, fish, and birds; they needed to learn how to grow corn, beans, and pumpkins—foods that they had never before tasted. Most had to learn how to hollow out the trunk of a tree so that grains of corn could be pounded into meal; others had to learn to mix those grains with water to create a bread. Even those who were not shoemakers learned how to make moccasins out of the skins of wild animals; even those who were not carpenters learned how to shape necessary objects out of wood, particularly canoes. Perhaps most significantly of all, the early settlers learned the virtues of becoming undifferentiated and unspecialized from the natives that they so quickly and unfairly displaced. They learned that survival in the New World required men and women to be adept at many skills: jacks-and-jills-of-all-trades, albeit masters and mistresses of none. And when they passed on that knowledge and those skills to their children, they created what subsequently came to be called the culture of "Yankee ingenuity."

The Colonial Economy

The economy that the European settlers created in North America was an economy based on trade across the ocean. The settlements that eventually became the thirteen colonies were each founded in a different way, in a different time. In the seventeenth century, Virginia and Massachusetts were started by companies, groups of investors who pooled their resources in order to finance a group of settlers. New York began in the same way, except that the original investors and settlers were Dutch rather than English, but New York was turned into a proprietary colony (belonging to one individual who had his grant from the king) when the English ousted the Dutch in 1664. New Jersey, Pennsylvania, Connecticut, Delaware, and both Carolinas were also seventeenth-century proprietary colonies, and so was Maryland, created to provide a haven for English Catholics. New Hampshire and Rhode Island were created in the seventeenth century by emigrants from Massachusetts who found the religious intolerance of the original settlers a bit too much to bear, and Georgia, the very last (organized in 1732), was founded by a company with the express intention of turning a profit for the investors by providing jobs and homes for people who would otherwise have been imprisoned for debt.

Each of these colonies faced the same set of difficulties, whether founded to make a profit or to make new homes. Lacking the ability to manufacture the goods that they needed for survival (axes and salt, cooking pots and woolen cloth), the colonists desperately needed some commodity that they could sell on European markets in exchange for those goods. Even after several decades, even after the difficulties of first settlement had been overcome, both hard cash and human labor were still scarce in the colonies, and manufactured goods had to be imported from abroad. By the time the seventeenth century had turned into the eighteenth, Atlantic trade had become the backbone of the colonial economy; without it the entire enterprise of settlement everywhere in North America (including the French colonies, which eventually formed the basis of Canada) would have failed—and failed miserably.

From Jamestown in 1610, from Massachusetts Bay in 1630, from Georgia a century later, the first products sent back to Europe after the settlers arrived were those that were very nearly raw, products that were easiest to procure with the fewest tools and the least labor. In many places, particularly where the native inhabitants were willing to cooperate, this product was furs—pelts of the animals that inhabited the endless forests of the eastern coastal plain and the Appalachian highlands. Furs were close to a perfect product from the colonists' point of view. Relatively lightweight and pliable, they were easily packed for shipment across the ocean, where they commanded very high prices. Western Europe had been depleted of most of its furbearing animals for a very long time, but the demand for fur remained high since there was no product better able to protect Europeans from the ravages of the cold. Rather than hunt themselves (an enterprise at which they were not very skilled), the settlers preferred to obtain pelts from the local tribes, who were often quite willing to trade beaver skins and deer hides for axes, knives, cloth, mirrors, kettles, beer, rum, and other products of European civilization. Thus the labor costs involved in the fur trade were slight, as long as the labor of the natives was discounted. The fur trade continued to be a major sector of the colonial economy through-out the eighteenth century; 2,000 packhorses loaded with "Indian goods" made their way west from Augusta, Georgia, just in the year 1740 alone, returning with pelts which would bring between £20,000 and £30,000 on the London market. Within decades, the eastern forests were depleted of furbearing animals and trading outposts had to be established farther west—one of the reasons the French and the English were at each other's throats in the territory between the Mississippi and the Appalachians in the middle decades of the eighteenth century.

Aside from furs, the forests of the New World provided the earliest col-onizers with several other products, much in demand in Europe, that were relatively easy to extract: potash, pearl ash, pitch, tar, and resin. Unglam-orous as these may be, they were used in a host of manufacturing process-es; the first two were essential chemicals in the making of soap and glass, as well as the dying and shrinking of cloth; the latter three, in shipbuilding and other forms of carpentry. Wherever there were pine trees (in the Car-olinas, for example, or on the sandy soils of Cape Cod), pitch, tar, and resin (frequently lumped together under the heading "naval stores") could be collected by tapping and purified by boiling. Wherever land had to be cleared of trees, potash and pearl ash could be produced as a cash crop, by-products, as it were, of the process of clearing.

Potash was made by constructing a huge pile out of the trees that had been felled, burning it, collecting the resultant ash, loading it into a huge iron pot, and heating the ash until it was semimolten. The heating process caused carbon, sulfur, and a few other elements in the ash to be driven off (as noxious gases), and the gray caked material that resulted when the pot cooled was nearly pure potassium nitrate. Reheating this once again in the same pot could—if high enough heats could be achieved—break down this compound and produce almost pure potassium, called pearl ash by virtue of its lighter color. Hauled to the nearest merchant, these alkalis could be traded for the supplies needed to get through the first winter, before land could be plowed and the first crop raised. Naval stores and alkalis were one of several forest products without which the young American economy might not have survived.

Wood itself was another such commodity, whether in the form of whole logs, boards, shingles, or barrel staves. Western Europe had long been starved for wood: in all of England, for example, there was no stretch of forested land more than five miles square in the early seventeenth century, and in parts of southern England (the most populous part of the country), there was virtually no wood at all with the exception of hedges. The earli-est explorers and settlers were well aware of the potential value of the North American forests: on June 22, 1607, the Jamestown settlers wrote to the officers of the Virginia Company to tell them that "the soil is most fruit-ful, laden with good Oake, Ashe, Walnut trees, Poplar, Pine, sweet woods, Cedar, and others yet without names." A few trees had already been felled, and some boards sawed so as to construct a few houses, the colonists reported, and they had already sent back "a taste of clapboard." That taste must have whetted English appetites because it was not long before the garrison in the Jamestown block house was set "for their exercise" to the cutting down of trees and the making with hand tools "of Clapboard and Wainscot"—and not long either before a sawmill had been constructed further up the James River in Henrico.[3] The lumber industry in the colonies expanded rather quickly after that. By 1675 there were fifty sawmills operating just in the province of Massachusetts. Some communities built sawmills even before they built gristmills, which meant that they had to import flour from somewhere else.

Some of these early sawmills were tidal mills, one or two may have been windmills; the vast majority used river falls to run a waterwheel, which, through simple wooden gearing, powered an up-and-down saw blade. All were situated so that trees could be cut upstream and floated down to the mill and the sawn boards carried out on boats or barges. At least for the better part of the seventeenth and early eighteenth centuries, most of these mills were built and operated by men who were not English, lumber having been so scarce in England for so many centuries that sawmilling had not flourished there. Where sawmills were to be built and operated, Danish and Dutch emigration was encouraged by the investors—an early example of what later came to be called technology transfer through the emigration of skilled labor.

Lumbering also provides an example of the way in which an implement can be changed by anonymous craftsmen when a new environment creates new opportunities and new demands. The ax that Europeans brought with them when they first crossed the Atlantic consisted of a brittle iron head (the poll) weighing about four pounds onto which a very light steel blade (the bit) was welded, with the whole head inserted in a straight, clublike handle. Over the course of several decades, colonial craftsmen began to improve this instrument, which had remained unchanged in Europe over many centuries. The weight of the poll and bit was evened out so that the center of gravity of the head was located near the centerline of the handle; the handle itself was carved so that it curved, frequently precisely to fit both the height and the grip of the person who was going to wield it. Using such a felling ax an American axeman could fell three times as many trees as someone using the older, European type.

Between the ax and the sawmill, the eastern forests were very quickly transformed—and in some cases obliterated. The great white pines of the northern forests were cut down for shipmasts; the smaller pines became the millions of miles of boards from which houses, barns, forts, and mills were built all over North America, Europe, and the Caribbean. The white and red oaks were transformed into barrels, casks, and trunks for transporting wine, sugar, rum, and flour. The birches became canoes and shingles. The cherries and the maples became furniture, the willows became caning; the chestnuts and the cedars became fences. A good lot of this lumber was exported, but not all of it. Every structure in the colonies was built, wholly or in large part, out of wood; virtually every implement, from printing presses to rakes, was made out of wood; and nearly every household, from Maine to Georgia, burned wood every single day of the year for heating or for cooking until well into the early decades of the nineteenth century. A typical New England household consumed as much as thirty or forty cords of firewood a year, a stack of wood four feet high, four feet wide, and three hundred feet long—better than an acre of forest. Where the Europeans favored half-timbered construction, the Americans moved quickly to full-timbered styles; roofs that had been thatch or slate abroad were shingled in America. American lumbering practices were regarded by European travelers as incredibly wasteful: "an incredible amount of wood is really squandered in this country for fuel," a Swedish visitor remarked in the 1740s, "day and night all winter, or for nearly half of the year, in all rooms, a fire is kept going."[4] Trees were cut down helter-skelter, without any thought being given to the renewal of the forests—hardly a serious manner when the forests themselves seemed endless.

But the forests were not, in fact, endless. Within a few decades, some types of trees—the great white "mast" pines—had disappeared entirely from the forests of the coastal plain, and the balance between species had been profoundly altered. By the end of the eighteenth century, deforestation had transformed the ecology of whole districts, most especially in New England, New York, and Pennsylvania—many habitats for game had disappeared, watercourses had changed, seasonal flooding was more frequent and more devastating, the soil was less able to retain water, summers were hotter and winters colder than they previously had been. The landscape Americans now seek when they travel to Maine or New Hamphire or to the White Mountains, the Catskills, the Poconos, or the Blue Ridge for their vacations is not, by any stretch of the imagination, the original, densely forested landscape that the settlers encountered when they first arrived.

Aside from furs and forest products, the three items that formed the backbone of the American economy during the first decades of settlement were fish, tobacco, and wheat. Wheat did not grow well in New England, where it was subject to devastation by a fungal disease called "blast," but it did flourish farther south, in parts of New York (particularly Long Island and the Mohawk Valley) and in New Jersey, Delaware, Pennsylvania, and Virginia. Europe experienced periodic shortages of wheat; a drought here or a battle there could, and frequently did, threaten some Europeans with starvation every year. The farmers of New York, Pennsylvania, New Jersey, Delaware, Virginia, and Maryland were happy to provide their European cousins with the staff of life—for a price. They were also happy to provide their compatriots overseas with the "noxious weed" that the native Americans had taught the Europeans how to roll into tubes and smoke for relaxation.

By 1700, the middle and southern colonies were starting to convert from subsistence to commercial agriculture, with wheat and tobacco as their principal products. Commercial farming was a labor-intensive activity; both wheat and tobacco required considerably more care in plowing, cultivating, irrigating, harvesting, and fertilizing than corn does. For this reason, the middle and southern colonies imported more indentured servants than New England during the seventeenth and much of the eighteenth centuries—until, that is, some commercial farmers (particularly in the southern colonies) realized that slaves provided even less expensive labor than indentured servants.

The cities of the middle colonies—New York, Philadelphia, and Baltimore—became centers of merchant milling by the end of the seventeenth century. Merchant mills were somewhat different enterprises from local gristmills. Colonial farmers brought their grain to a local gristmill to have it ground for their own consumption; the price the miller exacted for this service was some fraction of the grain or some other farm product—cider, perhaps, or a side of pork. When farmers brought their grain to a merchant mill, they expect to sell it to the miller in exchange for cash. The miller then ground it and packed it into barrels for shipment and sale somewhere else. Merchant mills were very profitable in the colonies, and they were also, as we shall see in a later chapter, a locus for important forms of technological change.

So were fisheries or, more precisely, some of the industries that were stimulated by the massive colonial trade in dried fish, particularly cod. Meat was not a basic staple of European diets in the seventeenth century, but fish was very popular. Fish were relatively easy to catch, very easy to preserve (in salt), and relatively easy to transport (in barrels)—hence, fish were cheap. Fish was also, in those countries that continued to abide by the practices of the Catholic Church, ritually required as a food on some days of the week and for some weeks of the year. Having been worked over for centuries, the prime fishing areas along the European coast were not bearing very well by the seventeenth century—the reason Portuguese, Dutch, French, and English fishing fleets had been scurrying down the coast of Africa, in and around the Baltic, and all over Greenland in search of fish.

Eventually they were bound to try North America. Stretching in a long arc from Long Island to Newfoundland, cooled by an Arctic current, lay one of the finest fishing grounds in the world. In the early decades of the sixteenth century, French and Portuguese fishing fleets established fishing stations in the areas that are today Newfoundland. Roughly a century later, the English who settled in Plymouth and Massachusetts Bay, as well as the Dutch who settled New Amsterdam, also began mining the gold that came from the sea. The abundance and variety of fishes in the waters around Massachusetts astounded the earliest settlers. The Reverend Francis Higginson, who was the first minister of Salem, told friends and relatives in England: "The aboundance of Sea-Fish are almost beyond beleeving, and sure I should scarce have beleeved it except I had seene it with mine owne eyes." One settler reported that during a few hours of fishing he and his companions "had pestered our ship so with Cod fish, that we threw numbers of them over-boord againe."[5] By 1700, New Englanders were exporting over 10 million pounds of fish a year, surpassing the fleets of England in both quantity and quality. By 1770, 665 vessels were employed in the New England fisheries alone, and 4,405 men (a number larger than the total population of several substantial towns) were making their livings as fishermen.

Such extensive fishing (and also whaling) stimulated other colonial industries. Someone, after all, had to build all those boats carrying the whalers and the fisherman out to sea; someone had to weave the sailcloth; someone else had to make the rope that was used to keep the sails in place. The colonists quickly became adept at all of these endeavors; to the dismay of English craftsmen, colonial shipbuilders were soon able to turn out a product that was cheaper (since transportation costs on the raw material were lower) and better adapted to coastal sailing.

Furs, trees, fish, wheat, and tobacco—these were the humble materials on which firm colonial foundations were laid. Some of the colonists may have dreamed of living without religious persecution, some may have dreamed of a life unfettered by poverty, some may even have remotely sensed the potential for democratic government in the New World—but all realized that first they had to trade something in order to survive in the wilderness. Some made fortunes from furs, trees, fish, wheat, and tobacco (and, of course, land), but every single colonist depended on the trade in those commodities simply for survival. In some cases, that dependence was direct, as when colonists exchanged potash for plowshares so that they could plant the seeds that produced their food. In other cases, it was indirect, as when the mother country, anxious not to interrupt the flow of profitable goods, sent soldiers and armaments to defend the colonists against the ofttimes hostile native inhabitants. Either way—direct or indirect—furs, trees, fish, wheat, and tobacco kept both the colonists and the colonies alive.

Colonial Economic Policy and Technological Change

Most of the social and technological systems that both the native Americans and the European colonists developed in North America evolved in response to the pressures and opportunities created by new environments. The very first human settlers on the continent used stone and wood spears to kill polar bears, but they eventually figured out how to add leather and make harpoons, which could effectively kill the sea lions and whales that were so abundant in the arctic seas. Centuries later, their descendants living on the temperate plains and woodlands learned how to propel small spears with the tensile forces of bent wood and twisted animal gut, thus adding small game (rabbits, deer, squirrels, pheasants) to their diets. Later still, the Europeans arrived, and their traditional implements and skills began to change also. Unable to grow wheat easily at first, they learned to make a semileavened bread from corn. Unable to depend on the network of crafts they had been accustomed to in Europe, they learned how to do their own butchering, baking, and candlemaking—and how to repair their implements themselves. Their unwieldy axes, little used in Europe, evolved into elegant and efficient instruments in the New World, adapted to the needs of people who wanted to eliminate and clear the forests as quickly as possible.

But necessity and opportunity, although powerful, are not the only forces that direct and shape technological systems. The European colonists brought with them another social instrument—governmental policy— which can also have profound effects on technological change. The North American tribes, lacking complex governmental arrangements and bureaucrats who devote their lives to creating and implementing economic policies, may not have been aware of the power that governments can have over technology, but the European settlers in North America most certainly were.

In the seventeenth and eighteenth centuries, mercantilism was the economic theory favored by most European monarchs. It was a very simple theory: the function of the king's government was to make the king rich, for when the king was rich he could defend his nation and extend its power, thereby making his subjects rich, or at least comfortable. The best kind of riches, the mercantilists thought, were the kind the Spanish had taken from South America, gold and silver bullion. Unfortunately, not all countries were lucky enough to be able to tap into the silver mines of Mexico and the gold lodes of Peru. For these unfortunate countries, there was a second best solution—trade. And the very best kind of trade was with your very own countrymen, settled in colonies far distant from your homeland. Colonies could supply you with raw materials that would stimulate industry at home and, in turn, provide a ready market for home manufactures.

To keep the colonial system working to the advantage of the monarch, it was, however, necessary to ensure that all the goods being traded back and forth between colony and homeland were carried on vessels owned by loyal subjects, manned by loyal sailors, and unloaded only at royally approved ports—the better to ensure that in times of peace the regulations would be followed and that in times of war there would be plenty of ships and plenty of skilled sailors to draft into the king's service. Now all of this required something of a bureaucracy to enforce, but the cost of the bureaucracy would certainly be offset by the riches that would flow into the coffers of the king—or so the theory said.

In practice, of course, nothing worked out quite that simply, either for the colonies or for the mother country. Bureaucratic control turned out to be a rather difficult thing to manage three thousand miles across the ocean, and some of the colonists turned out to be quite adept at smuggling; the word "yankee" is actually a corruption of a Dutch word that means "smuggler." After a while, in any event, the rules and regulations started to get so complicated that even a perfect bureaucracy could not have enforced them. As trade between the colonies and the homeland picked up and as the colonial economies became more stable, difficult situations inevitably arose in which it proved impossible either to know whose interests it was best to serve or whether the regulations were accomplishing what they were intended to accomplish. What do you do, for example, when the colonists—who are, after all, loyal subjects—begin producing ships—very good ships as it happens—in direct competition with shipbuilders at home, who also happen to be loyal subjects?

With all the goodwill in the world, the kings of England and their financial advisors could not have resolved such disputes either equitably or simply—and, of course, neither the king nor the officials of his customs agency had all the goodwill (or efficiency) in the world. By the middle decades of the eighteenth century, the mercantile system had become hopelessly complicated, and it continued to become more complicated—and more of an irritant to the colonists—as King George III attempted in the years after 1763 to solve his financial troubles by exercising ever more control over colonial trade. The oft-amended Navigation Acts, the Currency Act, the Sugar Act, and the Stamp Act eventually provoked the colonists to revolt; they also had a crucial, although less spectactular, determining effect on the pattern of technological change in the colonies.

In capitalistic economies such as those of Europe in the seventeenth and eighteenth centuries, whereever an industry flourishes, technological change occurs. When many people are engaged in the same activity, someone will eventually figure out how to change an implement here, or reorganize a system there, so as to give his or her enterprise just a little competitive edge. And where the activity is proving profitable, someone else may come along and invest the funds that the innovator needs to experiment with the new implement or develop the reorganized system, producing not just an idea for technological change but technological change itself. The waterwheel was a European invention, but Americans were building a waterwheel on every stream that could tolerate one. Sooner or later some American was going to figure out how to build one just a little bit better, as we will see in a later chapter. The mills that were powered by waterwheels—the merchant mills that ground wheat and the sawmills that created boards—were extremely profitable ventures (in part because the goods they were producing were in demand overseas and were relatively unpenalized by tariffs), which meant that eventually some investor, while laying out money to build a new mill, was going to consider building it with one of those "just a little bit better" wheels—and the technology of waterpower was going to change.

Yet while mercantilism favored American milling, it did not similarly favor American weaving or, for that matter, American spinning. The rules and regulations of the English international economy were constructed to encourage export of flax, cotton, and wool from the colonies while prohibiting the export of cloth. Cloth had been one of England's principal export commodities for centuries, and according to mercantile theory, the English would have been foolish to allow a cloth-making industry to develop in their colonies. Americans were allowed to spin and weave materials for their own consumption, but not for sale overseas; whatever they needed that they could not supply with their own hands, they were expected to buy from producers in England. This meant that spinning and weaving were both relatively unpopular and relatively unprofitable occupations in the colonies. Thus in the last decades of the eighteenth century, the early years of industrialization, when spinning wheels and looms were changing with dizzying speed in Britain, they were hardly changing at all in the British colonies. The power of even inefficiently enforced governmental policy was such that those colonial industries that were repressed by regulations of the British Board of Trade (textile production was one such industry; iron and steel fabrication was another) did not undergo technological change until several decades after independence had been won.

Conclusion: Quickening the Pace of Technological Change

By 1750, the continuance of European settlement in North America seemed assured. Each of the thirteen colonies was, in its own way, flourishing. The growth of foreign trade had led to the development of cities all along the eastern seaboard, from Boston in the north to Charleston in the south. Frontier settlements were still being attacked by bands of native Americans, and French armies on the western frontiers were still a source of worry—but most of the time most of the colonists could go about their business peacefully. New immigrants arrived daily: from Scotland and Ireland, from Scandinavia and France, from the Germanies and from the Netherlands. The population was increasing by natural means as well; since colonial mothers were better fed than their European sisters, they tended to have more children, and those children were more likely to survive infancy.

At every turn, civilization was being "implanted" in the colonies and the wilderness was being "subdued." Fish, furs, trees, wheat, and tobacco still provided the basis for the colonial trading economy, but other products had been added as the colonists had pushed into new territory and as money was invested in new facilities. Wheat was still flowing in an endless stream from the middle colonies, but by the 1750s, more of it was leaving as flour and biscuit than as whole grains. Pig and bar iron was leaving the smelter's furnace and traveling across the seas, just as rum was being poured out of the distiller's vats and into barrels destined for the Caribbean. Virginia and Maryland were still producing tobacco, but North Carolina had converted to indigo and South Carolina to rice (each of them also labor-intensive crops). Naval stores remained an important export, but so too did increasingly large quantities of finished wood products: knocked-down houses (intended for quick construction in the Caribbean), clapboards, shingles, and barrel staves.

As the European population of the colonies expanded, so too did the level of domestic demand and the quality of the goods being produced to satisfy it. Craftsmen in Philadelphia were making furniture for houses in Lancaster, while iron furnaces in Lancaster were producing box stoves to heat houses in Philadelphia. Shoemakers in Massachusetts were making footwear that was being sold to plantation owners in North Carolina. Weavers in New York were making cloth that was purchased in Connecticut. Orchardmen in Connecticut were pressing cider that was being sold in Maryland. From primitive beginnings, the colonial economy had developed into a rather complex affair by 1750.

Only a century earlier, the forests and the beaches of the eastern coastal plain had been occupied by people who cut the bark from trees with stone cudgels, dug root crops out of the ground with sticks, and ground grain in hollowed-out logs. Now there were sawmills and distilleries, gristmills and brickworks. Just a short time earlier, there had been people living directly off the fruits of the ground; now there were people living directly off the fruits of foreign trade. Where once the Europeans had been frightened by terrifying forests and mysterious rivers, they now found cultivated fields and prospering cities.

For several millennia, technological change had occurred slowly in North America. People migrated and adapted to new environments, but there had never been enough of them to profoundly alter those environments, even if their technologies had been capable of it, which they weren't. New technologies had developed in response to new environmental conditions, but never at a pace or to an extent that had prevented the natural environment from repairing itself. Animals had been killed, but their populations had never been completely decimated so they were always able to reproduce themselves. Plots of agricultural land, once abandoned, had fairly quickly became indistinguishable from the surrounding forest.

With the advent of Europeans, the pace of change quickened—and the changes began to look both permanent and profound. The Europeans brought with them social and technological systems that had evolved slowly, over many millennia, somewhere else; but in North America, they were totally novel and had totally novel effects. In just over a century, the human population of the eastern coastal plain had increased markedly and its natural environment had been completely altered. Many of the forests had been cleared; most of the furbearing animals had been killed; some species of birds were almost extinct.

At first, the new settlers had had to borrow survival skills and locally adapted implements from the native peoples, but once they had managed to stabilize their colonies and establish regular trade with their home country, they were able to recreate a semblance of the social and technological world of Europe. Some new settlers adapted their old technologies to fit new conditions. Other settlers used the technologies at their disposal (and the support of their government) to alter their new land so that it would more closely resemble, and be of service to, the one from which they had come. Yet others used the same technologies—the books and the deeds, the guns and the axes, the wheat and the cattle—to exploit the abundant natural resources of the new land in new ways, at a new pace, and with new convictions about the righteousness of what they were doing.

Notes

1. From the diary of Thomas Morton [n.d.], as quoted in William Cronon, *Changes in the Land: Indians, Colonists and the Ecology of New England* (New York: Hill and Wang, 1983), p. 38.

2. John Winthrop, *Winthrop's Journal*, ed. James Kendall Hosmer (New York: Scribner, 1908), p. 294.

3. All quotes are from "Your Poore Friends to the Council of the Virginia Company," as quoted in Mary Newton Stanard, *The Story of Virginia's First Century* (Philadelphia: Lippincott, 1928), p. 40.

4. Peter Kalm, *Travels in North America* [1753–61], ed. Adolph B. Benson, Vol. I (New York: Dover, 1966), p. 239.

5. Francis Higginson, "New England's Plantation" [1630], *Massachusetts Historical Society Proceedings* 62 (1929): 311.

Suggestions for Further Reading

James Axtell, *The Invasion Within: The Contest of Cultures in Colonial North America* (New York, 1985).

William Cronon, *Changes in the Land: Indians, Colonists and the Ecology of New England* (New York, 1983).

Alfred E. Crosby, Jr., *The Columbian Exchange: Biological and Cultural Consequences of 1492* (New York, 1982).

James Deetz, *In Small Things Forgotten: An Archaeology of Early American Life*, rev. ed. (New York, 1996).

Peter Farb, *The Face of North America: The Natural History of a Continent* (New York, 1963).

Brooke Hindle, ed., *Material Culture of the Wooden Age* (Tarrytown, NY, 1981).

Karen O. Kupperman, *Settling with the Indians: The Meeting of English and Indian Cultures in America, 1580–1630* (London, 1980).

Eric Sloane, *A Reverence for Wood* (New York, 1965).

Wilcomb E. Washburn, *The Indian in America* (New York, 1975).

Jon Ewbank Manchip White, *Everyday Life of the North American Indian* (New York, 1979).

2

Husbandry and Huswifery
in the Colonies

THE CIVILIZATION THAT THE EUROPEAN SETTLERS CREATED in North America was agrarian, founded and focused on farming. Some people worked small plots of land that could only provide the bare necessities of life. Others owned huge plantations; they were "white glove" farmers, supervising and managing laborers without getting their own hands dirty. Some colonists supplemented farming with other occupations, such as carpentry, blacksmithing, shoemaking, and weaving. Many people—servants, slaves, and day laborers—farmed even though they did not own land and even though the label "farmer" (or, to be less anachronistic, "husbandman") would never have been applied to them. Women fell into this latter category also, especially the wives and daughters of farmers; they winnowed wheat, hoed gardens, milked cows, churned butter—farmers in reality, if not in name. By 1750, about 80 percent of the colonial economy was agricultural and about 90 percent of all adults living in the colonies farmed for at least part of the year. The new nation, when it came to be created at the end of the eighteenth century, was truly a nation of farmers.

This is, of course, no longer true. As a result of industrialization, the contribution of agriculture to American economic output is now 3 percent, not 80 percent, and the percentage of people engaged in agricultural pursuits is now 4 percent rather than 90 percent. Nonetheless, the fact that so many Americans farmed in 1776 remains politically important to us today. The governmental structure under, around, and through which we still continue to operate was created to ensure that the interests of farmers

would not be overpowered by the interests of merchants and other city dwellers. Perhaps even more important, our agricultural origins are reflected in our myths, the commonly held ideas that underlie whatever is unique about American culture: that special set of attitudes that distinguishes Americans—no matter what their ethnic origins may be—from citizens of other countries.

One such myth is the myth of "self-sufficiency." Open almost any American history textbook, and you will read something about the self-sufficient farmers of the colonies, who could produce everything that they needed for their own support and sustenance—all their bread, all their meat, every inch of cloth, every cord of wood—and who never, as a result, had to purchase anything. Listen to any Fourth of July oration, and you will likely hear something about the "independent" life of our foreparents, people who did not need to rely on welfare systems or paychecks but who could, and did, make their way on their own through dint of hard work, self-employed. Our politicians frequently applaud self-sufficiency, whether it be in energy policy, defense policy, health policy, or even farm policy. Independence is a concept to which Americans attach multiple emotional meanings, some of which derive from our sense of our national past, and some of which derive from our respect for that past. Every time it is said that someone doesn't want to take money from her parents or doesn't want to be a renter forever or hopes soon to start his own business, the rest of us respond sympathetically, at least in part because of what we were taught about the people who settled this country and shaped it in its early years, the farmers who "made this country great."

The word "myth" has two meanings. It can mean a story, or a concept, that many people know or share, and which gives them a sense of communal identity; this is the positive meaning of the word, the meaning that was used—implicitly—in the preceding paragraph. But there is also a negative meaning, that of a story, or a concept, which is not true. The history of technology teaches us that the myth of self-sufficiency is a myth in both senses. As we examine the equipment with which colonial farmers sustained themselves in the wilderness we will begin to see why this is so.

Types of Farms in the Colonial Period

Roughly speaking, there were four kinds of colonial farms: pioneer farms, large plantations, small "mixed" farms, and subsistence farms. Pioneer farms were enterprises that were just getting off (or into) the ground, which might take, depending on the skill, health, and size of the farming family—and the character of the plot—anywhere from five to twenty years. A pioneering family had to devote much of its time and energy to clearing the land: girdling trees, chopping them down, pulling out stumps, removing rocks and boulders. Until cleared, the land on a pioneer farm could not provide even a low level of subsistence for a family; pioneers either had to have cash savings or a handy first crop that they could exchange for essential supplies. Historians have estimated that in the 1780s at least £25 (roughly a year's wages for a farm laborer) was required to buy the land for a pioneer farm and acquire the tools and supplies that would be required to sustain life for the first few years.

The two most common first crops on a pioneer farm were potash and maple sugar; both could be obtained with minimal effort and little expenditure for tools. Potash, as we learned in the previous chapter, was produced by burning trees, leaching the resultant ash with water (to produce lye), and then boiling the lye until nothing was left except a dark crystalline residue; all that was required was a special double kettle, a strong back, and a lot of patience. For maple sugar production, one needed first sugar maples, then an awl (for making a hole in the trunk of the tree), some spouts (for conducting the syrup), some pails (for collecting it), and a large kettle (for boiling it down to sugar). Both products enjoyed a ready market, and both could be exchanged in market towns for sacks of flour or a scythe, a pig or a plow.

Until the farmland became fully productive, however, a pioneering family lived at what contemporaries regarded as a substandard level: in houses made of logs chinked with mud, without a proper hearth, perhaps lacking beds, certainly lacking appropriate changes of clothing and an adequate diet. "They erect a square building of poles, notched at the ends to keep them fast together," a visitor to pioneer settlements in New Hampshire reported. "The crevices are plaistered with clay or the stiffest earth which can be had, mixed with moss or straw. . . . The chimney a pile of stones; within which a fire is made on the ground, and a hole is left in the roof for the smoke to pass out. Another hole is made in the side of the house for a window, which is occasionally closed with a wooden shutter."[1] "In many places," reported another visitor who was touring western South Carolina, "they have nought but a Gourd to drink out of. Not a Place Knive or Spoon, a Glass, Cup, or anything. . . . It is well if they can get some Body Linen, and some have not even that."[2] The pioneering family, living at this substandard level, was far from self-sufficient, for it was unable to provide for itself without constant recourse to the marketplace, both for basic foodstuffs—the sacks of flour and the pigs—and for fundamental tools—the scythe and the plow.

The plantation farm was not independent either, yet for a different reason. Plantations were huge tracts of land, owned by one person but worked either by employees, slaves, or tenants. Most were located in the southern colonies (although there were a few in New York, especially in the Hudson River Valley) and almost all were devoted to cash commodities: tobacco, rice, and indigo in the South; wheat or occasionally cattle in the North. So much land was available and so many slaves or servants were employed that the plantation could grow all the food that was needed for its own sustenance, spin and weave its own cloth, make and repair its own tools, even in some instances mine and forge its own iron.

A large plantation might be a self-sufficient community, but the people who lived on it were not independent. The tenants, slaves, and servants were specialists; those who farmed depended on those who wove, and those who wove depended on those who farmed. The owners of the plantation were the most dependent of all, not only because they needed dozens, perhaps even hundreds, of people to support their style of life, but also because they were enmeshed in trade networks that spanned the globe. Instead of wearing the cloth that their employees wove, they imported elegant fabrics from Europe; they shopped for furniture and rugs and wines in the great markets of the world; even the food that they ate was enlivened by condiments from other continents. In any event, few plantations, even in the South, were wealthy enough (either in land or in slaves) to support self-sufficiency; most traded actively, exchanging their cash crops (or, more frequently, credit for their impending crops) for such essential but labor-intensive goods as cloth and salt pork, shoes and potatoes. Great wealth created the possibility of self-sufficiency, but rarely the reality.

Which leaves us with the last two categories: mixed farms and subsistence farms. These are the two kinds of farms that have been traditionally regarded as self-sufficient, producing all that a family needed for survival and then (in the case of mixed farms) a little more, so as to have a marketable crop that could generate a profit. A close examination of the farming practices on these relatively small plots of land will give us some idea of the effort that had to be expended to achieve economic independence, and the likelihood that our ancestors ever really achieved it.

The Technological Systems of Colonial Agriculture

Colonial farming was done with a set of very simple implements; a hoe, a plow, a harrow, and a scythe were all that was needed to plant and harvest the major colonial crops: corn, wheat, rye, oats, tobacco, rice, and indigo. These were the implements that the earliest settlers brought with them and also the ones their descendants were using several generations later. European farmers had been using implements such as these since at least the early Middle Ages; American farmers did not stop using them until the pace of technological change began to quicken, which was not until the middle of the nineteenth century.

In New England and the Middle Atlantic colonies, the growing season began in the spring when plows were used to break and turn over the soil, which had become compacted during the winter, or to uproot and turn over weeds that had developed on a field that was left fallow. The plow consisted of a heavy wooden beam to which several wooden and iron parts were attached: a ring, which allowed a draft animal (usually an ox) to be harnessed; the coulter, a sharp, vertical knife, which cut through the soil and sod; a share, a perpendicular piece of wood or iron, suspended just behind the coulter, which displaced soil, creating a furrow; a moldboard, just behind the share, which forced the displaced soil off to the side, creating a ridge as the plow moved through the field; finally a set of handles, by which the farmer guided the path of the plow as the oxen pulled it.

Plowing was arduous labor, for the farmer as well as for the ox; an able man working with a competent beast could not expect to plow more than an acre a day. (Indeed, in the Middle Ages, an acre of land had been defined as the amount that could be plowed with a team of oxen in one day; long before the founding of the English colonies, this had been legally codified at 4,840 square yards, which it still is today.) The colonists practiced extensive, rather than intensive agriculture, which means that they preferred to clear and plow fresh land, rather than manure, till, and fallow old land on which the yields were falling. Hence, in any given year, they tended to have more land under plow than European visitors, who were accustomed to intensive farming, thought appropriate. James T. Lemon, a historian who has carefully examined eighteenth-century Pennsylvania farm records, estimates that an average Pennsylvania farm consisted of approximately 125 acres, of which roughly 30 acres were under plow every spring—a whole month of very difficult work.

In the South, plows were rarely used during the colonial period, even on small farms, for in the South fields were rarely left to rest overwinter and were rarely left for fallow. At least two crops grown in the South, corn and tobacco, were hill crops, suitable for cultivation with various kinds of hoes: a grubbing hoe, which looked like a lightweight pickax set at right angles to a long handle, was used to break up the soil and then a hilling hoe, which had a flat blade, six to eight inches wide and ten or twelve inches long, was used to prepare the hills. Several seeds were carefully placed on top and the hoe was used to cover them. Rice and indigo were planted not in hills but in rows, yet even with these two crops southern farmers generally prepared the ground with hoes rather than plows. In the North when corn was planted after plow cultivation, hills were made by cross plowing the field and the seed was placed in the rise that was created at the juncture of two furrows. Hoes were used later in the season to cultivate the soil around the hills to control the growth of weeds, work that was, among some ethnic groups, performed by women and children.

If small grains were to be planted in the North—wheat, rye, oats, barley—a different procedure was used. The seed for these grains, as their name implies, is very small, and consequently it was sown broadcast, the farmer casting it aloft with a sweep of his arm, letting it fall wherever it might. To prepare a field for such a method of sowing it had to be harrowed after it was plowed. A harrow is a triangular wooden frame, set with many iron or hardwood tines, which is dragged across the earth (usually by an ox, guided by a man) so as to break up the clods that are created by plowing; a second harrowing, after the sowing, ensured that the seed would be lightly covered by soil.

Agricultural work did not cease after the seed was planted. During the growing season the crops had to be tended in various ways: tobacco had to be pinched, pruned at the very top, when it had grown sufficiently high; worms had to be picked off everything. Soil was hoed regularly to keep down the weeds; weeds were controlled in rice fields by flooding them periodically, just high enough to drown the weeds without damaging the rice.

At harvesting time, some form of sharp blade was utilized to cut down the crops: a scythe (a sharp blade set almost at right angles to a wooden handle) or a sickle (a curved blade set into a wooden handle) or some form of knife. Harvesting was arduous work, for it had to be done quickly—just when the crop was ripe—and it was done entirely by hand. Many hands were required: all those who resided on the farm and extra hands hired just for harvesting. In some New England villages, craftspeople were required by law to lend their labor to their neighbors at harvest time.

The labor did not cease after cutting because the crop could not be left simply lying in the fields, lest it rot. Either some form of upright storage had to be devised in the field (corn could be tied up in bunches, for example) or the crop had to be transported to shelter. Tobacco went to drying sheds, small grains to barns for threshing and winnowing, ears of corn to lofts where they could dry before being husked. Those shelters were as much a part of the technological system of colonial farming as plows and harrows were; tools to wrest resources from the land. The transportation of crops from fields to barns was done with rude wooden carts, often little more than a platform on wheels, drawn by oxen.

Plowing, planting, tending, and harvesting crops was heavy, time-consuming labor, but the hard work did not stop even after the crop was safely home, for none of these products could be consumed by human beings without additional processing. All the small grains grow on a stalk from which they must be separated (threshing and winnowing), and all are covered with a husk that must be either removed or broken before they can be cooked (milling). None of this was easy work. For threshing, the straw (dried stalks with kernels attached) was placed on a barn floor and either hit with a jointed device called a flail or trod by draft animals; this part of the work was most commonly done by men. Then the mixture was placed on a large cloth and women would repeatedly toss it into the air, so that the broken stalks (now called the chaff), being lighter than the kernels, would be blown away; this process, depending on the state of the wind on any given day, might take hours before any given batch was relatively clean of chaff.

Following this, the grain was carefully stuffed into sacks to await milling. Rice could be milled at home since its husk comes off rather easily. A rice mill consisted of two round wooden slabs, one of which was made to revolve atop the other by the operation of a hand crank; the rice was passed between the slabs and the friction removed the husk. (Rice also has a hard cuticle, which must be broken following milling; this was done, more laboriously, by pounding it with a mallet.) Wheat, being harder to mill, was most frequently hauled to a water-powered gristmill, a task that is easy to write about, but much more difficult to do, given the poor (or nonexistent) quality of colonial roads: "I had 14 miles to go in winter to mill with an ox team. . . . No roads were broken and no bridges built across streams. I had to wade two streams and carry the bags on my back. . . . I got only 7 miles the first night, and on the 2nd night I reached the mill"—a not unusual, four-day trip to provide three or four bushels of wheat, perhaps a months' supply.[3] Where no gristmill existed, colonial farmers either dispensed with wheat and rye, or they had to contrive a mill of their own with two heavy, polished stones and at least one horse or ox.

Corn, which was the grain most commonly used in the colonies, both North and South, was just as difficult to process. Some of it could be eaten fresh after harvesting, straight off the cob, but most of it was destined for consumption several months later, which meant that it had to be dried, husked, decobbed, and then ground. Husking was tedious but not strenuous work, usually done by men, who sometimes relieved the boredom of their task by turning it into a social event—a husking bee—replete with music and drink. Once husked, kernels were removed from the cob by the use of a simple slicing device, a sharp blade set, slightly at an angle, into a flat board. Dried, decobbed corn could be cooked up (with dried beans) over a slow fire for many hours, creating succotash, but in order to become bread in any form, it had to be ground, and grinding was not easy. Most colonial farmers preferred when possible to carry their corn to mill for grinding, but when this was impossible, corn could be prepared for cooking at home and by hand.

Gristmills were preferred, no matter how difficult of access, because home preparation, pounding, was not only time-consuming but also backbreaking; in colonial Maryland, fines were instituted to prevent parents from forcing their children to pound corn. With a hand quern (a slab of wood, rotated by hand), two or three hours were required to grind a day's supply; with a spring mill (a hand-guided mallet, attached to a green sapling, which was used as a spring), one day of pounding might supply a family for a week. Frederick Law Olmsted, while traveling in the West in the 1850s, observed a winch-operated device that may have had its origins in the colonial period: "Two boys were immediately set to work by their father at grinding corn. The task seemed their usual one, yet one very much too severe for their strength. Taking hold at opposite sides of the winch they ground away outside the door for more than an hour, constantly stopping to take breath."[4] This task had to be repeated every two days.

Yet the men and women of the American colonies did not live by bread alone. Hoes, plows, harrows, flails, scythes, and mills were necessary for the production of grains; for the two cash crops on which the South became rich—tobacco and indigo; and for the small quantity of garden crops—potatoes, cabbages, beans, pumpkins, herbs, onions—that the colonists also planted. Other tools were required, however, for other forms of farm production.

If a family wanted to keep domestic animals, for example, then it had to provide food for them in the winter. This meant the necessity of planting

In the absence of a gristmill these pioneer farmers had to pound corn with a spring mill. In the absence of a sawmill they built their cabin out of logs, not boards. Note also, in this early nineteenth century drawing, the simplicity of their clothing. (Courtesy Ward Melville Memorial Library, State University of New York at Stony Brook.)

several fields with English grasses (the native American grasses were inadequate winter feed for horses and cattle) and regularly cutting the grass during the summer months (with a sickle) and transporting it (with a cart) for storage (in a barn). James Lemon estimated that on a hypothetical Pennsylvania farm of 125 acres, at least 20 acres might be planted in grasses, each of which provided about 1½ tons of hay every year. This went to feed roughly seven head of cattle (three cows, one steer, three calves), three or four horses, eight pigs, and perhaps ten sheep. On top of this, the farm animals consumed roughly 215 bushels of grain (mostly corn and oats) during the year, or the total produce of 22 of the 30 acres planted in grain. The animals themselves and the food they ate, as well as the tools required to harvest their food and the barns in which that food was stored, were all part of a single technological system, intended, just like plowing and harrowing, to turn land into food, energy, and profit for the use of human beings

Furthermore, horses, oxen, and even cows could not be used as draft animals unless they were harnessed. Steers and pigs could not be turned into roasts, steaks, hams, and sausages without butchering knives—and considerable skill at using them. In addition, the household required barrels for salting away the meat and a hearth over which it could be hung for smoking, else how could the meat be preserved for future consumption? Collecting milk from cows required, at the very least, a pail and a stool; making butter meant a churn; making cheese, at least a strainer and maybe also a press (butter and cheese, when salted, were convenient ways to preserve milk after it left the cow). Even wild game, such as deer, pheasant, pigeons, rabbits, turkeys, and squirrels, could not be acquired without guns or traps, and fish could not be caught without nets or scoops or weirs (underwater traps that could be left unattended).

Along with meat and bread, a family also required something to drink. Apples were grown all over the eastern seaboard, most especially in the North, not so much for eating off the tree or for pies but for hard cider (apple juice that had been left in casks to ferment), which was the universal colonial beverage. But juice could not be made without a press, which was an enormous contraption made of heavy hardwood beams: the apples were placed on a grooved bed, then a heavy slab of wood was lowered onto them by tightening a screw.

According to Lemon's estimates, that hypothetical Pennsylvania farm of thirty acres under plow and twenty devoted to grass also had two acres of orchards, roughly enough to produce 800 bushels of apples and, later, 50 barrels of cider. Those 50 barrels were, however, the product of very hard and tedious work, which began when the apples were collected (imagine picking 800 bushel baskets of apples!). Following this they had to be hauled to the press and dumped into it; the press had to be cleaned of cores and skin after each batch was processed (the cider remains were fed to the pigs). In terms of human labor (and capital investment), cider was not cheap, but then neither were its closest competitors, beer and milk. Even water, were it to be collected for drinking, washing, cooking, or any other household chore, could not be obtained without pails and barrels or, in more elaborate installations, a well, a bucket, and a rope.

In addition to food and shelter, farming families needed clothing. In the colonial period, cloth could not be produced without yet a different set of specialized tools and a different crop. Some of the colonists' clothing was made of leather (moccasins, for example, or vests and leggings), but making leather goods from the hides of your own animals required (in addition to skill) frames for stretching and drying the hide, special knives for cleansing it of fat and hair, special chemicals for tanning and vats in which to tan it, and special needles (called awls) for stitching it together.

Wool and flax were the other two substances most commonly found in the bodices and skirts, the aprons and kerchiefs, the shirts and trousers, the leggings and petticoats, the mantles and capes that the colonists wore—not to speak of the pillow covers and bedsheets on which they slept, the table linen (if they were fortunate) off which they ate, and the rags that they stitched into rugs or with which they diapered their babies. Wool came from one set of domestic animals, sheep, but a self-sufficient farm would have needed more than sheep in order to obtain woolen thread for knitting or woolen cloth for sewing. Special shears (and a good deal of patience) were needed to remove the wool from the animal and special combs (called cards) were required to clean and straighten it. Spinning wheels turned it into thread, and looms wove the thread into cloth.

Linen, which was derived from the fibrous stalk of the flax plant, required yet another set of tools. Flax grew exceptionally well in the colonies; from two acres planted in flax, a farmer might have been able to derive 300 pounds of plants at harvest time, as well as ten bushels of seeds (which, when pressed, yielded another valuable colonial commodity, linseed oil). Flax did not, however, yield up its thread easily; the plant had to be broken apart (which required a special instrument called a brake, a heavy log set into frame in such a way as to crush the stem of the plant when lowered) and then the fibrous inner material had to be combed carefully with a special device called a hackle—all this before any spinning could begin, or any weaving, for that matter.

Spinning and weaving were among the most time-consuming, and tedious, of all occupations. Spinning was women's work, so universally and symbolically related to femininity that our language still contains the colloquial expression "the distaff side" to describe the female partner to a marriage; distaffs were the implements (sometimes nothing more than a split stick) that held the wool or flax fibers before they were spun. After the wool or flax had been suitably prepared for spinning (the chores of washing and carding and hackling were often assigned to children), the women of the household would turn the short fibers into continuous threads by twisting them together; the wheel, which was nothing more than a spindle turned by a crank, aided the twisting process by providing a continuous source of tension on the developing thread.

Weaving, which was sometimes done by men and other times by women, was performed on looms of various sizes, some small enough to be placed on a lap, others large enough to fill an entire room. All colonial looms, no matter what their size, were rectangular frames that held a set of threads (called the warp threads), all running in the same direction. Warping the loom, tying down these threads at just the right tension and spacing, was a meticulous enterprise, requiring considerable dexterity and a great deal of experience. Once warping had been completed, the operator proceeded to weave, which meant that he or she passed a continuous thread over and under, under and over, at right angles to the threads of the warp. Large looms for wider fabrics had warps threaded on two frames, each of which could be alternately raised and lowered by the operation of a foot pedal. The raising and lowering eliminated the need to carefully run the weft (this was the name given to the continuously weaving thread) needle over and under every single warp thread thousands of times, but looms that had such a double frame were exceedingly heavy, which meant that operating them was arduous work nonetheless. Smaller looms were less ponderous but slower; they also produced narrower fabric, which was not suited to such items as bedsheets.

In farm households, the months from December through May were largely devoted to spinning and weaving. We can derive a rough gauge of

Making cloth involved some of the most complex, and expensive, preindustrial
tools—and the labor of all members of a family was involved. In this late
eighteenth century depiction the man is weaving, the woman is spinning, and the
children are using flax stems to weave mats. (Courtesy Ward Melville Memorial
Library, State University of New York at Stony Brook.)

what a household required from the diary of Elizabeth Fuller, a teenage
girl, who in the 1790s recorded having spent the months of January and
February spinning and the months of March, April, and May weaving, with
the resultant production of 176 yards of cloth. She finished on June 1 and
exulted in her diary: "Welcome sweet Liberty."[5] If it was of sufficient
width, 176 yards of cloth was probably all that her family needed for the
year (although it was unlikely to be of the right density for outerclothing),
but the sewing—of shirts and sheets, of petticoats and breeches—was, of
course, yet to come.

We have slowly been developing the list of technological systems required to maintain and sustain a family farm. We need add only one more:
the fuel system. Most colonial farmhouses were not large structures; the
stately colonial homes of such notables as Washington and Jefferson, which
have been preserved as part of our national heritage, bear about as much
resemblance to an ordinary colonial residence as a Model T bears to a Rolls
Royce. An ordinary farmhouse essentially consisted of one room, ofttimes
called the hall. If the ceiling were high enough in the hall, a loft might have
been added or some partitions might have been built to create one or two

new rooms at one end or a lean-to may have been constructed at the back, with a door cut through the rear wall of the house. The hall was the busiest room of the house. Here the women of the family cooked, the family sat down to meals, and some members of the family slept. The loft and side rooms were usually used either for storage (bolts of cloth, bags of corn-meal, ropes of onions) or for sleeping. The lean-to, if there was one, was used for storing milk or cider, also for churning butter.

The most striking feature of the hall (in fact, its only feature) was the fireplace, inserted in one wall of the room, framed in brick or stone, topped by a wooden beam, which served as the mantle, with a separate small en-closure (the baking oven) off to one side. The fire burning on this hearth rarely was allowed to go out: during the cold months of the year, it was the only source of heat for the house; every day of the year, it was the place at which cooked meals were prepared. Its furnishings were few, but each was essential: andirons were used to keep the logs off the floor of the hearth and improve air circulation; bellows, to encourage the fire; a set of tongs or a fork, to manipulate the logs; a shovel, to remove or rearrange ashes. On a swivelling iron bar inserted in the side wall (called a crane) were hung heavy kettles; perhaps another iron bar was set further up in the chimney for hanging meats to be smoked. The cooking equipment was also mod-est: one or two hanging kettles, perhaps a footed pot with a cover, a skil-let or two, and a few wooden spoons of various sizes completed the house-wife's set of tools. Many high school graduates go off to college today with more cooking equipment than an eighteenth-century farmer's wife was likely to see in her lifetime.

Keeping that hearth fire burning was no easy business. Open hearths are notoriously inefficient. Most of the heat goes straight up the chimney and whatever radiates laterally into the room does not radiate very far; anyone who has tried it can report that when it is cold outside you can sit two feet from a fireplace flushed with heat in front and freezing behind. To supply an average colonial farmhouse for a year, more than an acre of woodlot had to be cut down (although farmers sometimes killed their trees by girdling and waited until the weather brought them down), hauled to the farmyard, chopped and split to fit into the fireplace. Kindling for stimulating the fire when it had flagged or for starting a fire in the baking oven also had to be collected and stored. The tools involved were simple—several axes of dif-ferent sizes, a sturdy cart, a good ox, a wedge, and a mallet—but the work was hard, and unending. Everyone in the household participated, but the heavy part—the cutting, hauling, and chopping—was usually reserved for men; roughly one third of a man's time during the year was occupied in fuel-related chores.

Conclusion: The Myth of Self-Sufficiency

Most colonial farming families who had passed through the stage of pio-neering had a very simple standard of living, not impoverished, but hardly luxurious either. Meals were probably nutritious (we guess this was so in part because the birth rate was much higher in the colonies than it was in Europe and people lived considerably longer), but they were very simple and exceedingly monotonous: perhaps porridge in the morning, some cheese and bread at night, and, as the main midday meal, some combination of meat, grain, and vegetables that had cooked all morning. A Swedish traveler in the 1750s reported that farmers in New York "[for breakfast] drank tea . . . and with the tea they ate bread and butter and radishes. . . . At noon they had a regular meal, meat served with turnips and cabbage. In the evening they made a porridge of corn . . . into which they poured fresh milk. . . . This was their supper nearly every evening."[6]

Wardrobes were simple, too, and so were house furnishings. Each member of the family might have one pair of heavy and one pair of lightweight shoes; children whose feet were still growing went barefoot or wore moccasins most of the time. A petticoat, a skirt, a bodice, one dress, two or three aprons and neckerchiefs, knitted stockings, and a heavy cape for cold weather were thought adequate for a grown woman; girls' apparel was made loose fitting so that it could be easily altered as they grew. Boys and men wore stockings, trousers, and very plain shirts, and for them, two changes plus outerwear (also some form of cape) were thought adequate. Underwear was unheard of and we have little record of sleepwear either; people may have slept in their bodices and shirts.

Houses came equipped with a large table for eating, a small number of chairs (children probably ate standing up), perhaps a bench or two, some chests and boxes, some bedsteads (wooden frames, laced with rope), featherbeds (large pillows, stuffed with feathers, which served as mattresses), some sheets and quilts—and little else. Especially comfortable households might have had a bed for each member; otherwise, people slept on the floor, as near as possible to the hearth. For eating utensils, they had wooden bowls (trenchers) and wooden spoons; as things improved, families added some pewter mugs and platters.

Had the average farming family, engaged either in mixed farming or in subsistence farming, tried to maintain even this relatively simple standard of living entirely with its own labor, a large stock of tools would have been needed: cider presses and plows, looms and harrows, harnesses and spinning wheels, knives and candle molds. As simple as these tools appear to us today, they were expensive in their day. Complex wooden tools, such as presses and double looms, had to be built by skilled craftsmen. Metal parts not only had to be created by skilled workmen but also had to be transported great distances (at considerable expense) since few colonial farmers were located within easy distance of either an iron mine or a foundry. Rough estimates, derived from probate inventories (careful lists of possessions and their value that were, and still are, made at the time of someone's death) suggest that, in the middle of the eighteenth century, a self-sufficient farm, large enough to support a family of seven (two adults and five children), would have cost about £150, or roughly fifteen times the annual wages of an indentured servant—far beyond the means of most farming families.

On top of the expense, self-sufficiency would have required enormous amounts of time, many hands, and extraordinary levels of skill. The simplest job, the description of which might take no more than two sentences in this text, could take hours, days, weeks or even months to perform. If a housewife made everything from scratch, she had to combine the skills of more than a dozen crafts: flax and wool spinster, weaver, dyer, fuller, tailor, seamstress, knitter, baker, gardener, brewer, dairymaid, chandler (someone who makes candles), and soapmaker—all this on top of cooking, tending the sick, caring for infants and small children, winnowing, carrying water, and, occasionally, laundering. Her husband likewise had to be competent at plowing, reaping, carpentry, wood carving, leatherworking, butchery, stonemasonry, woodchopping, carting, milling, brewing, not to speak of a bit of blacksmithing and shoemaking. Had a young person been set to an apprenticeship at any one of these crafts, he or she would have been expected to take five years, at the very least, for adequate training.

Yankee farmers and their wives were famous for their versatility, but no one could have performed competently all these tasks at one time; the quality of farm handwork was not high. Worse still, lack of skill in any of these endeavors might occasionally be life threatening: a cart, poorly loaded, could dump its sacks of grain into a river and a family's food supply for a month might float downstream; a fire, improperly banked, could spark a conflagration; some herbal dosages administered in the wrong amounts could kill. In the face of all this—hard work, great expense, the need to acquire complex skills, and the need to face not inconsiderable risks—why were so many American colonists enamored of self-sufficient farming and the standard of living that went with it? The answer is, they weren't.

Self-sufficiency was never the dominant economic model in the southern colonies. Having found three exceedingly profitable cash crops—tobacco, rice, and indigo—plantation owners in the South preferred to import the commodities that their households needed for sustenance. Cloth and hats, shoes and furniture, candles and soap, salted beef and casks of flour all arrived at the wharves of southern plantations, located, as so many of them were, at tide's or river's edge.

Probate inventories indicate that the vast majority of small farms in the New England and Mid-Atlantic colonies simply did not possess enough tools to be self-sufficient. An inventory might reveal, for example, that a farmer owned four cows, but nary a pail for carrying milk nor a churn for making butter nor a strainer for pressing cheese. Another might show that at the time of her death a widow had seventy batches of linen thread in her loft but no loom. Only 27 percent of all the inventories made in Suffolk County, Massachusetts, in 1774 included spinning wheels, and only 2 percent of the households opened up for the court's inspection had wheels, looms, sheep, and flax listed together. In the Massachusetts countryside in that year, 71.4 percent of all the inventories listed cows and milkpails, but only 44 percent had cheese and butter equipment. Even more startling, despite the widespread use of beer and cider in Massachusetts, only 5.6 percent of the inventories included either presses or barrels or anything else needed for home brewing. In Somerset County, Maryland, where there were many small farms, only about half the households inventoried in the 1720s had looms, and only 10 percent had cooking equipment beyond the rudimentary pots, kettles, and skillets.

Colonial farmers specialized. Some owned plows but hired neighbors to do their harrowing. Others grew flax but did not chop their own wood. A farmer who was a reasonably good carpenter might make a chair for another farmhouse and then trade it for the pig that he had not had the time to raise himself. A housewife might keep chickens and trade the eggs with her counterpart three miles away in exchange for the butter that she had neither the equipment nor the time to churn. Some people fattened pigs and steers for eating but sold the meat on the hoof, only to buy it back again salted away in barrels. Some people raised sheep but traded the wool to others who had wheels and looms, perhaps in exchange for cider or cheese. The market economy flourished in the colonial countryside even where there were no merchants or even storekeepers to manage it.

Supplementing this barter economy were the itinerant craftsmen who plied colonial roads, supplying skills and tools that farm households needed. There were, for example, roving shoemakers who would stay for a week and turn the family's hides into boots; neighborhood seamstresses who would come in to do the sewing; midwives who supervised childbirth and cared for the mother and newborn child; knife sharpeners who carried their flintstones on their backs; plowmen who assisted when schedules were tight; stonemasons who mended chimneys or built new ones. Farm men and women may have been jacks-and-jills-of-all-trades, but they clearly preferred, whenever such preferences could be exercised, to have specialists take over some of their work.

Not surprisingly, the wealthier a household was, the more tools it contained and the larger the number of technological systems in which it was embedded. In addition, the wealthier a household was, the larger the number of waged servants, indentured servants, and slaves that appear on the inventory (colonial courts were very thorough; slaves were part of an estate, and wages due a servant were part of its debts). Put another way, this means that complete self-sufficiency was possible only for those who were rich enough to own a great deal of land and many tools; they also had to be rich enough to control the labor of a great many people, whether those people were family members, tenants, employees, indentured servants, or slaves. Only the very wealthy, such as Jefferson and Washington, could afford to have their own gristmills, built with their own funds, located on their own lands, operated by their own millers. Everyone else had to pay (in cash or in kind) for the services of a miller, which means that everyone else was dependent in some sense on the market economy—even for so basic a commodity as flour.

Economic independence—complete household self-sufficiency—was thus the province of the very rich. Thomas Jefferson deeply believed in the virtues of self-sufficient farming; as a framer of the Constitution, he tried to ensure not only that the interests of the small farmer would be respected but also that the culture of self-sufficiency would become embedded in American political processes. Unfortunately no political contrivance—representative legislatures, strong state governments, or restriction of the franchise to landowners—could ever have produced the ideal society that Jefferson sought: a nation of truly independent small farmers. The technological systems of the day simply were not up to the challenge; in the world of hoes and harrows, looms and spinning wheels, economic independence for individuals was simply out of the question. With all the best intentions, even with horses and oxen to help, a colonial family simply could not have produced all that it needed for its own sustenance; the tools at its command were not sufficiently productive. Self-sufficiency was a myth—even in its own day.

And it became even more mythical as time wore on, increasing in potency as the process of industrialization carried us further from the world in which the idea had been conceived. Even Jefferson's contemporaries seem to have been aware of the contradiction between what he said and what he did. Many wished to be spared, as Jefferson was spared, the long hours and back-breaking labor of farming. Many wished to have, as Jefferson had, more comfortable clothing than that which could be spun, woven, and stitched at home. Many hungered after the varied diet that became possible when there was cash to pay for coffee from Brazil or sugar from the Caribbean or wine from Madeira. Those wishes, for what we would now call a higher standard of living and a more comfortable style of life, stimulated the most profound economic and technological transformation of all time: the industrial revolution. Self-sufficiency is attractive to those of us who are nostalgic for what we think were simpler, preindustrial times, but it didn't seem particularly attractive to most of the people who lived in those times. They understood precisely how much hard work, and how much discomfort, was really required.

Notes

1. Jeremy Belknap, *The History of New Hampshire*, Vol. III (Boston, 1792), p. 258, as quoted in Percy Wells Bidwell and John I. Falconer, *History of Agriculture in the Northern United States, 1620–1860* (Washington, DC: Carnegie Institution, 1925), pp. 81–82.

2. Richard J. Hooker, ed., *The Carolina Backcountry on the Eve of the Revolution: The Journal and Other Writings of Charles Woodmason, Anglican Itinerant* (Chapel Hill: University of North Carolina Press, 1953), p. 39.

3. W. J. McKnight, *History of Northwestern Pennsylvania* (Philadelphia, 1905), p. 596, as quoted in Louis C. Hunter, *A History of Industrial Power in the*

United States, 1780–1930 (Charlottesville: University Press of Virginia, 1979), p. 13.

4. Frederick Law Olmsted, *Journey through Texas: A Saddle Trip on the Southwestern Frontier* [1857], as quoted in Hunter, *History of Industrial Power*, p. 12.

5. Diary of Elizabeth Fuller, as quoted in Mary Beth Norton, *Liberty's Daughters: The Revolutionary Experience of American Women, 1750–1800* (Boston: Little, Brown, 1980), p. 17.

6. Peter Kalm, *Travels in North America* [1753–61], ed. Adolph B. Benson, Vol. I (New York: Dover, 1966), p. 602.

Suggestions for Further Reading

Percy Wells Bidwell and John I. Falconer, *History of Agriculture in the Northern United States, 1620–1860* (Washington, DC, 1925).

Lois Green Carr and Lorena S. Walsh, "The Planter's Wife: The Experience of White Women in Seventeenth-Century Maryland," *William and Mary Quarterly, 24* (1977): 542–71.

John Demos, *A Little Commonwealth: Family Life in Plymouth Colony* (New York, 1970).

Lewis Cecil Gray, *History of Agriculture in the Southern United States to 1860* (Washington, DC, 1932).

James T. Lemon, *The Best Poor Man's Country: A Geographical Study of Early Southeastern Pennsylvania* (Baltimore, 1972).

Gloria Main, *Tobacco Colony: Life in Early Maryland, 1650–1720* (Princeton, 1982).

Judith McGaw, ed., *Early American Technology: Making and Doing Things from the Colonial Era to 1850* (Chapel Hill, NC, 1994).

Michael Merrill, "Cash Is Good to Eat: Self-Sufficiency and Exchange in the Rural Economy of the United States," *Radical History Review, 3* (1977): 42–71.

Darrett B. Rutman, *Husbandmen of Plymouth: Farms and Villages in the Old Colony, 1620–1629* (Boston, 1967).

Carole Shammas, "How Self-Sufficient Was Early America?" *Journal of Interdisciplinary History, 13, no. 2* (Autumn 1982): 247–72.

Laurel Thatcher Ulrich, *Good Wives: Image and Reality in the Lives of Women in Northern New England, 1650–1750* (New York, 1980).

3

Colonial Artisans

BY THE MIDDLE OF THE EIGHTEENTH CENTURY, roughly one out of every ten residents of the colonies was an artisan, a person trained in one or the other of the crafts: blacksmiths and printers, cordwainers (people who made shoes) and seamstresses, coopers (people who made barrels) and shipwrights, midwives and tanners, joiners (people who made furniture) and millwrights (people who constructed water mills), apothecaries and iron smelters. Although small in numbers, these artisans were big in impact. Many artisans manufactured the tools that farmers needed to work the land; they also provided the services and skills that were essential for the growth of cities. The political activities of artisans helped to foment the Revolution; the economic activities of artisans laid the groundwork for industrialization.

Some artisans were rural, living and working in the small villages that dotted the countryside from Maine to Georgia. Many of these rural craftspeople were indistinguishable from the farmers who were their neighbors and on whose business they depended; many owned plots of land and grew some of the crops that they needed for their own sustenance, acquiring the rest in trade for their services: a barrel of salted pork in exchange for mending a clock, five pounds of butter in trade for two yards of cloth. Rural artisans frequently practiced several trades at once, specializing in none: one Long Island craftsman, for example, advertised himself as a wheelwright, clockmaker, carpenter, cabinetmaker, toolmaker, and a repairer of spinning and weaving equipment as well as guns—all the while collecting fees for the pasturing of other people's cows. These jacks-of-all-trades were versatile but not innovative; they supplied their customers with sturdy, durable pieces of familiar equipment.

Artisans who lived in cities were, however, in a different situation. The vast majority of colonial craftspeople were city dwellers and the vast majority of city dwellers were craftspeople and their families: silversmiths and pewterers, candlemakers and bakers, printers and carpenters, masons and coopers. Colonial cities were transportation hubs, ports from which ships plied the coastlines and the oceans, the ultimate end points for what few major roads there were. During the colonial period transportation was very expensive; artisans needed to be located in cities so that they could minimize their transportation costs, both on the items they needed to buy (such as iron and wood or wool and hides) and on the items they intended to sell (such as nails and guns or carpets and shoes).

Colonial cities were also demographic hubs: centers of population, constantly expanding, concentrations of people who lived more or less within walking distance of each other and who had cash with which to purchase goods and services. In 1690, the total population of the five leading towns of the colonies was only about 19,000 people, but by 1776, less than a hundred years later, this had multiplied to 108,000—a spectacular increase, indicating not only an expansion of the crafts but also an expansion of the market for the products of the crafts. Artisans congregated in Boston and New York, Newport and Philadelphia, Charleston and Baltimore—competing with each other, learning from each other, joining together in organizations intended to advance their common interests.

In colonial cities, artisans could specialize so as to become perfectionists in just one branch of a trade. They could also subdivide their work by hiring several assistants so as to become more productive. The silver and pewter bowls, the dining tables and armchairs, the embroidered fabrics and building facades that are preserved in museums as examples of colonial creativity and elegant design are all products of America's early urban artisans.

The Apprenticeship System and Labor Scarcity

There were only two routes by which a person could enter the crafts in the colonies: emigration and apprenticeship. Artisans were relatively well paid in Europe, which meant that, except for those who had suffered religious persecution, they had no particular incentive to brave the fearsome oceans. In the seventeenth century, those few artisans who did emigrate were almost instantaneously converted into farmers as soon as they disembarked; the pressure for survival in a wilderness was such that every man, woman, and child had to turn his or her energy toward wresting food and shelter from the land. Among the early founders of Plymouth Bay, there were, for example, three carpenters, two tanners, three tailors, and a blacksmith, but on arrival, they transmuted into farmers, one and all—a decade later, they still were not practicing the crafts in which they had been trained.

Yet the colonies desperately needed artisans, people who could work the abundant deposits of iron, people who could operate sawmills, people who could compose type and publish newspapers. As a consequence, communities often offered special inducements for artisans to emigrate. In 1656, for example, the town of Scituate, Massachusetts, offered a free mill site (in Europe these would have been very expensive pieces of property), as well as free use of lumber on the common lands of the town, to any individual who would build and maintain a sawmill there. Seventy-five years later and many miles away, the government of South Carolina offered £175 to any individual who would set up a printshop and publish a newspaper in Charleston. Over the years, thousands of people responded to these and similar inducements, bringing German ironworkers to Pennsylvania and New Jersey, Irish weavers to Massachusetts, Danish and Swedish sawyers to Delaware and New York, French seamstresses to Virginia. Despite the inducements, the supply of emigrant artisans was never equal to the demand for colonial skilled labor; European artisans, by and large, preferred to remain at home.

As a result, by the end of the eighteenth century, most colonial artisans were native-born men and women who had learned their trades by serving an apprenticeship. An apprentice was a young girl or boy being trained to practice a skilled craft; sometimes, especially with young girls, an apprentice might just be called a servant. Informal apprenticeships occurred when the son or daughter of an artisan was taught, so to speak at the parent's knee or when a young child, often a girl, was sent to work for wages in someone else's house. Formal apprenticeships depended on a contract between the parents and the master, the articles of indenture. According to the terms of an apprenticeship contract, the parents gave up their rights to the child's labor; the child in question may have been as young as six or as old as thirteen at the time. In exchange, the master promised to provide food, shelter, clothing, religious instruction, and, of course, training in a craft. In effect, the master received seven to ten years of very inexpensive labor from the child, while the child presumably received seven to ten years of instruction from the master.

The apprenticeship system had developed in western Europe during the Middle Ages. By the time that Europeans began settling in America, it was an ingrained part of their culture; all the residents of the colonies understood how it worked and accepted its basic premises. One of those basic premises was that the child was going to have to work, and work hard: ten or twelve hour days, six days a week. Younger apprentices were not expected to manage particularly heavy jobs (such as raising and lowering printing presses), but they were expected to handle everything that was routine and monotonous in the shop and to act as if it were a privilege to be allowed to do so.

Which is precisely what apprenticeships had been in Europe for most of the seventeenth and eighteenth centuries—a privilege. In Europe, where prime agricultural land was scarce, there was always an oversupply of labor on the land: more people than could be adequately fed, more children than most households could support. Landowning or land-working parents sought apprenticeships for their younger sons and daughters as a way of keeping them out of poverty: the sons who would not be able to inherit the land, the daughters who could not be provided with an adequate dowry. If these children had stayed with their parents, they would have been expected to work hard, so it came as no surprise to anyone that as apprentices they also worked hard. If they had stayed with their parents, they might also have starved or been poorly nourished or become transient laborers working for a pittance. As apprentices or servants, their futures looked somewhat brighter, and in gratitude for this change in their fortunes, they were expected to respond willingly to the demands of their masters and mistresses.

In Europe ever since the Middle Ages, the formal apprenticeship system had been regulated by guilds, voluntary but formal associations of master craftspeople. The guilds performed many functions: setting prices for certain goods and services, setting standards for the quality of products, representing the interests of their members in royal courts and parliamentary proceedings, providing insurance pools for their members, arranging special feasts and religious services on holidays. Regulation of the apprenticeship system was one of the guild's most important functions. Master craftsmen liked to take on apprentices because they were a source of exceedingly cheap labor. Guilds acted to prevent an oversupply of skilled artisans by limiting the numbers of apprentices that any member could indenture and by setting fixed periods for the term of apprenticeship.

A European guild was something of a cross between our modern labor unions and our modern trade associations. Artisanal shops were small businesses. Everyone who worked in them (the masters, the apprentices, even the wage laborers, who were called journeymen if they had finished serving their apprenticeships) was supervised and regulated by the guild, although the guild's policies were developed by, and in the interests of, the masters. By the fifteenth and sixteenth centuries, guilds had become very wealthy institutions, and their policies had an enormous impact on the growth of European cities. Guildhalls, the places where guild meetings were held, were imposing structures, many of which still stand—reminders of a world long lost.

In the American colonies, everything was different. Colonial artisans tried to re-create the guild and apprenticeship systems, but with little success. European guilds could not extend their authority across the ocean, and there weren't enough masters in any craft in any colony to underwrite the expenses that organization of a new guild would require. To make matters worse, colonial parents had no particular reason to want to set their children to apprenticeships.

The land itself was the source of the problem. In Europe, there was usually a buyer's market for apprentices because there was a paucity of land. In the colonies, this was reversed: the market for apprentices was a seller's market because land was plentiful and cheap. Many parents owned parcels large enough to divide among their children and their children's children for several generations. Other parents knew that, when their children reached maturity, just a small inheritance, or even a small gift, would purchase a suitable parcel of land—just a little bit farther west. Under such circumstances, parents had no incentive to set their children to apprenticeships, and those few young people who were indentured to colonial masters rarely served their full term.

Apprentices were important in the organization of work in a craft shop because they provided extremely cheap but extremely necessary manual labor. Colonial master craftsmen tried all kinds of devices for solving their endemic problem of labor scarcity: some tried to import European journeymen (people who had finished an apprenticeship but had not yet set themselves up as masters); some tried to indenture young people who needed to pay off the cost of their passage; others tried using slave labor. But the labor shortage did not improve. Imported journeymen were likely to notice the attractions of cheap land themselves, and failing that, they were likely to notice that, where there is a labor shortage, higher wages can easily be demanded. Indentured servants had a penchant for running away before completing their terms of service; they, too, were attracted to cheap land on the frontier, and they learned after a while that the arm of the colonial law was not long enough to reach them there. Even slaves were not a perfect solution since a slave who had learned a trade was expensive, and a slave who had not learned a trade had precious little reason to become adept at one.

The net result was that throughout the entire colonial period, from the first decades of the seventeenth century to the last decades of the eighteenth, there were never enough apprentices and journeymen to keep colonial masters happy—and never enough masters to supply the manufactured goods that an expanding population needed. "They have their Cloathing of all sorts from England," the Virginian Robert Beverley complained of his fellow colonists, "as Linnen, Woolens, Silks, Hats and Leather. Yet Fleece and Hemp grow nowhere in the World, better than there; their Sheep yield a mighty Increase and bear good Fleece, but they shear them only to cool them. Tho' their country be overrun with Wood, yet they have and bear all their Wooden-Ware from England; their Cabinets, Chairs, Tables, Stools, Chests, Boxes and Birchen Brooms, even so much as their Bowls, to the eternal reproach of their Laziness."[1]

But laziness was not the problem; labor economics was. Where land was cheap, craft labor was bound to be scarce; and where craft labor was scarce, the crafts were slow to develop. However many artisans there were in the colonies, the number was never great enough to keep the colonists from importing more than half of the manufactured goods that they needed; never enough to keep communities from having to advertise and offer inducements to artisans whose skills they required; and never enough to support a network of guilds that might have advanced the interests of their members.

Despite their scarcity, American artisans were the people most responsible for technical change during the colonial period, and it was they, and their descendants, who eventually put the new nation on the road to industrialization. Hence we need to know something more about them and about the conditions under which they worked. To this end, for the remainder of this chapter we will explore three crafts—printing, milling, and iron smelting—that were particularly critical to the growth of the colonial economy and particularly illustrative of different kinds of colonial technological systems and work processes.

Printshops and Printers

Every colony needed at least one printer, probably more. Printers published newspapers; they broadcasted the acts of colonial legislatures; they created the legal and business forms that people needed; they recorded minister's sermons for posterity. The potential for profit as a printer was high, but establishing a printing business in the colonies was not an easy matter. Printing required the construction and/or importation of some fairly complex, fairly expensive pieces of equipment; it also required the services of people who were literate.

Few master printers emigrated. The first press to operate in the colonies was operated by a locksmith. The Reverend Jose Glover, a representative of the Corporation for Propagating the Gospel in New England, purchased a press and its associated equipment and embarked for Massachusetts Bay in July 1638; he must have intended to use the press to publish missionary pamphlets in the Indian languages. Unfortunately, Reverend Glover died aboard ship. The press became part of his wife's property, and she fulfilled the contractual arrangements that her husband had made. She provided a house for Stephen Daye, a locksmith, who had emigrated with the Glovers and who had agreed to operate the press. The first publications to emanate from this press, located in Cambridge, Massachusetts, were a copy of "The Freeman's Oath," a form needed by colonial officials, and an almanac (for 1639). Virginia did not get its first press until 1682, almost sixty years after the founding of the colony; New York lacked one until 1693, over seventy years after the Dutch first settled there; even Massachusetts did not get a second press until 1674, close to forty years after Mrs. Glover and Mr. Daye began their operation.

By the middle of the eighteenth century, the various obstacles had been overcome and print shops abounded. Almost all eighteenth-century print shops were urban, and almost all were owned by the craftsmen who operated them. Some master printers had silent partners who were not trained in the craft, but who had put up capital either to buy new equipment or to pay off old debts. Some print shops were owned and operated by the wife or the daughter of a deceased master printer. In most cases, however, there was no distinction between the person who owned the establishment and

In this drawing of a printer's shop, from Edward Hazen, *A Panorama of Professions and Trades* (1837), the man on the left is composing: choosing type and placing it in a composing stick. The man on the right, with the help of an apprentice, is sliding a galley full of type onto the bed of the screw press, prior to printing. (Courtesy National Museum of American History.)

the person who worked there; virtually all owners were workers, and all members of the owner's family, male and female, were expected to participate in some way in the business.

An ordinary colonial printing shop contained one or two presses: huge, heavy contraptions, made of four-by-four beams, designed to exert pressure by means of a lever. Until the middle of the eighteenth century, presses had to be imported from England, as no colonial carpenter was capable of producing such a complex device. Close by the presses were wooden boxes filled with type, small pieces of metal, each with the raised form of a letter on it. Composing—slipping pieces of type into rectangular wooden forms so as to create words and sentences as prescribed by a manuscript—was relatively light work, which young boys and girls could easily learn to do if they were literate. Once the forms were filled with type, they were placed on a table and locked into a frame called a galley, which was then carried (this was very heavy labor) to the bed of the press. The galley was then inked by being beaten with two soaked leather balls (work for an apprentice) and a piece of paper was placed on top of it. An impression was made on the paper by lowering a very heavy wooden block onto the galley, and then moving it across the galley on a carriage. When the press was raised, the paper could be pulled; then the galley was inked again and the whole process was repeated.

Only a master printer or a journeyman was able to operate the press, the hardest and the most highly paid work in the shop. Two good pressmen working together and, alternating stints, could print 240 sheets of paper in an hour, or 2,400 in an average day. The work was difficult because considerable upper-body strength was required as well as considerable skill in appropriately adjusting various parts of the press; printers were known for their special gait, a result of muscular overdevelopment on the right side of their bodies, the side they used when applying weight to the lever. In printing, terms of apprenticeship tended to last for eight or nine years; Benjamin Franklin began serving when he was nine; John Peter Zenger when he was ten. A young apprentice printer began as a go-fer: responsible for sweeping out the shop, making the leather balls (by soaking goat's skin in urine), stirring the vats in which ink was made (by boiling lamblack in oil), hanging newly printed sheets to dry, or just generally running errands. He might then graduate to composing type and cleaning galleys when they were ready to be broken up. Only when he was sufficiently grown might the apprentice be put to work on the press itself; some apprentices never went this far, either because their masters decided not to let them or because they never developed sufficient strength.

Huge quantities of type had to be readily at hand for composing every galley: an average printshop required about 1,200 pounds of type, or roughly 1,200 individual pieces. Sets of type, called fonts, were very valuable. Each piece was used and reused until it was too worn to make a good impression, and then the metal was melted down and recast. Type was not made by the printer himself, but rather by an especially skilled craftsman, called a type founder. Every font was difficult and time-consuming to prepare, as it required the creation in metal of a master form for each letter, a reverse form for each letter, the smelting of special alloys for the type, and hours and hours of metal filing. Type was always in short supply in the colonies; until 1768, when Abel Buell of Killingworth, Connecticut, entered into the business of type founding, every font had to be imported from England.

In addition to presses, galleys, tables, frames, and type, a printer needed ink and paper. In the eighteenth century, ink was made from lampblack, the almost pure carbon residue that results from the incomplete combustion of wood or candlewax, dissolved in linseed oil. Colonial printers did not make their own linseed oil or their own lampblack, but several were owners or partial owners of lampblack houses, and many mixed their own inks by combining the two raw materials.

Colonial paper was made from rags. In the early years, almost all paper had to be imported from England, but paper mills sprang up in various places in the colonies during the eighteenth century. Not suprisingly, several printers, Benjamin Franklin among them, had part ownerships in those mills. In a paper mill, linen rags were soaked in water and beaten (often by a trip-hammer operating off a waterwheel) until they formed a pulp; this pulp was then spread out very thinly on rectangular frames, the bottom of which were closely set with wires, rather like a sieve. When the pulp was partially dry, it was turned off the frame onto a felt mat; piles of interlaced felt and semidry paper would then be placed into a press to squeeze out the remaining water. Paper was not cheap, but printers had to have considerable supplies on hand in order to ply their trade effectively.

All in all, a colonial printer could not set himself up in business on even a modest scale without an investment of something on the order of £75 to £150. The printing house of Franklin & Hall in Philadelphia, one of the most impressive in the colonies, was valued at £184 in 1766. Had sums of this dimension been invested in the land, colonial printers would have been very comfortable farmers, not among the richest of their contemporaries but far from the poorest. In this they were akin to most other urban craft masters. Some (such as the silversmiths and the carpet weavers) may have required a greater expense to set themselves up in business; others (such as the carpenters and the tailors) required considerably less. All, if they were masters or journeymen, were persons of middling economic status, neither very rich nor grindingly poor. Benjamin Franklin paid his journeymen 6 pence (a pence was one twelfth of a shilling, which was one twentieth of a pound) for every thousand letters that they composed and 12 pence for every token (240 pages) that they printed at the press. This comes out to roughly four times the daily wages of an unskilled laborer.

The printshop, like all other artisanal shops in the colonial period—and, indeed, like colonial farms as well—was as much a living space as it was a working space. The weaver's shed was a lean-to attached to his cottage; the fire in which the silversmith heated his ingots was the same one in which his wife cooked porridge; the printer's family usually slept in a room adjacent to the press. Business roles and domestic roles were interchangeable: being the wife of an artisan meant feeding the apprentices as well as your family; training your children to your trade was considered part of being a parent; working very hard in the family business was what children were expected to do. Many of the people who worked in these shops but were not, by blood, members of the family were treated as if they were members of the family anyway: apprentices were fed and clothed by the master's wife; indentured servants slept in quarters provided by their employers; at midday everyone in the shop ate together; not infrequently they also went to church together.

Such close relations between employers and employees were not always mutually satisfactory; like all close relations, even today, they had their ups and their downs. Sometimes a young apprentice was able to report to his parents, as Elisha Waldo wrote to his father in 1785, a few weeks after beginning an apprenticeship to the Massachusetts printer Isaiah Thomas, that he liked his trade "exceedingly well" and that his master and mistress were "very kind."[2] Other times the results were not so positive. Benjamin Franklin's master (who was his older brother) beat and mistreated him so regularly that he ran away (fleeing from Boston to Philadelphia) before the legal end of his apprenticeship. John Fitch, who was apprenticed to a clock maker in 1762, recalled with more than a little bitterness that on one occasion his master's wife boiled up a mutton and bean broth, but "when it came to be about one week old I began to grow tired of eating it constant twice a day and frequantly three times and began to complain. To which she found an immediate remedy by adding water." Fitch's apprenticeship was so unsatisfactory that he pleaded with one of his brothers to buy out the remainder of his indenture, and his pleas finally succeeded a few months before his twenty-first birthday. As he traveled home, contemplating his future—his master had steadfastly refused to teach him how to make clocks—he "cryed the whole distance."[3]

A colonial printshop is an example of a preindustrial urban workplace. Like printers, most urban artisans were literate; the terms of their apprenticeships had required either that they attend school or that the master (or his wife) see to their instruction. Like printers, most urban artisans were also businessmen or businessmen in training; they worried about setting fees, about competition, about tariffs and taxes. Some crafts were more respectable than others; printing and silversmithing were at the top of the craft hierarchy (masters in these trades charged fees to parents who wanted to set their children to apprenticeships) and tailoring, shoemaking, and candlemaking were at the bottom (as they could be undertaken with a minimal investment, their ranks were filled, which meant that they were not as profitable), but the social status of the urban artisan was easily the equivalent of the rural small farmer.

Silversmiths, milliners, carpenters, and masons produced items very different from the newspapers and business forms that printers made, but the organization of work in their shops was very similar. All were based on the artisanal or craft (apprentice-master-household) system of production that had originated in European cities. In American cities, however, this system lacked two of its crucial European features: a large supply of potential laborers and a powerful, supervisory social institution (the guild). As a consequence, as we shall see in a later chapter, American industrialization was very different from its European and English counterparts.

Mills, Millwrights, and Millers

A printing press was, for its day, an impressive piece of equipment, but it paled by comparison to the waterwheel. By the middle of the eighteenth century, virtually every city and town contained at least one water-powered mill and usually more: a gristmill that ground grain for the rural population; a merchant mill that ground grain for overseas shipment; a sawmill for turning logs into planks; a carding mill for untangling newly shorn wool; a fulling mill for pounding wet cloth so that it would shrink; a tanning mill for grinding the oakbark powder that made leather goods supple.

Every resident of the colonies was probably familiar with these huge, noisy contraptions, the most complex technological systems of their day. Some colonial mills were powered by the wind and some by the tides, but by far the majority—as suited the geographical character of the eastern seaboard—were powered by water that flowed through or fell down on a wheel. Located on the banks of a stream toward the bottom of a downward slope, the noise of an operating watermill could be heard for great distances as its cumbersome wooden parts creaked and groaned. Few residents of the colonies were likely to observe a printing press in operation, but virtually all were familiar with the impressive insides of a mill.

The most primitive kind of mill was a tub wheel: rather like a turbine, this was akin to a thick wagon wheel lying horizontally in a stream, its spokes replaced by broad wooden paddles. Tub mills tended to be small (wheel diameters ranged between three and six feet) and were fairly easy to build since they consisted of little more than a trough (for directing the flow of water toward the paddles), the wheel assembly, and a hut to protect the assembly from the elements. Sometimes called democratic or egalitarian, tub mills sufficed for small rural communities or for individual family farms. Tub wheels were used to power rural gristmills (the vertical shaft of such a wheel could be easily attached to grindstones, which themselves turned horizontally), the ubiquitous mills in which the grain eaten by farm families was ground.

All other waterwheels used during the colonial period were placed vertically in the water, with horizontal shafts. These installations (the ones that have survived in many historic restorations today) were considerably more complex than the tub wheel. A vertical waterwheel required a dam, a barrier erected across the stream with its ends anchored in the banks on either side. A dam could increase the height of fall (since the water traveled over its top edge instead of through the streambed); it could also create a millpond (a reservoir of water that could be drawn on in times of drought) or could allow the miller to control the fall of water (so that springtime floods might not wash away his capital investment). Frequently a vertical waterwheel also had a millrace, an artificial watercourse made from hollowed out tree trunks that conducted the water from the pond to the wheel—and provided the miller with even more opportunities for control: records exist of millraces more than a mile in length.

Vertical wheels can be divided into three types: undershot wheels (most of the wheel was above water level and the racing stream hit only the submerged paddles), overshot wheels (the millrace ended in a penstock, which dropped water onto the topmost paddles), and breast wheels (half submerged, with the lower half enclosed in a wooden "breast" and the water applied midway along the wheels circumference). When properly built, an undershot wheel could utilize between 15 and 30 percent of the energy in falling water, an overshot wheel, between 50 and 70 percent. Both types of wheel had capacities of somewhere between one and ten horsepower. Breast wheels, the most efficient but also the most difficult to build, were rarely used during the colonial period. They became popular in the decades just after the Revolution as the waterwheels used to operate the new factories (also often called mills) in which thread was being spun and textiles woven.

Millwrights, the artisans who specialized in mill construction, were probably the most mechanically skilled craftsmen of their day. Trained through apprenticeship, lacking what we would refer to as formal education, pursuing a craft that made him itinerant for a good part of each year, the millwright was responsible for deciding which type of wheel was best suited to each locale (overshot wheels, for example, were better suited to places in which the water supply was low for part of the year) and each job (sawmilling, for example, required a different number of revolutions per minute than gristmilling) also for deciding what size the wheel would be, how the paddles or buckets would be shaped, how best the dam and millraces might be constructed, and how best to gear the waterwheel into the other parts of the operating machinery. Having made all those decisions, he was then responsible for overseeing the construction of the mill; a good millwright combined the skills of the carpenter, the joiner, the mason, the stonecutter, the blacksmith, the wheelwright, and the surveyor.

Many millers owned their mills as well as the mill seat, the piece of land on which the mill was located plus rights to the use of the watercourse in both directions. Like the millwright, a miller was possessed of many skills: he had to know how to pursue his particular trade, how to maintain his equipment once it was constructed, and how in general to conduct a complex business enterprise. Maintaining the equipment was not an insignificant matter and may have taken up as much of the miller's time as the actual work from which his name derived. Mill machinery—wheels, gears, and shafts—was made almost entirely out of wood, huge pieces of it; shafts were commonly two feet and vertical wheels eighteen to twenty feet in diameter. Under the best operating conditions, the joints holding such pieces of wood together were bound to come loose, and the wet and damp atmosphere of a watermill was hardly the best operating condition. In grinding mills, the stones (which might weigh more than a ton a piece) needed regular dressing, which was akin to resharpening; in sawmills, the iron or steel blades required constant attention. Fire was a constant worry (sawdust and flour dust are both highly flammable), and some reconstruction was almost always required each year after the spring floods had wreaked their havoc. As businessmen, millers had to set prices, or tolls, on their services, make decisions about extending credit to their customers, manage the labor of numerous other persons, and represent themselves before legal bodies of various kinds.

Millwrights and millers were, as one can easily imagine, especially valued as colonists. Europeans, whose culture had depended on such mills for four or five centuries, had some not unreasonable doubts about whether survival would be possible without them; mills were constructed in many communities before there were churches, schools, or courthouses. All kinds of special inducements were created to interest millers in emigration: free mill seats or the right to condemn land upstream and down was offered, special town funds were allocated to underwrite the costs of mill construction, or free labor might be offered to help the miller construct his equipment. Each new community was eager to avail itself of a miller's services; the deficiency of a mill was regarded, as one Vermonter put it, as "inconsistent with the existence of civilized life."[4]

A gristmill freed the pioneer family from the labor of grinding its own grain, and a sawmill freed it from the labor of preparing lumber for its own home. Lacking sawmills, colonists were forced to live in log houses with dirt floors or log and stucco houses with thatched roofs; however nostalgically we may look back on them, these were drafty and leaky and muddy and generally a threat to the health of the people who had to live in them. Unfortunately, no farm family could manage the labor needed to clear and plant the land as well as that needed to cut, plane, and shave either the boards that might go into a wooden house or the shingles that might create a better roof—not, that is, until there was a sawmill in the vicinity. Tanning mills, carding mills, and fulling mills all made it possible for the colonists to supply themselves with inexpensive and durable clothing, which, like meal for bread and boards for roofing, were essential items of comfort and survival. The miller's fee was not small: in return for his services, a farm family might have to give up half to two thirds of the leather that it brought to a tanning mill, perhaps a quarter of its grain, and an equal amount of its logs. Mute testimony to the colonist's reliance on these mills lies in the fact that few controversies about the fees ever erupted into the pages of court records.

Part mechanic, part businessman, the miller's services were essential to the livelihood of his neighbors. A watermill in operation may have been annoyingly loud, destroying the peace of the forests, and the by-products of the mills may have been offensively noisome (tannery chemicals befouled the rivers and sawyer's logs made boat travel and fishing impossible at some seasons of the year), but the colonists did not seem to mind. They understood full well that water-powered mills saved them from backbreaking labor and at the same time improved the standard of living of every community. As creators, owners, and managers of the most complex technological systems of their day, millwrights and millers were, as a consequence, among the most respected and valued members of their communities.

Iron Foundries and Iron Workers

Colonial iron foundries differed from printing shops and gristmills; these manufacturing establishments were not owned by any of the skilled craftsmen who worked there. Ironworkers were artisans who were employees and who had very little hope of ever ceasing to be employees. Only a very few colonial artisans were in that kind of social situation—shipwrights worked under similar conditions, so did the millers, bakers, and coopers employed at large merchant flour mills—but their situation has considerable historical significance. During industrialization the unusual pattern of their employment became much more widespread; the earliest factory owners fashioned the new kinds of workplaces that they were creating on the model of traditional iron foundries, merchant mills, and shipyards.

Ironworkers, shipwrights, and merchant millers worked with extremely expensive pieces of equipment, which is why most were not owners, or at least not sole owners, of the businesses in which they worked. In the middle of the eighteenth century, the start-up cost of an iron foundry was roughly £3500—forty-six times what was required to become a master printer. The apprentice printer or the journeyman mason could have expected eventually to become a master owner, but such an expectation was essentially out of the question for the ship's carpenter or—to draw from the example we are going to describe in detail—the furnaceman.

The element iron is never found naturally in a pure state, but rather as one of several forms of iron oxide mixed with other minerals; the process by which the oxides are reduced and the impurities driven from the ore is called smelting. Europeans were aware of two ways to smelt iron: one was to put it in a hearth with some charcoal, ignite the charcoal, and increase the heat on the hearth with a hand-operated bellows. As the heat rose, the oxygen in the ore combined with the carbon in the charcoal to form carbon dioxide (the iron oxide was reduced). The resulting black pasty material was removed from the hearth to an iron block—an anvil—on which it was beaten while it cooled, driving out some additional impurities. The black pasty material was called a bloom; its color provided a name for the person who did this sort of work, a blacksmith, and its name became the name of the place in which the work was done, a bloomery forge.

In colonial America, bloomery forges served the needs of rural populations; they were small craft shops precisely akin to printing houses in their form of work organization. Bloomery forges produced most of the horseshoes, ax blades, and plowshares that colonial farmers needed. Bloomery forges did not, however, produce the iron that other craftspeople (wheelwrights, for example, or gunsmiths) needed, and they did not supply the iron shipped overseas.

That was pig iron, or bar iron. It was smelted in blast furnaces (large structures of stone or brick) that were located on huge tracts of land, so large that they were called plantations. An iron plantation was an almost self-sufficient community, consisting of an iron deposit ready to be mined; a furnace, a forge, and a waterwheel; acres of forest to provide the charcoal for the furnace; housing for all the ironworkers and their families; and fields on which food for the workers could be grown. Most plantations were located on navigable rivers so that the pig iron, which was very heavy, could be transported to market at the lowest possible cost.

In order to start production at an iron plantation, the land had to be purchased and the houses, furnace, forge, and wheel all had to be constructed. A plantation usually belonged to a group of entrepreneurs who pooled their resources to finance all this activity. The first colonial plantation was organized just outside of Jamestown early in the 1620s, but it was not successful. A group of Indians attacked and massacred the ironworkers just after the furnace had been "put into blast" for the first time; we can only speculate upon what the natives thought the colonists were doing with a fire that lit up the skies of the forest night. Almost twenty years passed before another group of entrepreneurs proved willing to try again, but this and succeeding trials were more successful: productive blast furnaces were established in Taunton, Massachusetts, in the 1640s; New London, Connecticut, and Providence, Rhode Island, in the 1650s; and Shrewsbury, New Jersey, in the 1670s.

Iron making was a difficult enterprise in the colonies throughout most of the seventeenth century, however: the capital partners (who were usually far distant from the plantation) fought with each other, and very few skilled workmen could be located. In addition, the export market for American iron was not stable and there was as yet no secondary market in the colonies since colonial blacksmiths operated their own bloomery forges. The English economy began to expand in the early decades of the eighteenth century, however, and one of the commodities that it craved was unfinished iron bars. At the same time, rich lodes of iron ore were discovered all over the western frontiers of the mid-Atlantic states and trading networks between the mother country and her colonies became firmly established. All of this gave a tremendous boost to the colonial iron industry. Plantations sprang up all across the countryside, and by the middle of the century, pig and bar iron had become the third leading colonial export—behind only wheat and timber.

The affairs of an eighteenth-century iron plantation were managed by one individual, the ironmaster, who was sometimes a skilled worker who had accumulated enough money over the years to become one of the partners in the enterprise and sometimes a complete stranger to the business. The ironmaster exercised patriarchal authority over the entire plantation: he hired and fired workers; supervised all aspects of the production process; determined when the furnace would be "in" blast and when it would be "out"; decided what crops would be grown in the fields and what prices would be set for them in the company store; he decided what religious services, if any, would be conducted on the plantation and kept an eye on the behavior of all the workers and the members of their families. The plantation was essentially a world unto itself, usually far distant from any center of population. The work was arduous for all parties, and heavy drinking was endemic. All this meant that the ironmaster, like the captain of a ship, had to be a production supervisor and a police force rolled into one. Ironmasters had a difficult row to hoe—and so did their wives, who found themselves doubly isolated, first from settled communities and second from friendly contact with the workforce of the plantation.

Some of the members of that workforce were unskilled and more or less transient laborers (unless they were slaves or indentured servants): they mined ore with pickax and shovel or managed the carts and wagons with which it was hauled to the furnace. Others had served lengthy apprentice-ships in their crafts: the colliers, who knew how to make charcoal from wood; the furnacemen, who knew how to build, maintain, and operate the furnace; the chafers, who knew how to refine the product of the furnace.

The colonial iron industry, like almost all industries in the colonies, was plagued by labor scarcity. Ironwork was hard labor and the iron plantations were not regarded as pleasant places to live; hence few colonial parents saw much reason to set their children to apprenticeships there. As a result, wages were high, and those workers who chose to stay in the trade were constantly moving from plantation to plantation on the offer of better wages or better living conditions. Ironmasters tried to solve their labor problems by importing ironworkers from Europe, but that was rarely a suc-cessful strategy since the immigrant workers began demanding higher wages soon after their arrival. "They pretended to have their wages raised," one ironmaster complained about a dozen Germans he had recently hired, "which I refused. They made bad work; I complained and reprimanded them; they told me they could not make better work at such low wages; and, if they did not please me, I might dismiss them. I was, therefore, oblig-ed to submit, for it had cost me a prodigious expense to transport them from Germany; and, had I dismissed them, I must have lost these dis-bursements, and could get no good workmen in their stead."[5] Indentured servants and slaves did not solve the problem either; once trained, they could run away easily since most ironworks were located in isolated places. Throughout the colonial period, labor problems remained the ironmaster's greatest difficulty.

Smelting did not commence with the stacking of the furnace, but rather with the production of charcoal, the fuel for the furnace; an average blast furnace consumed fifty to one hundred acres of wood a year. Charcoal is partially burned wood. To make it, hundreds of logs were carefully laid, one atop the other (sometimes the resulting stack was beehive shaped, sometimes rectangular) so that as little air as possible could penetrate the inner layers of the stack. A mixture of mud and leaves was laid over the stack and then ignited; over the course of several days, if the stack was care-fully monitored, the inner logs turned to charcoal while the outer logs were reduced to ash. Charcoal had to be made continually when a furnace was in blast because it could not be stored; since it could not be made success-fully when there was rain or wind, the weather limited its production to certain seasons and, therefore, also limited the operation of the furnace.

A blast furnace was a stone tower between twenty-five and forty feet high, shaped like a truncated pyramid, and usually set into the side of a hill to make loading (which was done from the top) easier. When a blast was about to begin, the foreman would organize the stacking of the furnace with alternate layers of ore, charcoal, and limestone (which was also mined on the premises). Once the stack had been ignited, its internal heat had to be increased by successive blasts of cold air from a huge leather bellows (operated by a waterwheel) until the ore was molten.

During smelting, some of the impurities in the ore were driven off as volatile gases, creating a flame that could be seen for miles around, which the foreman monitored to determine how the blasts should be regulated. Most of the remaining impurities in the ore combined with the limestone to form a mixture, slag, which floated on top of the liquefied iron. After several days in blast, the foreman decided that the smelting was completed and a special hatch at the bottom of the furnace was opened, allowing the molten iron to run out into sand troughs that formed the hearth in front of the furnace. Usually there was one central trough with many offshoots, resembling (or so some ironworker must once have thought) a sow with nursing piglets, and so the iron bars formed in the troughs when the molten material cooled were forever after called pigs. Other times the sand could be shaped into rectangular molds, so that stove plates (or other simple objects) could be cast.

Pig iron had a certain amount of carbon dissolved in it and a certain amount of slag; as a result it was both harder and less malleable than bloomery iron. In order to make it more useful on the commercial market, it was refined at a forge before it left the plantation. A forge consisted of a raised hearth and a very large trip-hammer (which often operated off the same waterwheel that ran the bellows for the furnace). A block of pig iron was heated on the hearth until it was red and almost molten, then it was lifted with special tongs, placed on an anvil and beaten with the trip-hammer, thus altering its crystalline structure, driving off some additional carbon, and redistributing other impurities. The end product, bar iron, was more ductile and less likely to fracture than pig iron; it was the material that plantations offered most frequently for sale to secondary processors. Bar iron might be purchased by the owners of rolling mills (where it would be squeezed into thin sheets), or it might be cut into strips for subsequent shaping and welding by blacksmiths.

In all its forms, iron production was labor intensive. Had the average blast furnace (and its associated forges) been kept in operation twelve months of the year, a colonial plantation might have produced between 1,000 and 2,000 tons of iron a year (roughly what a modern blast furnace produces in a few hours), but the furnaces were rarely in continuous operation; one blast period lasted from October to December, another from late April to July or August. When the furnaces were shut down, some workers were assigned to repairing and rebuilding them, but others were sent either into the fields for plowing and harvesting or into the woods for chopping and hauling.

Wage arrangements varied from plantation to plantation, and even from individual to individual on any one plantation. Some ironworkers were paid by a piece rate when they were working iron and by a daily rate when they were chopping wood or plowing fields; others were paid an annual wage, which was partially in kind and covered all the varied activities that might be demanded. "Agreement [from the New Pine Forge, in New Jersey] with John Shaw, July 23rd. 1761 to stock the upper forge and at any time to as-

An eighteenth-century forge in operation. The waterwheel shaft and trip hammer are in the center; the hammerhead was a heavy piece of wrought or cast iron, weighing several hundred pounds; the wheel had to be at least twenty-five feet in diameter to raise the hammer. Pig iron bars were heated on the hearth to the left, and then refined into an "ancony" bar by the hammer. (Adapted from a drawing, "Forging an Ancony at Refinery Forge," from Arthur Cecil Bining, *Pennsylvania Iron Manufacture in the Eighteenth Century* published by the Pennsylvania Historical Museum Commission and used with its permission.)

sist in stocking at any of the other two forges when he has not stocking to do at the said upper forge. The said Shaw is to be paid for the faithful performance of the above agreement eighteen pounds and a pair of shoes if he does not get drunk above once in three months, a pair of stockings and his diet."[6]

Some foremen were subcontractors, receiving monies from the ironmaster, which they then distributed as they saw fit to the men who worked under them. Some ironworkers received their housing free of apparent charge; others paid a fee, which was deducted from their wages. Wives of ironworkers and their children were paid a daily wage when they worked in the fields. Apprentices, indentured servants, and slaves received nothing but their sustenance, some clothing, and whatever training was contractually required. All this was determined by the discretion of the ironmaster in negotiation with the individual worker (or, in the case of apprentices, with the young man's parents); there was no guild, or any other kind of organization, to protect the worker's interests.

And as it was on the iron plantation, so it was in some of the other colonial workplaces that were not small craft shops: the shipyards that employed shipwrights, caulkers, cordwainers, sailmakers, and carpenters, as well as day laborers; the merchant flour mills that might be the workplace for several millers, coopers, and bakers; the occasional weaving or spinning establishment that employed a dozen or so workers to operate the equipment belonging either to a wealthy master or to a group of entrepreneurs. We have no way to estimate what portion of colonial artisans worked under these more centralized conditions, employed for a wage, operating equipment that they did not own, and, in some cases, living in houses that belonged to their employers. They were, however, a large enough group to be noticeable, enough so that the owners of the first full-scale factories of the nineteenth century had no difficulty copying the employment practices of their eighteenth-century predecessors.

Conclusion: Reasons for the Slow Pace of Technological Change

During the colonial period, the pace of technological change was as slow as the proverbial tortoise. American craftspeople adopted the technologies with which they had been familiar in Europe, making only minor modifications. Technological change is not an independent variable, a self-propelled social engine; quite the reverse, it is a dependent variable propelled by a host of other factors. Many of those factors were missing in the North American colonies.

The English government had no interest in helping colonial artisans become more productive; indeed, quite the reverse was generally the case. Mercantilist doctrine dictated that a colony should supply raw materials that could be worked up into finished products in the home country. The English government had only a few mechanisms for encouraging technological change: issuing patents, awarding prizes for innovations, creating protective tariffs. It chose to use these mechanisms very rarely, governments being less attuned to such issues in the seventeenth and eighteenth centuries than they are today. Because of mercantilist theory, these mechanisms were used in support of technological change only at home, not in the colonies.

On top of this, the density of artisans in the colonies was too low to encourage technological change. Such changes tended to proceed fastest, as the creators of industrial-research parks clearly know, when and where there is a high density of skilled people willing and able to talk together, compare notes on how they are solving the problems in their work, and collaborate with each other. Separated by long distances, lacking a guild or any other social institution that could facilitate communication between them, oftimes farming for part of the year, always in short supply—colonial artisans were unlikely to feel much impulse toward changing the tools with which they did their work.

Indeed, many of the tools that an artisan needed could be constructed only by another artisan. The spinner needed a wheelwright to construct the spinning wheel; the weaver needed a carpenter to construct a loom; the printer needed a joiner to make a press and a founder to make type. Since artisans of all types were in short supply, spinners were grateful when they could find a wheelwright, and printers were grateful when a new font finally arrived by ship from England. Under such circumstances, artisans do not require innovation from each other, and there is very limited competition between artisans to supply the producer's market. The net result was that colonial artisans tended, on the whole, to re-create that with which they were already familiar.

The anonymous craftspeople of the colonies did, nonetheless, accomplish a good deal. They laid the foundations for a flourishing economy. They built the sturdy, durable tools—the cast-iron pots, the barrels, the wagon wheels, the woolen and linen cloth, the needles and the nails—with which their neighbors conquered the wilderness. Some of these artisans even managed against all odds to be innovative; the perfectly balanced American ax, the elegant products of Boston silversmiths and New York furniture makers—these are all testimony to the skill and to the creativity of many an American artisan.

The printers and sawyers, the blacksmiths and clock makers, the foundrymen and the hatters were important to the future of their countrymen as well, for it was their political activity that helped foment the Revolution. Urban artisans differed from the rest of the colonists because their fortunes, like the fortunes of the merchants, were tied to the politics of trade rather than the politics of landholding. Land prices and land taxes worried those who practiced husbandry; but export quotas and import duties, arrangements for credit, and the stability of currency worried those who practiced the crafts.

After 1760, it was precisely these matters that disturbed the relations between the home country and its colonies. Artisans, like merchants, found their business hampered first by the Currency Act (1764), which reduced the value of the paper money that had been issued by colonial legislatures, and then by the Stamp Act (also 1764), which increased the cost of the documents necessary to conduct business. The Townshend duties (1767) led merchants to organize a boycott of imported British goods, a boycott supported, in their own self-interest, by American craftspeople. In the early 1770s, in Philadelphia, Boston, and New York, master printers and carpenters and smiths were happy to supply the crowds of apprentices and journeymen that radical politicians needed to harass British troops and customs officials. Once Parliament passed the so-called Coercive and Intolerable Acts (1774), urban artisans were enthusiastic once again in their support for the non-importation movement organized by the first Continental Congress. Artisans did not occupy positions of political leadership in the colonies, but as valued and respected members of the communities in which they lived, their views carried considerable weight. Having objected for decades to the mercantilist structure of the British tariffs and duties,

American artisans understood that they had many things to gain and very little to lose from independence.

Once independence came, once colonial artisans were freed from the shackles of mercantilist policy, the rate of American technological change accelerated. Although artisans were small in number, even after the Revolution, their impact on the future would be great. The skills that they possessed—the mechanical aptitudes of the millwrights, the business acumen of the sawmillers, the organizational insights of the ironmasters—were the skills that the new nation was going to need as it began to compete on world markets with the nations of Europe. In addition, the work processes to which at least some artisans had become accustomed—working in places that were not their own homes and for an employer rather than themselves—would become the social foundation on which in the nineteenth century a mighty industrial economy would be built.

Notes

1. Robert Beverley, *History and Present State of Virginia* [c. 1696], ed. Louis B. Wright (Chapel Hill: University of North Carolina Press, 1946), p. 295.

2. From a manuscript letter, as cited in W. J. Rorabaugh, *The Craft Apprentice: From Franklin to the Machine Age in America* (New York: Oxford University Press, 1986), p. 28.

3. John Fitch, *Autobiography,* ed. Frank D. Prager (Philadelphia: American Philosophical Society, 1976), pp. 41–42.

4. From *Vermont Historical Gazetteer*, as cited in Louis C. Hunter, *Waterpower in the Century of the Steam Engine* (Charlottesville: University of Virginia Press, 1979), p. 30.

5. Peter Hasenclever, *The Remarkable Case of Peter Hasenclever, Merchant* (London, 1773), p. 9.

6. Ledgers of New Pine Forge, as quoted in Arthur Cecil Bining, *Pennsylvania Iron Manufacture in the Eighteenth Century* (Harrisburg: Pennsylvania Historical Commission, 1938), p. 118.

Suggestions for Further Reading

Arthur Cecil Bining, *Pennsylvania Iron Manufacture in the Eighteenth Century* (Harrisburg, PA, 1938).

Carl Bridenbaugh, *The Colonial Craftsman* (Chicago, 1980).

Charles F. Hummel, *With Hammer in Hand: The Dominy Craftsmen of East Hampton, New York* (Charlottesville, VA, 1968).

Louis C. Hunter, *Waterpower in the Century of the Steam Engine* (Charlottesville, VA, 1979).

W. David Lewis, *Iron and Steel in America* (Greenville, DE, 1976).

Paul F. Paskoff, *Industrial Evolution: Organization, Structure and Growth of the Pennsylvania Iron Industry, 1750–1860* (Baltimore, 1983).

Terry S. Reynolds, *Stranger than a Hundred Men: A History of the Vertical Water Wheel* (Baltimore, 1982).

W. J. Rorabaugh, *The Craft Apprentice: From Franklin to the Machine Age in America* (New York, 1986).

Rollo G. Silver, *The American Printer, 1787–1825* (Charlottesville, VA, 1968).

John Storck and William Teague, *Flour for Man's Bread: A History of Milling* (Minneapolis, 1952).

Lawrence C. Wroth, *The Colonial Printer* [1938] (Charlottesville, VA, 1964).

II

INDUSTRIALIZATION

THE WORD *INDUSTRIALIZATION* IS A SEVEN-SYLLABLE TONGUE TIER, and rightly so; the complex process that it denotes is the most profound historical change of modern times. To most people the word conjures up images of factories and machinery, railroads and smokestacks. Industrialization was more than that, however. Every nook and cranny of social and economic life was implicated in the process and affected by it, which is why this second section is so much longer than the first and also why it discusses so many different kinds of Americans: young and old, men and women, rich and poor, immigrant and native born.

Industrialization was a complex process that took a very long time to complete, which is why we no longer refer to it as a revolution. The beginning and end of a historical processes is not as easy to locate as, say, the beginning and end of a war or the beginning and end of an election campaign. Because industrialization is a process and because it did not unfold quickly, this section considers people who lived and things that happened as early as 1790 and as late as 1930.

Chapter 4 focuses on the lives of three innovators: Oliver Evans, Eli Whitney, and Samuel Slater. These three were not the only movers and shakers in the early years of industrialization, but their stories reveal an enormous amount about the general character of industrialization and the unique character of American industrialization. Chapter 5 deals with transportation, an often neglected aspect of industrialization. New roads, canals, steamboats, and later railroads were vitally important in unifying the new nation and creating an enormous national market in which new industrial

67

products could be sold at reasonable prices to millions of people. This chapter also demonstrates the crucial role that governments—federal, state, and local—played in encouraging industrialization to proceed.

Chapter 6 is a joint biography of some of the people—inventors, engineers, and entrepreneurs—who built the industrial economy. Of necessity the chapter focuses on the lives of just a few of those people, but the point that it makes is somewhat at odds with that focus: each part of the process of innovation—invention, development, diffusion—is collaborative. No one, not even the greatest of the inventors and entrepreneurs—neither Edison nor Carnegie, neither Bell nor Rockefeller—was able to do what he did alone.

Chapter 7 is the keystone of the section. It explores the concept of a technological system in some detail by focusing on five of the systems that developed during the nineteenth century—the telegraph, railroad, telephone, petroleum, and electric systems. The intention of the chapter is to explain what makes industrial societies different from earlier societies, and by focusing on those five systems it becomes possible to understand why, in industrial societies, people have become *more* rather than *less* dependent on technology and, at the same time, *more* rather than *less* dependent on each other.

Chapter 8 discusses workers, the men and women who used their hands and minds to run the machines and provide the services that created the industrial economy. In assessing the impact of industrialization on these people, we have to remember that they were very different from each other: unskilled workers were unlike skilled ones, and domestic servants differed from machine tenders. Thus the impact of industrialization on ordinary people is as complex and as complicated as people themselves. Chapter 9 makes a similar point about ideas. Many scholars say that we Americans have had a very special romance with technology, that more than any other Western culture, the United States is somehow uniquely technological. In fact, this romance has been, if anything, a love–hate relationship. Some people have thought that industrialization was the best thing that has happened in our history; others have thought it the worst.

Once we fully understand the process of industrialization, we can more fully understand not only that love–hate relationship but also why Americans sometimes feel so nostalgic about preindustrial times and preindustrial products. With part of our national soul we love technology, because industrialization has led to affluence; today even the poorest American has a higher standard of living than most of the people who have ever lived. With another part of our soul, however, we hate technology, because industrialization has made us dependent on technological systems; as a nation devoted to individualism, we are inclined to rebel against anything that compromises our independence.

4

The Early Decades
of Industrialization

THE MOMENTOUS DECADES IN WHICH THE UNITED STATES WAS BORN
and its national character determined were also the decades in which American industrialization began: the forty years between 1780 and 1820. Before independence was achieved, many colonial leaders, particularly merchants, had business connections with England; many had made the trip
across the Atlantic and had spent months, sometimes even years, on King
George's turf. The pace of economic and social change was quickening in
England; industrialization had, by the 1770s, already begun there. Some
Americans did not like what they saw, but others were favorably impressed.
Those who were impressed believed that their newly created country
would never be able to succeed economically unless it could beat England
at its own game, exploiting the United States's vast, untapped reservoir of
natural resources to manufacture goods that could compete successfully
against English goods on the world's markets.

American industrialization followed quickly on the heels of English industrialization, but American industries were not carbon copies of their
English cousins. American industrialization had its own special character,
dictated partly by the unique complement of resources that were available
in North America, partly by the nature of the American labor force, and
partly by the unique skills and interests of the men who were in one form
or another industrial leaders. We can best gauge the special character of
American industrialization by examining the lives and the contributions of
some of those men as well as the social and economic constraints that
shaped their labors.

On October 19, 1781, General Cornwallis, who was in charge of the British campaign against the southern colonies, surrendered at Yorktown, Virginia, and the Revolutionary War ended. Almost nine months later to the day, several hundred miles to the north in Delaware, Oliver Evans and two of his brothers purchased a parcel of land—300 acres in all—from their father. By September 3, 1783, when Benjamin Franklin, John Jay, and John Adams signed the Treaty of Paris, granting Americans unconditional independence from Britain, Evans had already started to alter the old gristmill on what had once been his father's property. In September 1786, as delegates from the state legislatures met in Annapolis, Maryland, to discuss whether the thirteen states could agree on a unified commercial policy that would stimulate what were then called "domestic manufactures," the Evans mill had been operating successfully for more than a year.

Oliver Evans was born in 1755, one of twelve children of a reasonably successful farmer; early in his teenage years, he was apprenticed to a wheelwright and learned how to build both with wood and with metal. At the beginning of the 1780s, when he and one of his brothers owned a village store, Evans carefully observed the operations of several small gristmills. Confident of his mechanical ability, he thought of several devices that would make milling more efficient—it was these devices that he intended to install in the mill he now owned in Delaware.

Evans called the first of his new devices a "grain elevator" and the second a "hopper boy." The elevator was a continuous leather belt holding a series of small buckets; pulleys attached to the shaft of a waterwheel kept the belt in motion. The grain elevator could haul, or elevate, 300 bushels of grain per hour, work that had previously been done by hand by several strong men. The hopper boy was a large revolving rake, twelve feet long, attached to a vertical drive shaft, which was also connected to the main shaft of the mill. The rake spread the ground meal evenly on the floor of the mill so that it could dry, and then guided it gradually to a chute, which led to a hopper, in which the meal was sifted. This work had previously been done by a young boy, hence its name.

Both of Evans's inventions had the potential for increasing the miller's profits by decreasing his labor bill. As Evans put it, the requisite work could be done by machines "without the aid of manual labor."[1] When machines substitute for manual labor *mechanization* occurs; Oliver Evans's singular contribution to the industrialization of the United States was to demonstrate how one of the country's most crucial businesses—flour milling—could be more fully mechanized.

In 1786 and 1787, Evans appealed to the legislatures of Delaware, Maryland, Pennsylvania, and New Hampshire to grant him exclusive monopolies on the manufacture of these implements and the right to require licensing fees from people who used them for a period of fifteen years. Evans wanted to be compensated for the value of his inventions; he also wanted to support his family while continuing to invent. All four legislatures agreed. The General Assembly of Delaware remarked, in January 1787,

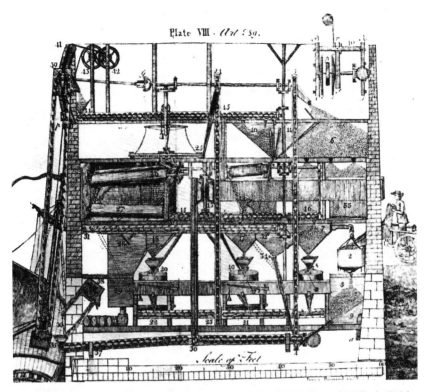

Evans's drawing of his mill from his book, *Young Mill Wright and Miller's Guide* (1795). No laborers are shown inside the mill; the man on the right is pouring grain into the mill from a wagon. The grain elevator is shown on the right, raising grain from a ship (*39*) and then dropping the grain onto an Archimedean screw (*44*), which moves it horizontally. The hopper boy is to the left of 25. There are three millstones in operation (at *8, 19,* and *20*). (Courtesy Ward Melville Memorial Library, State University of New York at Stony Brook.)

that the new devices would provide "a means of carrying on the business of Milling to a greater advantage than has heretofore been done, by causing it to be transacted and carryed on with less labour and costs."[2]

In the eighteenth century, and in the first half of the nineteenth, merchant milling was the economic mainstay of the middle Atlantic states as well as most of the new midwestern states. Millers bought grain from farmers, ground it into meal and flour, and then sold it to wholesalers, who shipped it overseas or to domestic retailers. Evans had, quite deliberately, chosen to put his mechanical abilities to work where they might do both him and the economy of the new country the most good. When the United States Congress passed the very first Patent Law (in 1790), Evans was quick to apply; he forfeited (as the law required) the individual monopolies granted to him in the several states in return for the sole right to exploit his inventions in all the states. Evans received United States Patent #3 for his improvements in milling.

Unfortunately, as Evans discovered to his dismay, new inventions were difficult to market and patent rights were not easy to enforce. Millers were conservative; they already had profitable enterprises, so why should they invest capital in new, unfamiliar equipment that they did not know how to install, operate, or maintain? Evans discovered that being a manufacturer of milling equipment meant not just building his own forges and shops, not just supervising a workforce of artisans, but also writing pamphlets that would advertise and explain his equipment, traveling to distant places to install it, then answering mail from millers who had maintenance problems. Within a few years, he was employing several traveling salesmen (including his brothers), who also had the task of collecting the annual licensing fees. In return, Evans was supposed to receive a small percentage of the profits of each mill that bought and installed his equipment, but the millers were not always prompt in their payments. Even more annoying, sometimes millers had hired millwrights to copy Evans's inventions; then he had to hire lawyers to sue for infringement of his rights. When his patent rights expired (in 1804), he had to lobby Congress (a four-year effort) to renew them. As a result of all this, Evans and his family lived in considerable poverty for about a decade, despite the fact that his inventions were being adopted in mills all over the country and were transforming the most important industry in the new nation. "The introduction of them into general use was," as he put it mildly, "extremely laborious."[3]

Whatever problems Evans may have faced in being compensated for his inventions, the machines that he developed for making milling more efficient played a significant role in the growth—and the industrialization—of milling in the first half of the nineteenth century. Old mills were converted to accept his machinery; virtually all new mills were built to his designs. By 1840, roughly 24,000 gristmills were operating in the United States, and a very large number of them had incorporated one or another of Evans's laborsaving devices. The total value of the flour produced in such mills had been $14 million in 1810; by 1860, that figure had increased almost 200 percent, to roughly $250 million. In that year, reckoned in terms of dollar value, flour milling was the country's leading industry, although the number of people employed in the industry was fairly small: roughly 27,000 compared, for example, to 114,000 employed in the manufacture of cotton cloth or 123,000 making boots and shoes.

Skilled labor had been scarce in the colonial period, and it continued to be scarce during the early decades of nationhood. As long as land remained cheap, many people, male and female, preferred working their own land to working for someone else. Where labor was scarce, it was also expensive, which meant that laborsaving devices increased profits. This is part of the reason why so many American industries subsequently trod the path of mechanization, the path Evans had pioneered in milling, substituting machines for workers.

Oliver Evans, Steam Engines, and Machine Shops

Oliver Evans's mechanical ingenuity was not, however, limited to flour-milling machinery. In the 1780s, he was also experimenting with steam engines, having read about the devices with separate condensers that James Watt had patented in Britain a decade earlier. Unlike Watt, who was interested in building engines that would stand still (for example, for pumping water out of mines), Evans thought that the most interesting use of steam power would be in transportation. For this purpose, he realized that smaller, lighter, and more powerful engines would be required (the Watt engines were so heavy they could not have been moved by the power that they generated) and that it would take many experiments conducted over the course of many years to develop one. In 1787, he asked the Delaware legislature for patent protection on an invention that was still in its planning stages: a "land Carriage" propelled by "the Power of Steam and the Pressure of the Atmosphere."[4]

Since Evans had not yet built a working engine, or even a model of one, the legislature refused. Evans persevered . . . and then persevered some more. In 1793, he moved from Delaware to Philadelphia to be at the center of what was then the most "improvement minded" of all American cities. In 1801, he built a small, stationary, high-pressure steam engine and used it to power a grinding mill designed to pulverize plaster of Paris. In February 1804, he received a United States patent on the design for that engine. This Evans engine had two features that made it both pathbreaking and radically different: first, it had no condenser (the steam was vented directly to the atmosphere) and second, both the cylinder and the crankshaft of the engine were placed at the same end of the engine beam. This high-pressure engine was smaller and lighter than a Watt engine.

In September 1804, Evans proposed in a letter to the president of the Lancaster Turnpike Company (see Chapter 5) that the company finance the construction of several "steam carriages" that would carry freight at the then unheard of rate of forty miles per day; the proposal was rejected. By December, he had had another idea for utilizing his new engine; he proposed that the Philadelphia Board of Health build an enormous, steam-powered shovel for dredging Philadelphia harbor, a shovel that could move from place to place, on land and in the water, under the power of its own high-pressure steam engine. Apparently the Board of Health was more adventurous than the Turnpike Company. Evans's proposal was accepted, and in July 1805, "Orukter Amphibolus, or Amphibious Digger" as Evans called his monstrous machine (it was twelve feet wide by thirty feet long) lumbered first from Evans's workshop to the river, and then several miles up the river to the harbor, where it was employed, not altogether successfully, for several months.

In 1806, an Evans high-pressure steam engine was installed to power a gristmill in Lexington, Kentucky. When that machine proved to be, in Evans's words, "simple, cheap, and durable," he joined several other investors in constructing a steam flour mill in Pittsburgh, Pennsylvania. These were two of the country's first manufacturing plants to be powered by neither waterwheels nor draft animals but by steam. By 1810, Evans's workshop in Philadelphia was producing several high-pressure steam engines a year, some of which were being used in mills, some in pumping stations (the Philadelphia Water Works employed several), and some in steamboats. In 1811, he invested in the construction of another workshop further west in Pittsburgh; the first of the many Evans steamboats to sail on western rivers was launched from there in 1817.

Thus, when Evans died in 1819, part of the legacy that he left not only to his family but also to his country were two sophisticated machine shops and several dozen highly skilled machinists. In 1811, his Mars Works in Philadelphia consisted of an iron foundry with two furnaces, a blacksmith's shop, a millstone manufacturing plant (producing patented grindstones), and of course a steam engine workshop—thirty-five people were employed there. In the first decade of the new century, two of his sons-in-law (James Rush and John Muhlenberg) became active participants in his business, and they inherited control of it on his death. The steam flour mill and the steam engine workshop in Pittsburgh had long before been placed in the skilled hands of his oldest son, George, whom he had also trained to follow in his footsteps.

Evans's shops were not unique; in the first decades of the nineteenth century, machine shops, large and small, were being established in cities and towns all over the settled parts of the country. During the eighteenth century, there had been blacksmiths and wheelwrights and coppersmiths and carpenters in the colonies, but no one who was really skilled enough to make or even to repair a complex piece of equipment such as a steam engine. The very first steam engine used in North America was imported from England in 1753; the owner of the machine ordered that its parts be supplied in triplicate, under the assumption that no craftsman in the colonies would be able to fabricate them. Fifty years later, when Robert Fulton and Robert Livingston were laying plans to construct a steamboat that would ply the Hudson River (see Chapter 5), they also decided to have the engine imported since they couldn't be confident of getting precisely what they wanted in the United States. Fulton and Livingston also paid the cross-Atlantic expenses of an English mechanic to build the machine and an English engineer to operate it.

By 1820, thanks to Evans and others like him, such transatlantic dependency was declining—by 1840, it was pretty much defunct. Grievously, the names and accomplishments of many of the country's earliest engine builders have disappeared from the historical record, but we do know that there were successful machine shops in several East Coast cities before 1820. In each of these shops—and many others—native-born metalworkers and machine builders and draftsmen were being trained, sometimes by native-born masters, sometimes by skilled artisans who had migrated from England. Having completed their apprenticeships, several of the most en-

In nineteenth-century machine shops the engine parts that needed strength in compression—the parts that would support a lot of weight—were cast in foundries by pouring molten iron into sand molds inside large wooden boxes. This drawing of the foundry of the Norris Locomotive Works in the 1850s is from "A Visit to the Norris Locomotive Works," *United States Magazine*, October, 1855. (Courtesy Smithsonian Institution.)

trepreneurial of these young Americans went into business for themselves (as did a goodly number of the English emigrants)—and the American machine tool industry began to expand.

By 1820, the popularity of the steamboat had created a considerable domestic demand for steam engines, and much of that demand was being met by American machine builders. The advent of the railroads in the 1830s and 1840s increased the demand even more. When Robert Stevens decided to build the first steam railroad in the United States in 1830, he went to England to choose and purchase a locomotive, but a decade later, as the result of such inventive mechanic-entrepreneurs as Matthais Baldwin, George Escol Sellers, and William Norris, Philadelphia had become a prolific producer of locomotives for the burgeoning American market. Evans may not have realized it at the time, but his modest machine shop in Philadelphia had started his country down the road to mechanical preeminence.

Machine shops such as Evans's make producer's goods, the things that people buy in order to produce other things: hopper boys and grindstones for making flour, spinning mules and power looms for making cloth, pumps for getting water out of mines, steam engines for powering boats and railroad trains, lathes for cutting sofa legs and gun stocks, borers for smoothing gun barrels and engine valves. Improved producer's goods have a multiplier effect on an economy; because they get used in multiple ways, they have multiple amplifying effects. Evans's grain elevator, for example, helped to lower the price of flour because it lowered the production costs of flour. Then his steam engines helped to lower the price even more by lowering the cost of transporting the flour from mill to market. As the price got lower, more and more people could afford fine white flour. Demand increased, which means that millers were encouraged to expand their facilities and thus to buy more hopper boys and grain elevators and steam engines—the multiplier effect.

The pace of industrialization in the United States was exceptionally rapid in part because of the strength of its machine tool industry and because of the effects that such an industry can have in many and diverse corners of the economy. Grievously, Evans died before the fruits could ripen on the several kinds of trees that he had planted, but in succeeding generations, the harvest turned out to be very abundant indeed.

Eli Whitney and the Cotton Gin

One of the people who reaped the benefit of that harvest was Eli Whitney. Like Oliver Evans, Whitney was the son of a prosperous farmer, and, also like Evans, he demonstrated both entrepreneurial and mechanical ingenuity when still a fairly young man. Whitney was born in Massachusetts in 1765; he was in his teens during the Revolutionary War. When imports from Britain were curtailed during the hostilities, the price of various crucial items rose. Whitney had the clever idea of earning money by installing a forge in his father's workshop so as to produce and sell nails. The enterprise succeeded; in fact, demand was so great that Whitney, no more than sixteen at the time, had to hire an assistant so that he could keep up with orders.

Neither farming nor nail making could, however, satisfy Whitney for very long. When he was nineteen, he made up his mind to acquire an education. As he lacked preparation for college, and as his father could not afford to send him to a preparatory school, Whitney took a job as a schoolteacher to earn the tuition that he needed. In 1789, at the age of twenty-four, he was able to pass the entrance examinations for Yale College. Whitney enjoyed his years at Yale, except for his frequent need to take odd jobs to supplement the meager funds sent by his father. He hoped to become a lawyer after graduating, but finding himself rather deeply in debt, decided to take a job, just for a few years, as a tutor to the children of a wealthy southern plantation owner.

As things turned out, Whitney never arrived at the plantation. While traveling south, he met Catherine Greene, widow of the Revolutionary War general Nathaniel Greene, who invited him to spend a few weeks on her plantation in Georgia before taking up his tutoring position. Greene was trying to rescue her plantation from a crushing load of debt; her plantation manager, Phineas Miller, was experimenting with new crops. One of these new crops was upland (short fibered) cotton. Upland cotton was easy to grow in Georgia, but profits were elusive because the seeds of the plant were difficult to separate from its fibers. On learning about Whitney's mechanical ability, Greene suggested that he try his hand at inventing a machine that could do the job. Miller outfitted a small workroom and offered to bear the expense involved in developing the machine in return for a share of the profits that sales of such a machine would surely generate.

Within ten days, Whitney produced a small model of a machine, which he dubbed a "cotton gin" (a contraction of "cotton engine"); within a few weeks, he had built a full-sized model. Whitney, Miller, and Greene were all delighted by the capabilities of his invention; they all believed that their various financial travails had come to an end. In a letter to his father, Whitney exclaimed that with a hand-cranked gin, "one man will clean ten times as much cotton as he can in any other way before known and also clean it much better than in the usual mode." Turned either by horse or by waterpower, Whitney said, "it makes the labor fifty times less, without throwing any class of People out of business."[5]

One of the gin's virtues was its simplicity—little more than ordinary carpentry skills were required to build and maintain one. For Greene and other cotton planters, this simplicity was a blessing, but for Whitney and Miller, it turned out to be a curse. By the time they succeeded in obtaining a patent on the gin (March 1794—eighteen months after the patent application was filed), dozens, perhaps hundreds, of copies had been made by carpenters working for plantation owners. Neither Whitney nor Miller had the resources to begin suing such a large number of people for patent infringement. To make matters worse, after building a few gins of his own and setting up a few ginning mills in Georgia, Whitney had returned to Connecticut to establish a factory for the manufacture of gins, only to discover that the costs of production would be higher than he had calculated. Miller's pockets were not deep enough to meet these additional expenses; as a result, Whitney was forced to borrow money from old friends and acquaintances. In the spring of 1795, the already troubled enterprise suffered yet another setback: the Connecticut workshop burned to the ground and had to be completely rebuilt. Whitney was mortified to discover that he could not pay his creditors.

The undertaking that had started out so optimistically a few years earlier (Yale had even awarded Whitney an honorary master's degree in acknowledgment of his achievement) was fast becoming a profound disappointment. When a Georgia judge, persuaded by a group of planters who were angered by what they considered the excessive ginning fees charged in Whitney and Miller's mills, voided an infringement suit that the firm had brought, Whitney almost gave up altogether. Ironically, the machine that produced enormous profits for the southern cotton industry produced nothing but heartache for its inventor.

The Armament Industry and the American System
of Manufacture

There is, however, an even greater irony in the story of the cotton gin. In the winter of 1797–98, Whitney was distraught about his financial failure; he was indebted to so many people in New Haven that he was afraid to take a walk for fear of meeting his creditors on the street. He needed to create an enterprise that would bring in money quickly and, having learned a sad lesson with the gin, was determined not to make the same set of mistakes twice. He needed to invent something for a customer who could afford, as Phineas Miller could not, to pay the setup costs of an enterprise. Only one such customer came immediately to mind: the United States government.

In the winter of 1798, many people feared that the United States would shortly go to war with France. Little more than eight years old, and just fifteen years past its first major encounter with a European power, the young country was not equipped to mount another military campaign. Its supply of both arms and warships was negligible and neither of the two federal armories (at Springfield, Massachusetts, and Harper's Ferry, Virginia) was capable of producing the number of muskets and bayonets that would be needed to defend a coastline that stretched from Maine to Georgia. The War Department was going to have to turn to private contractors.

Whitney already knew how to build a factory, and he already had trained a workforce. Having succeeded in developing machinery for making some of the component parts of gins, why couldn't he develop machinery for some of the component parts of muskets? "I am persuaded," he wrote to Oliver Wolcott, Secretary of the Treasury, "that Machinery moved by water adapted to this Business would greatly diminish the labor and facilitate the Manufacture of this Article. Machines for forging, rolling, floating, boreing, Grinding, Polishing etc may all be made use of to advantage."[6]

Whitney was proposing to mechanize arms manufacture, proposing, essentially, that he could substitute machinery for the work of highly skilled artisans and, in the process, increase the speed with which guns could be produced. His timing was perfect; Congress had just appropriated $800,000 for the procurement of arms, and Wolcott was not sure where, or how, to spend it. Within three weeks, Whitney had signed a contract obligating him to what was then considered an extraordinary delivery schedule: 10,000 muskets in two years' time. The government, in its turn, agreed to pay *in advance* at various stages of the contract. "Bankruptcy and ruin" Whitney wrote to a friend, "were constantly staring me in the face. By this contract I obtained some thousands of Dollars in advance which saved me from ruin."[7]

As things turned out, the United States did not go to war with France—which was fortunate, because Whitney did not deliver the muskets on time. Although he did build a factory and did hire a workforce, much of his attention in the next several years was devoted to the continuing legal battle over cotton gin royalties rather than to the construction of special purpose machinery. To cover the costs of his enterprise, Whitney had to keep asking for payments from the federal government, but he also had to offer some explanation of why the shipments of finished muskets were delayed.

Whitney repeatedly told Wolcott that the delay was being caused by his effort to develop new kinds of machines, machines that would not only radically transform the way in which arms were produced but also radically transform the quality of the arms he would deliver. In 1799, in one of a series of apologetic letters to Wolcott, he explained what he called his new principle:

> One of my primary objects is to form the tools so the tools themselves shall fashion the work and give to every part its just proportion—which when once accomplished, will give expedition, uniformity, and exactness to the whole. . . . In short, the tools which I contemplate are similar to an engraving on a copper plate from which may be taken a great number of impressions, perceptibly alike.[8]

In those three words, "expedition, uniformity, and exactness," Whitney summed up the *American system of manufacture*. By expedition, Whitney meant what he had meant when he had first secured the contract: mechanization would increase the efficiency of production. Uniformity and exactness, on the other hand, were quite novel claims. By uniformity Whitney meant that all the muskets coming out of his factory would be identical to one another. By exactness he meant something more: that an assembled musket would be composed of parts (the gunlock alone had more than a dozen separate parts) finished to such exact specifications that they would be identical to the parts of every other musket. Whitney was suggesting, in short, that he was developing machinery to produce *interchangeable parts*.

Whitney was not the only manufacturer who was trying to satisfy the government's desire to purchase a large quantity of armaments with interchangeable parts. In the end, as it happens, he wasn't even particularly successful at it. Although his armament business flourished, none of the muskets that he finally delivered (in 1809, nine years late) had uniform parts and no records exist to indicate that he had developed new machinery with which to mechanize their production. But his experience with the cotton gin had taught Whitney a valuable lesson: invention is difficult, but development is even more difficult. To think of a new way to mill flour or produce gunlocks requires creativity and ingenuity. To make a profit off the invention, or actually produce the new devices, takes perseverance, patience, political acumen, and lots of money: not just any kind of money, but up-front money—what is sometimes called investment capital—to sustain an enterprise for months and years before anything is ready to be sold. Eli Whitney had the wit to realize that the federal government was one of the very few potential sources of up-front money, and he also had the political acumen to understand precisely what it was that the federal government wanted from the arms manufacturers in the early decades of the nineteenth century.

Officials of the War Department wanted to be independent of foreign suppliers of armaments; they also wanted to be able, on short notice, to obtain cannons, muskets, pistols, and bayonets. In addition, they wanted battlefield equipment that could be repaired on the battlefield, instead of having to be discarded and replaced. The United States military was willing to pay the very high costs involved because it wanted the upper hand on the battlefield.

The manufacturing system that the War Department required took a fairly long time—almost two decades—to develop. In its full-blown form, the new system was complex, involving not just the design of new machines but also the creation of new ways to hire, train, manage, and pay workers. Whitney took the first steps when he tried to convince the War Department that specially designed machinery was required. Another step was taken by his Connecticut neighbor Simeon North, who finally succeeded in building such a machine. North had been a manufacturer of scythes and plowshares, but in 1798, he had also obtained a federal contract (North's was for pistols) and had promised, like Whitney, to deliver a large number in a very short period of time. At first, North tried doing this by the division of labor (so that some employees made only the body of the pistol, for example, while others made only the firing mechanism), but by 1816, he had invented the first milling machine. This machine was a special piece of power-driven equipment, which could finish a metal part by passing a specially shaped, hardened steel cutter over it, thus mechanizing what had once been done laboriously by hand.

The War Department also instructed the superintendents of the Springfield Armory to find ways to increase the rate at which they were producing armaments. Two successive superintendents, Henry Morgan (1802–5) and Benjamin Prescott (1805–13), tried to make the traditional form of craft production more efficient. Morgan introduced, as North had, a division of labor. Prescott's innovation was a system of payment by piece rates (payment for the number of parts produced rather than for the amount of time spent on the job).

Perhaps because both of those innovations angered the master armorers employed at Springfield, neither Morgan nor Prescott did much to increase the speed of production, so a subsequent superintendent, Roswell Lee (1813–33), tried a different tack. In 1819, a local mechanic, Thomas Blanchard, patented a special purpose machine (a lathe) that could cut irregular wooden shapes (which is what gunstocks are) by following the outlines of a model placed in the machine. Lee allowed Blanchard to set up shop as a subcontractor in the armory. By 1826, Blanchard—in addition to turning out thousands of precisely identical gunstocks—had created a series of machines that increased the efficiency of gunstocking while eliminating the need for skilled laborers.

At about the same time, John H. Hall, who had started his career as a cabinetmaker and shipbuilder in Portland, Maine, became a subcontractor to the other federal armory, the one in Harper's Ferry. In 1811, Hall had patented a breech-loading rifle. (A breechloader is a rifle that receives bullets and gunpowder near the firing mechanism.) Sometime between 1811 and 1820, Hall decided that he could produce this new rifle with interchangeable parts, partly by designing special purpose machines, partly by designing special purpose fixtures (the vises that hold the separate parts in the machine as they are being worked), and partly by developing a series of gauges that could be used to test each part for uniformity. His 1820 contract with the War Department allowed him to establish a separate factory adjacent to the federal armory in Harper's Ferry—by 1822, he was able to report with considerable glee that he had reached the goal that Whitney and others had sought: "I have succeeded in establishing methods for fabricating arms exactly alike, & with perfect economy, by the hands of common workmen."[9]

"Armory practice"—the use of special purpose machines, division of labor, and an unskilled workforce—spread quickly to civilian industries: clock making, for example. Traditional clock makers, who worked by hand in both wood and brass, could make four or five clocks a year. As a result, clocks were very expensive, costing $50 if they had brass wheels and gears, about half that if the workings were wooden. In 1802, Eli Terry, a master clock maker, built a factory in Connecticut in which he connected some of the clock maker's traditional tools to a waterwheel, enabling production to increase to 200 a year. Within a few years, Terry had simplified the design of the clock (so that special purpose machinery could be built to produce some of the parts) and had licensed his design to another manufacturer, Seth Thomas. Production rates increased tenfold, and prices fell accordingly. By the 1830s, there were several dozen factories producing machine-made wooden clocks in Connecticut and Massachusetts, and many of the parts of those clocks were interchangeable. When high-quality sheet brass began being produced in Connecticut (beginning in 1837), many of the wooden clock manufacturers began making brass wheels for their wooden clocks, which led to the development of increasingly sophisticated die-stamping machinery. By 1850, an average clock factory was turning out 130,000 to 150,000 clocks annually, and the price of a clock had fallen to $1.50.

In the decade before the Civil War, sewing machines also began to be produced by the techniques of armory practice. The Wheeler and Wilson Manufacturing Company was the country's largest manufacturer of sewing machines in the decades before and after the Civil War. When they opened for business in 1851, Wheeler and Wilson used traditional hand methods to build the machines, turning out about 700 to 800 a year; each cost about $125. However, in 1856, William H. Perry became superintendent of the Wheeler and Wilson factory. Perry had learned armory practice when he was employed by Samuel Colt's small arms (pistols and revolvers) factory in Hartford, Connecticut, and he was determined to introduce special purpose power machinery in the production of sewing machines. The owners of the company were willing to invest the required capital: by 1859, production had soared to 21,000 machines; in 1872, the company's best year, 174,000 machines were produced.

In 1854, the British Parliament assembled an investigatory commission to advise Her Majesty's government on the best possible ways to manufacture small arms. Several witnesses before the commission reported that American armorers had found ways to produce more uniform products, more efficiently, and at a lower labor cost. For some reason, the commissioners decided to coin a new term for what had previously been called armory practice: the *American system of manufacture*. The name was appropriate. The new method of production that Eli Whitney had imagined and dozens of his successors had created was uniquely American; by 1854, it had spread far beyond the armories and was complex enough to be called a system. Whether in the armories or in the clock factories, the basic elements were always the same: relatively unskilled workers utilized special purpose, power-driven machinery and standardized gauges to create uniform parts for specially designed products intended to meet either very high (in the case of consumer products) or very exacting (in the case of military products) levels of demand.

The American system of manufacture was the foundation of what would subsequently be called mass production for a mass market. In the years between the end of the Revolution and the beginning of the Civil War, the U.S. government, through the War Department, had provided the funds without which the development of the system could not have proceeded. The historical situation in which that government found itself in the very first years of its existence—mounting defenses against wealthier, more experienced armies and navies—had been a crucial motivating factor for that expenditure, encouraging the increased productivity that resulted from the division of labor. The uniquely American balance of resources—cheap land, expensive labor, an expanding population—had shaped the system in its developmental stages, favoring techniques that would substitute unskilled for skilled workers. Eli Whitney, Simeon North, Thomas Blanchard, John H. Hall, and a host of other inventors and manufacturers had contributed mechanical ingenuity, entrepreneurial skill, and—perhaps most crucial of all—dogged persistence.

Samuel Slater and the Factory System

One consequence of industrialization was the growth of factories. During the colonial period, only a tiny fraction of Americans worked in places that employed dozens of people at once, but in the early decades of the nineteenth century, the number of such "manufactories" began to grow. One of the first men to develop techniques for designing, building, and operating factories was Samuel Slater.

Unlike Oliver Evans and Eli Whitney, Samuel Slater was an immigrant. Born into a fairly prosperous merchant family in Belper, England, in 1768, Slater had been apprenticed when he was fourteen years old to Jedediah

Strutt, who owned and managed several of England's largest textile factories. Strutt had trained Slater in both technical and managerial skills with the intention that he would become a mill superintendent, someone midway between the owner of the business and the shop floor overseers. In 1789, having learned that some of the American states were paying bounties to British citizens who knew how to build and maintain cotton spinning machinery, Slater emigrated.

Before the Revolution, there had been no textile industry to speak of in the colonies. Many farm women spun wool and flax; sometimes they or their husbands wove the resultant thread into cloth, but more frequently they sold or traded it with other farm families who did handloom weaving. Cloth had been one of England's most important export products for centuries; many of the Crown's mercantile policies had been intended to protect its domestic industry from competition. As a result, the volume of cloth production in the American colonies had been very low; much of it was of a fairly coarse grade, intended for local consumption by people who could not afford more luxurious imports.

In the years before Slater emigrated, several pioneering inventions had transformed the English textile industry. First there was the jenny, which had several spindles and a reciprocating action, driven by a hand powered spinning wheel. The jenny was followed by the water frame, which used rollers to spin and which was designed to be driven not by hand but by a waterwheel. Finally there was the mule, a hybrid machine, which combined the reciprocating action of the jenny with the power drive of the water frame. Together the jenny, the water frame, and the mule had mechanized the production of English yarn.

The British government, recognizing the crucial economic role of these new inventions, tried to keep the machinery from being exported to other countries: anyone who was caught trying to leave Britain with the machinery faced imprisonment, a fine, and confiscation of the equipment. There were, however, numerous ways to get around the regulations—and many people had succeeded in one way or the other. Samuel Slater exemplified one of those ways: following in the footsteps of many of his countrymen, he had simply lied about his occupation (he had told officials that he was a farmer) and had emigrated to New York with the design of the machines in his head.

Unhappy in New York, Slater heard a rumor that a firm of merchants in Rhode Island—the firm's name was Almy and Brown—had purchased some spinning machines and was looking for a skilled mechanic to set them up and keep them running. Given both his background and his experience, Slater was not interested in just becoming someone's employee; in return for setting up the equipment and constructing new equipment based on more recent British models, he wanted a share of the business. William Almy and Smith Brown were reluctant to agree as, following common practice, they believed that business partnerships worked best if they were restricted to relatives. Slater, however, was operating in a seller's market:

the firm couldn't find anyone else capable of doing the work they wanted done. After a few months of hard negotiating, Almy and Brown caved in. Slater promised to build the equipment and a mill, to supervise day-to-day production at the mill, and to pay half the expenses. His partners agreed to purchase the raw material, to lend Slater the money, to take the yarn from Slater's mill and have it woven into cloth on handlooms in their own workshops, to sell the finished cloth, and to pay the other half of the expenses.

The spinning mill that Slater built in Pawtucket, Rhode Island, early in the 1790s is widely regarded as the first textile factory in the United States. From the start, there was a good deal of friction between Slater and his partners, partly because they really didn't want to share control with someone who was not kin but also partly because the mill was organized and managed in a fashion with which the Americans, lacking Slater's experience with factories, were totally unfamiliar. Almy and Brown had thought that they were going to go into business as "putters out": giving wool and flax and hand-powered jennies to spinners who would work in their own homes, taking the yarn and giving it to weavers who would work on handlooms located in small shops that belonged to the firm, selling the resultant cloth to retail merchants. Slater had something else in mind altogether.

Slater's mill was working with cotton, not wool or flax. Since cotton was not grown in Rhode Island, the raw material had to be purchased from wholesalers, not from local farmers. The mill, furthermore, was completely mechanized. It had carding machines (for preparing the raw cotton) and water frames (for spinning it). Children between the ages of seven and twelve were hired for wages, not piece rates, to operate the machines. Almy and Brown simply could not accustom themselves to the demands of this kind of operation. They were not, for example, in the habit of purchasing raw materials regularly: "The machinery is now principally stopped for want of cotton wool," Slater regularly complained. Necessary supplies, which Slater had to purchase through Almy and Brown, were frequently not forthcoming. "Brushes much wanted!!!" Slater once wrote, "none to sweep the mill with"—a serious matter since accumulations of lint were a fire hazard.[10] Worst of all, Slater sometimes ran out of both cash and supplies with which to pay his help. Merchants by training, Almy and Brown understood what it meant to retrieve yarn or cloth from workers in exchange for payment in cash or in kind, but they did not understand what it meant to pay wages and to pay them with some regularity. In February 1796, Slater almost ran out of patience with his partners, asking them to send

a little money if not I must unavoidably stop the mill after this week. It is now going on four weeks since I recd 15–20. Can you imagine that upwards of 30 people can be supplyed with necessary articles . . . with that. . . . Or, do not you imagine anything about it. This is the 3d and last time I means to write until a new supply is arrived.[11]

Eventually Slater left Almy and Brown and built mills of his own. By the time he died in 1835, he either owned or had an interest in at least thirteen textile mills and two machine shops, and each of his sons and one of his brothers had also been brought into the business. In addition, by 1835 the factory system had almost completely supplanted the household system and the putting-out system. Slater's mill was the first, but by 1805 there were at least twelve more spinning mills in Rhode Island and southern Massachusetts. After Jefferson's embargo of English goods in 1808, the number grew even faster: between 1808 and 1812, thirty-six cotton mills and forty-one woolen mills were started, all making yarn.

These mills were all rural, located in small villages, often built by the mill owners themselves. By today's standards the mill buildings were rather small: two-story structures, built by the side of a small river or a large stream, with the waterwheel mechanism on the lower floor and the machinery above. Each mill employed thirty to eighty people, most of whom were children—boys and girls between the ages of seven and fourteen who lived in the villages with their parents. Usually the fathers of those children were employed in some other way by the mill owners (they might be working on farms that belonged to the mill, repairing machinery, weaving on handlooms, or supervising the employees). Their mothers, in time-honored fashion, kept house: preparing meals, nursing the sick, making and mending clothes.

Slater had begun in Pawtucket by employing the children of poverty-stricken local farmers, but as his workforce grew, he had trouble locating a sufficient supply of such children. Following English practice he next tried employing pauper apprentices, orphaned or abandoned children who were "put out" for a year at a time by the governmental authorities responsible for their care. Slater wasn't satisfied with the pauper children system either because it turned out to be very expensive. Slater had to pay local families to board the children, and he was required, by the terms of his contract with the local authorities, to provide a basic education for them at his own expense. In the end, he settled on the family labor system as the best solution to his problem. "Slater's system," as it came to be called, was used in many of the earliest factories.

Housing was built adjacent to the mill site and rented to families. The older children were immediately employed in the mills; the younger, when they reached the appropriate age or the family's finances required it. Fathers who deemed mill work socially unacceptable were employed in other ways, often as handloom weavers (power looms did not begin supplanting handlooms in these villages until the 1820s). Mill owners saw to it that churches and schools and shops were built in each village. Mill families were frequently required to attend church and to shop only in the village. Children's employment contracts made specific provisions for the months during which they would be free to attend school.

The whole system was profoundly patriarchal. Mills were usually owned

Webster, Massachusetts, a typical rural mill village in the early nineteenth century. The river, which powered the waterwheel, runs down the center of the painting; the mill building is the large two-story building (*left, center*). The smaller two-story buildings may have been homes for the mill owners; the one-story buildings homes for mill employees. (Courtesy Smithsonian Institution.)

either by one individual or (in a pattern that Slater adopted from Almy and Brown) by a partnership of family members. At least one member of those partnerships lived in the mill village, ran the factory, and controlled most of the activities in the factory and out of it, including the retail shops in which workers purchased the food and supplies that they needed. Fathers (and sometimes mothers, especially if they were widows) exercised some control over their children's labor and their children's discipline, bargaining with owners for improvements in wages or withdrawing their children from the workforce until unsatisfactory conditions were improved.

As the years went on, however, the textile industry became, at one and the same time, larger and more impersonal; the Slater system was replaced by what came to be called the "Lowell system." Francis Cabot Lowell, like Almy and Brown, was part of a family that had been very successful merchants for generations. On a trip to England as a young man, Lowell had visited textile factories and had seen power looms in operation; on his return, he had described the basic features of those looms to a Boston mechanic who succeeded in building workable duplicates. Lowell then solicited investments from several family members and business associates to create the Boston Manufacturing Company, which was a limited liability corporation, not a partnership. The capital investment that Lowell had in mind for the kind of factory he intended to build was more than any one individual, or even two or three individuals, could have contemplated. In addition, a limited liability corporation, as its name suggested, limited the risk for each participant to the amount of money invested. If the corporation became bankrupt, only *its* assets could be sold to pay off creditors, not the assets of the individual investors, a very attractive feature.

The Boston Manufacturing Company opened its first mill in Waltham, Massachusetts, about ten miles west of Boston, in 1814. When this proved financially successful, the investors went about looking for a site for another, finally settling on an uninhabited site on the Merrimack River, twenty-five miles north of Boston, which eventually became the city of Lowell, Massachusetts. After several years of construction work, the first mills opened there in 1822.

What made the Lowell system different from the Slater system? First, in the Lowell mills both spinning and weaving (and such subsidiary processes as dying, fulling, and bleaching) were completely integrated: raw cotton came in at one end of the mill complex, and finished cloth went out the other end. Second, in the Lowell mills every process that could be mechanized had been. These mills were four stories high: The waterwheel was in the basement; raw cotton was picked and carded on the first floor, spun on the third, and woven on the fourth; all the machinery was connected to the waterwheel by complicated systems of shafts and belts. Third, the mills were operated, day in and day out, not by the owners of the company, but by managers who were employees, chosen for their organizational rather than their technical or financial skills. Fourth, the workforce was not composed of families or even of children who continued to reside at home with their parents; adjacent to the mills, Lowell and his company had built boardinghouses for their workers, most of whom were unmarried young women, relatively unskilled, recruited from the farms of the surrounding countryside. The company owned the houses and employed respectable widows to operate them, with room and board fees deducted from the worker's pay; the whole community was governed by a strict set of moral rules, established by the company and enforced both by the factory overseers and boardinghouse keepers. Fifth, the business was owned by a corporation, which meant that if the owners wanted to expand, all they needed to do was to sell new stock, attracting additional investors. Sixth, as a consequence of corporate ownership, the initial factories and villages of the Lowell system were very large, and they could grow even larger over time. By 1855, there were fifty-two mills in Lowell, employing over 12,000 people.

In its early years, Lowell was organized just as patriarchally as Slater's mill villages had been, but the corporate form of ownership, the development of a class of nonowner managers, the fact that the workers were (at least at first) living away from their families and their hometowns, and the growing size (and later on, diversity) of the workforce essentially made the continuance of patriarchal arrangements difficult, if not impossible. Parents could not control their children's labor, and they could not negotiate with their children's employers. As the workforce got larger and larger, managers stopped being able to control where their workers lived and how frequently they went to church. The Lowell system created a new kind of wage laborer: a person who contracted individually with her or his employer and who lacked either an older family member or any other form of communal support system to help with the negotiations.

Whether organized under the Slater system or the Lowell system, textile manufacturing was extraordinarily profitable, at least in the decades before the Civil War. Samuel Slater had crossed the Atlantic with virtually no capital at all in 1789, but when he died in 1835, he left an estate of $690,000. In the ten years between 1816 and 1826, the Boston Manufacturing Company returned an average annual dividend of just under 19 percent. In the 1850s, as the business became increasingly competitive, profits fell somewhat, but the Merrimack Company (which operated some of the Lowell mills) averaged 12 percent annually for most of the decade.

The factory system spread quickly to other industries in which it could be appropriately applied. Not surprisingly, the Lowell system, with its potential for almost infinite expansion, eventually supplanted the Slater system. The boot and shoe industry, which had been organized under the putting-out system for more than a century, became a factory-based enterprise in the 1830s and 1840s and was almost completely mechanized in the 1850s and 1860s. Papermaking, which had once been a handicraft, was mechanized and organized in factories during the 1840s. In some localities, dairy products such as butter and cheese (which had previously been made in people's homes) were being made in factories, by water- or steam-powered machines tended by unskilled laborers, in the 1880s. Meatpacking, the slaughtering of animals and the processing of their carcasses, was industrialized by the 1870s. In these industries the rise of the wage laborer was accompanied by the demise of household production.

The story of the Slater and Lowell systems suggests two additional characteristics of American industrialization. First, as a nation of immigrants, the United States was able to take advantage of the technological traditions of many countries. In the early decades of the nineteenth century, a good deal of what is now called technology transfer was accomplished through immigration. Some enterprising Americans actually made their livings by canvassing local manufacturers to discover their employment requirements and then traveling to Britain and Europe to induce trained people to emigrate. Others would meet ships newly arrived in port, interviewing all the immigrants on board, offering jobs to those whose skills might turn out to be valuable. City and state governments sometimes placed advertisements in European newspapers; manufacturers sometimes sent their sons abroad, not just to observe how things were being done in other places but also to offer cash bonuses to artisans who might be willing to pull up stakes and resettle. We do not have reliable statistics on how many actors in the drama of American industrialization were actually, like Samuel Slater, immigrants, but every machine shop for which records still exist seems to have employed a strikingly large number of Englishmen, Scots, and Germans and an occasional Dutchman.

Second, American industrialization remained a rural phenomenon for many decades. In part this was an accident of geography: the eastern half of the United States—most particularly the eastern coastal plain, which stretches from the Appalachian range to the Atlantic Ocean—is unusually well endowed with mill sites, places in which falling or moving water can be used to operate waterwheels, and in which water systems can be used to transport both raw materials and finished goods. Thus, when entrepreneurs began building factories in the United States, they almost invariably chose waterpower over steam power, which meant that they almost invariably chose rural rather than urban locations for their mills.

Such a choice was also favored by American ideological preferences for rural, rather than urban, culture. By 1820, the evils of British industrial cities were well known to Americans, and many politicians were determined not to allow the same overcrowded, unhealthy, exploitative conditions to exist on these shores. Some industrial villages eventually grew to the size of cities (Lowell, Massachusetts, is one such example; so is Paterson, New Jersey), but in the decades before the Civil War, most still retained their rural character. Population densities were, relatively speaking, low, and perhaps even more crucially, the people who lived in these villages retained familial connections with agricultural land and with agricultural practices. The single women who worked in Lowell were the daughters of farming families, and many returned to farming when they married. Many mill villages were surrounded by large plots of land, which factory families tended and on which they grazed animals. Some factories employed transient individuals and families (many of them immigrants) who intended to work in the mills for a few years in order to accumulate enough money to buy land further west. Other factories employed men and women who were supplementing the income from their farms with seasonal work indoors. Industrial workers were often dissatisfied with the conditions under which they labored—so dissatisfied that they periodically went out on strike. Despite this, long-lasting industrial labor unions were slow to form in the United States, at least in part because in the beginning unskilled industrial laborers did not see themselves as likely to be permanently employed in the mills. Although the United States industrialized quickly, it still understood itself to be predominantly an agricultural society. The process that Samuel Slater began in the 1790s was in some ways quick—and in other ways slow—to bear fruit.

Conclusion: The Unique Character of American Industrialization

"He that studies and writes on the improvement of the arts and sciences labours to benefit generations yet unborn," Oliver Evans wrote to one of his sons, "but it is not probable that his contemporaries will pay any attention to him."[12] Evans was right in one sense—his accomplishments were largely unsung not only in his own day but also in ours—but wrong in another. Imitation can often be the highest form of compliment, and Evans's work was imitated repeatedly; the fruits of his labors multiplied, during his own lifetime and certainly thereafter, in ways even he, farsighted as he was, could not have imagined. In the 1760s, when Evans was still a boy, England was beginning to industrialize, building on a substantial manufacturing capability that had been expanding for the previous two centuries. Contrarily, when the United States began to industrialize in the 1780s, it had virtually no manufacturing base to speak of, except for the iron furnaces in which pig and bar iron were produced—its former mother country, mercantilist to the core, had done everything it possibly could to stymie colonial industry. In the space of the forty years between 1780 and 1820, the country went from being one of the world's weakest economies to being potentially one of the strongest—forty years after that, it was second only to Britain both in its manufacturing output and in the monetary value of its exports. The rapidity of American industrialization depended on the skills, the patience, the farsightedness, and the persistence of pioneers like Evans, Whitney, Slater, Hall, Blanchard, and Lowell.

But industrial technology, like all technology, is shaped by the contexts in which it is developed, applied, and diffused. Rapidity is only one of the special characteristics of American industrialization; some of the others, as we have seen, depend as much on the unique characteristics of American geography, economy, and society as they do on the unique characteristics of American innovators. Because of the availability of cheap land, American skilled workers tended to be both scarce and expensive: for this reason, innovators like Whitney, Evans, and Hall tended to create machines that could replace workers or that could be operated by workers with little or no training. Cheap land also meant that a permanent class of industrial workers was slower to develop in the United States than, for example, in either England or Germany; in the early decades of American industrialization, factory workers—whether they were children, women, or men—tended to think of themselves as transient workers, earning some money before starting up or returning to a farm. Because the American military needed large numbers of armaments quickly in the first decades after independence, it was willing to pay large sums of money for the development of new forms of special purpose machinery; manufacturers in civilian sectors of the economy were subsequently able to benefit from what had been learned in the armories. Because of the availability of mill sites, American factories were powered by water in rural settings far longer than their European counterparts, and American industrial cities took longer to become overcrowded. And, finally, because its expanding population meant an expanding domestic market, American manufacturers were more willing than many of their European and English counterparts to invest capital in new equipment—those same special purpose machines—with which to lower costs and increase production capacity. American industrialization was indeed spurred by the skills, the patience, the farsightedness, and the persistence of American innovators, but it was also shaped by the unique characteristics of the United States in the early decades of its independent nationhood.

Notes

1. This quote comes from a pamphlet that Evans wrote in 1814, as quoted in Greville Bathe and Dorothy Bathe, *Oliver Evans: A Chronicle of Early American Engineering* [1935] (New York: Arno Press, 1972), p. 215.

2. Quoted from the Proceedings of the General Assembly of Delaware, in Bathe and Bathe, *Oliver Evans*, p. 18.

3. Patrick N. I. Elisha [pseud. Oliver Evans], *Patent Right Oppression Exposed* [1813], as quoted in Bathe and Bathe, *Oliver Evans*, p. 11.

4. These terms come from the Proceedings of the General Assembly of Delaware, as quoted in Bathe and Bathe, *Oliver Evans*, p. 19.

5. Quoted without attribution in Constance M. Green, *Eli Whitney and the Birth of American Technology* (Boston: Little, Brown, 1956), p. 46.

6. Quoted from correspondence between Wolcott and Whitney in Green, *Eli Whitney*, p. 102.

7. From a letter from Whitney to Josiah Stebbins, quoted in Greene, *Eli Whitney*, p. 110.

8. As quoted in Merritt Roe Smith, "Eli Whitney and the American System of Manufacturing," in Carroll W. Pursell, ed., *Technology in America: Individuals and Ideas* (Cambridge: MIT Press, 1980), p. 51.

9. From a letter from Hall to John C. Calhoun, as quoted in Merritt Roe Smith, *Harper's Ferry Armory and the New Technology: The Challenge of Change* (Ithaca, NY: Cornell University Press, 1977), p. 199.

10. These quotations are from letters that Slater sent to Almy and Brown. All are quoted in Barbara M. Tucker, *Samuel Slater and the Origins of the American Textile Industry, 1790–1860* (Ithaca, NY: Cornell University Press, 1984), p. 54.

11. Tucker, *Samuel Slater*, p. 84.

12. This is part of what was found, written in Evans's own hand, on the back page of the copy of his book, *The Abortion of the Young Steam Engineer's Guide* (1805), which he willed to one of his sons on his death; see Bathe and Bathe, *Oliver Evans*, p. iv.

Suggestions for Further Reading

Greville Bathe and Dorothy Bathe, *Oliver Evans: A Chronicle of Early American Engineering* [1935] (New York: 1972).

Thomas C. Cochran, *Frontiers of Change: Early Industrialism in America* (New York, 1981).

Carolyn Cooper, *Shaping Invention: Thomas Blanchard's Machinery and Patent Management in Nineteenth-Century America* (New York, 1991).

Felicia Johnson Deyrup, *Arms Makers of the Connecticut Valley: A Regional Study of the Econoomic Development of the Small Arms Industry, 1798–1870* (Northhampton, MA, 1948).

Eugene S. Ferguson, ed., *Early Engineering Reminiscences (1815–1840) of George Escol Sellers* (Washington, DC, 1965).

Constance M. Green, *Eli Whitney and the Birth of American Technology* (New York, 1956).

Brooke Hindle, *Emulation and Innovation* (New York, 1981).

Brooke Hindle and Steven Lubar, *Engines of Change: The American Industrial Revolution, 1790–1860* (Washington, DC, 1986).

David A. Hounshell, *From the American System to Mass Production: The Development of Manufacturing Technology in the United States* (Baltimore, 1984).

Steven Lubar, *The Philosophy of Manufactures: Early Debates over Industrialization in the United States* (Cambridge, MA, 1982).

Judith A. McGaw, *Most Wondrous Machine: Mechanization and Social Change in Berkshire Papermaking, 1801–1885* (Princeton, 1987).

Carroll Pursell, *Stationary Steam Engines in America Before the Civil War* (Berkeley, 1963).

Philip Scranton, *Proprietary Capitalism: The Textile Manufacture at Philadelphia, 1800–1885* (Cambridge, 1983).

Merritt Roe Smith, *Harper's Ferry Armory and the New Technology: The Challenge of Change* (Ithaca, NY, 1977).

Barbara M. Tucker, *Samuel Slater and the Origins of the American Textile Industry, 1790–1860* (Ithaca, NY, 1984).

Anthony F. C. Wallace, *Rockdale: The Growth of an American Village in the Early Industrial Revolution* (New York, 1978).

5

Transportation Revolutions

ON MARCH 2, 1807, THE UNITED STATES SENATE asked Secretary of the Treasury Albert Gallatin to prepare a plan "for the application of such means as are within the power of Congress, to the purpose of opening roads and making canals."[1] A year later, after making a thorough investigation, Gallatin presented his report.

Today, most Americans probably think it was inevitable that the United States would succeed as a nation "indivisible," but in Gallatin's day that outcome appeared by no means certain. The country was surrounded by potential enemies—the Spanish and the French to the south and the west, the British to the north—and there were large numbers of potentially hostile people within its borders. In addition, the federal government was deeply in debt both to its own citizens and to foreigners for the expenses involved in fighting the Revolutionary War.

To make matters worse, the several states were jealous of their own authority and were frequently unwilling to cooperate either with each other or with the federal government. In this regard the newly created western states—Ohio, Kentucky, and Tennessee—were the most worrisome. These states had no Atlantic coastline and therefore no trade with European and Caribbean ports. As a result, they lacked both economic and political ties to the older states, those that had been developing for the previous hundred and fifty years and that had fought the Revolution together. "Numerous have been the speculations on the duration of our union," Robert Fulton admitted in a technical appendix to Gallatin's report,

93

and intrigues have been practised to sever the western from the eastern states. The opinion endeavored to be inculcated, was, that the inhabitants beyond the mountains were cut off from the market of the Atlantic states; that consequently they had a separate interest . . . that remote from the seat of government they could not enjoy their portion of advantages arising from the union, and that sooner or later they must separate and govern for themselves.[2]

Since the demise of Roman Empire in the fifth and sixth centuries, no government had succeeded in unifying a territory the size of the United States for more than a few years. The classically educated founders of the American republic knew this very well. They also knew something about how the Romans had succeeded in holding their territories together: partly by creating a republican form of government, partly by making citizens out of the inhabitants of the territories they conquered, and partly by building excellent roads, roads that led from the conquered territories to the capital of the republic. Fulton, Gallatin, and President Thomas Jefferson hoped that the formula that had worked for the Romans—republicanism, widespread citizenship, excellent transportation—would work for the Americans as well. In the domain of transportation, as Gallatin's report to the Senate made clear, they hoped that a national transportation system, a system of roads and canals stretching from Maine to Georgia and from the Mississippi to the Atlantic, would unite the diverse parts of the country, making secession of any of the states, but most particularly the western states, both economically and politically unthinkable.

Transportation Difficulties

The difficulties that Americans faced when they attempted to transport themselves or their goods or their mail from one place to another at the turn of the nineteenth century are difficult for their late twentieth-century descendants to imagine. Overland, there were three modes of transport: on foot, on a horse, or in a wheeled vehicle; in the water, there were only sailing ships and boats, pole driven barges, rowboats, and canoes. Whether on land or on water (most trips required a combination of both), traveling was an arduous and expensive enterprise.

In 1801, when Jefferson became president and Gallatin took over the treasury, the roads of the new nation were in a deplorable state. Most were little more than clearings in the woods and trodden-down places in the fields, paths that had been made over the centuries by the native peoples and by animals. Many of these roads were so narrow that they were suitable only for foot traffic, or an occasional horse and rider. Virtually none were paved; when it rained, horses couldn't gallop, and carts and carriages were in danger of sinking up to their axles. Bridges were few and far between; where the roads met streams and rivers, fording was necessary. In 1800, the overland trip from Boston to Washington—over some of the best of these terrible roads—required ten days, fifteen to sixteen hours a day.

Road repair had been, and continued to be, haphazard. The colonial governments had generally adopted the English system for road planning and maintenance. Local governments—town and county councils—had been given the right to decide where roads would go and the obligation to maintain them. Lacking revenues, these local governments made roadwork a condition of property ownership: every property owner was responsible for a certain number of days a year, at the bidding of the supervisor of the roads, of clearing underbrush, moving boulders, cutting limbs from trees, laying logs in swampy places, and possibly even building or repairing bridges. The local authorities had always varied in the attention that they paid to the roads—at some times a great deal, at other times not a bit; in some places very careful, in other places profoundly sloppy—and the local citizenry had been just as varied in its willingness to respond. In the years since the Revolution, while few state governments had done anything either to alter or to enforce these arrangements, the federal government had had other, more pressing expenses.

"From Philadelphia to Newport [Delaware]," one traveler between New York and Washington reported in 1796,

> the roads were tolerable. On quitting the latter place, they became frightful. A little beyond Havre-de-Grace [Maryland], the axle tree of the carriage broke through . . . as no other coach could be had, the travelers and their baggage were stowed into a wagon, and driven to the next stage, after being forced to walk several miles on foot before even a wagon could be found. On the evening . . . another carriage overset with them. Some of the ladies, and other people in it were dangerously hurt. The company passed the greater part of the night in a wood, where they contrived to kindle a fire, and where they were during the whole time in the midst of a shower of rain and snow. . . . The passengers were, besides the waste of useful time, put to considerable expense by staying almost five days on the road. . . . Through a large proportion of the United States, travelling is alike perilous.[3]

For people to the west of the Appalachians, the situation was even worse since they had a mountainous terrain to contend with and only very widely scattered communities willing or able to maintain the roads. In the autumn, after the crops had been harvested, those communities would organize trains of pack animals to carry produce to Baltimore or Pittsburgh or some other potential market. The paths over the mountains led through gorges, valleys, and precipices. The pack trains moved in single file; one man led the way, another brought up the rear. In this manner, two men could transport roughly a dozen animals, each loaded with about two hundred pounds of freight.

Only the most valuable produce was worth transporting this way, lest the cost of the transportation increase the price of the goods beyond most customers' ability to pay. A western commercial product such as wheat was only viable if it had first been compacted for transport by being turned into liquor or if it was literally on the hoof, in the form of pigs that were driven to market by drovers. The animal drovers and the pack train drivers were the only people, aside from an occasional representative or senator making his way back from a session of Congress, who frequented these Appalachian roads. A trip to visit old friends and relations, or travel for any other kind of business or pleasure, was simply out of the question.

In those years, given the difficulties involved in moving on land, much of the nation's commerce moved by water. At the time of the Revolution, virtually all settlements of any size were either on, or within easy overland travel distance of the coastline or a river that ran to the coastline. Sailing ships carried goods in coastwise trade—from Portsmouth, New Hampshire, to New York City, for example, or from Charleston to Baltimore—and larger ships could carry the same freight from the Atlantic ports to the Caribbean and beyond.

Both large and small ships could travel on stretches of a few broad rivers—from New York to Albany along the Hudson; up the Delaware from Wilmington to Philadelphia and Trenton; along the James from Norfolk to Richmond—but in general, inland transportation by water was not easy. Many rivers were interrupted by falls and rapids and even sometimes beaver dams. Some had shallow places or fallen trees or giant boulders and any variety of other hazards. Some flooded in the spring and fall; others were blocked by ice in the winter; yet others became shallow or dried up altogether in the summer. Dense forests broke the wind, thus necessitating barge traffic; but poling a barge, which was easy enough going downstream, with the current, was murderously difficult going upstream against the current and thus rarely undertaken either with passengers or with large quantities of freight. Western settlers faced a particular problem: none of their arterial rivers could provide clear sailing either to the Mississippi or to the Atlantic.

Even the coastwise trade was hampered by numerous treacherous places, which meant that ships could not travel direct routes and many coastal voyages were longer and more expensive than they might have been. On top of this, variations in wind patterns meant that ships could not travel to a schedule. No shipper could be precisely sure when a shipment would either arrive or depart, and travelers might be becalmed for days on end.

As the eighteenth century turned into the nineteenth and more and more people settled more and more land more and more distant from navigable waterways, the economic problems created by the transportation system, such as it was, intensified. A sack of wheat might double in price as it was hauled overland from western New York to the Hudson River at Albany, and it might double in price again while it waited for a thaw that would permit shipment to Manhattan. The same thing might happen to salt or maple syrup or preserved meat that settlers in Ohio and Kentucky might want to send across the Appalachians for sale in Baltimore or Philadelphia.

Many national leaders, Jefferson and Gallatin among them, understood that vast stretches of the new nation were underpopulated because farming at such great distances from the centers of commerce was doomed to be unprofitable. And as long as those places were underpopulated, they were vulnerable to attack—a fact that became even more troubling in 1803 when, as a result of the Louisiana Purchase, the size of the new nation almost doubled. Jefferson and Gallatin believed that for the economic and political health of the expanding nation, the transportation logjam simply had to be broken.

The *Report on the Subject of Public Roads and Canals* was a transportation master plan for the new country. The federal government, Gallatin proposed, should spend $20 million over ten years: $3 million for canals; a little under $5 million for a "great turnpike road, from Maine to Georgia, along the whole extent of the Atlantic sea coast"; $1.5 million to improve the navigability of the rivers that run from the Appalachians into the Atlantic; another $2.8 million for "four first rate turnpike roads from those rivers across the mountains, to the four corresponding western rivers"; a canal should be built around the falls of the Ohio River; and, finally, the roads to Detroit, St. Louis, and New Orleans would be "improved."

Gallatin's plan was eventually realized—in fact, surpassed. The problem of transporting goods and people across the width and breadth of the United States, thereby creating both national unity and a national market, was solved in the decades between 1800 and 1870. The still new country managed to build what was widely acknowledged to be one of, if not *the*, most advanced transportation system in the world. That system was not, as Gallatin had imagined it would be, entirely the result of federal initiatives. Rather, as we shall see, it arose from a uniquely American partnership between the federal government, local governments, and private business.

Toll Roads and Entrepreneurs

Gallatin and Jefferson were actually latecomers. In the movement for transportation reform, the first competitors off the starting block were several of the states rather than the federal government; they had been moved to act not by farsighted political figures but by profit-seeking entrepreneurs. Once the turmoil of the Revolutionary period was over, merchants in cities and towns began to press their local and state governments to do something about the deplorable condition of the roads on which their goods had to travel. Since public ownership and maintenance had clearly failed either to build or to maintain good roads, they argued, why not turn the task over to businessmen?

In many state capitols, legislation was passed to create joint stock companies (in those days, every such company required its own piece of enabling legislation) for the purpose of building turnpikes, paved roads on which tolls were charged. Each company was given well-defined stretches of public property on which to build, as well as the right to sell stock so as to raise capital, which would pay for the widening, straightening, and paving of those roads. The companies were also given the right to charge tolls on the roads; these tolls would be used to maintain the roads and to generate profits so as to compensate the original stockholders for their investment.

Pennsylvania was the first state to charter a turnpike company. The Philadelphia and Lancaster Turnpike Company was formed in the early months of 1792. Its stock went on the market on June 4; by noon, the offering of 1,000 shares at $300 each was oversubscribed. The road was completed, sixty-two miles of it, two years later at a total cost of $465,000, or $7,500 a mile: "a masterpiece of its kind . . . paved with stone the whole way, and overlaid with gravel, so that it is never obstructed during the most severe season."[4]

Other entrepreneurs quickly learned the lesson of the Lancaster Pike. By 1821, the Pennsylvania legislature had authorized the creation of 146 turnpike companies; 84 of them succeeded in raising the money they needed and 1,807 miles of paved roads had been constructed. In 1806, in a significant move, the legislature passed an act allowing the state itself to buy stock in these companies. Henceforth, the taxpayers of Pennsylvania would subsidize the construction of a privately owned, toll-charging road system. Maryland, perceiving that it would have to compete with Pennsylvania's roads for the trans-Appalachian trade, chartered its first turnpike company in 1796; New York, content at first to have its western produce move on the Mohawk River, waited until 1798; New Jersey got into the act in 1801.

Between these toll-bearing turnpikes and the older, unpaved, free state roads, a continuous highway of sorts had come into existence between Maine and Georgia and between the western and eastern borders of the mid-Atlantic states by 1812. The war that broke out in that year led to a British blockade of the Atlantic coast, which put enormous pressure on land transportation, both for moving troops and for moving the freight on which the economic life of the country depended.

After the war was over, many national legislators, reflecting on this experience and recalling Gallatin's proposal of 1808, began to consider the possibility of building a national road system. They imagined a system that would be paid for and owned by the national government, providing effectively for the defense of outlying regions, including roads where there were not yet states or where settlement was too sparse for a stock company either to raise subscriptions or to turn a profit.

Construction on the so-called Cumberland Road, an east–west national road that was intended to become just the first of many federally financed "internal improvements," was under way by 1815. The first interstate road, this turnpike was routed from Cumberland in western Maryland, through southwestern Pennsylvania, to Wheeling, on the Ohio River in western Virginia. The road builders had reached Wheeling by 1818, but thereafter the funding for the road was mired in political controversy.

Remembering how much the colonists had hated the English monarchy, many national legislators were wary of giving the federal government too much power; others, concerned about the problems of unification and integration, were wary of giving it too little. Strictly interpreted, the Constitution did not allow the federal government to engage in internal improvements. Broadly interpreted such activities could be justified under the provisions that allowed the national government to "provide for the common defense," and "establish . . . post roads." The net result was that various schemes for extending the National Road were proposed by one president or Congress and then vetoed by another. Having reached Wheeling by 1818, the National Road did not get to Columbus, Ohio, until 1833, and didn't begin to approach the Mississippi until the 1850s (it is now the basis for U.S. Route 40).

In technical terms, none of these roads was particularly innovative; the fundamental road-building techniques had been pioneered centuries before by the Romans. Using simple hand tools and a great deal of muscle, the roadbed would be cleared of such obstructions as tree stumps and boulders. Then teams of shovel-bearing laborers would dig and haul so as to shape a bed that was slightly higher in its center than at its edges to permit drainage. Stones were chipped and laid to create a foundation for the road and to line the drainage ditches on both sides; gravel was then raked over the road bed to ensure that water would not remain on its surface. In some places, the turnpike companies built new wooden bridges; in other places, they enlarged and strengthened bridges that already existed. In hilly and mountainous terrain, repetitive sharp curves were carefully designed so that carts, wagons, and carriages would not overturn in negotiating them, and there were some places in which the twists and turns of old roads could be straightened by hacking out new paths, marked by surveyors, through woods and fields.

By the time the turnpike boom ended, just before the Civil War, many thousands of miles of hard-surfaced roads had been built in the United States. After the war, most of those roads fell into disrepair; by 1880, the vast majority of the road companies were bankrupt and the roads they had once so proudly planned and built had reverted to state ownership. As long as there were no other alternatives, the turnpikes were a boon for emigrants traveling west and for farmers and merchants who wanted to haul freight a short distance. But in fairly short order there were other alternatives. The turnpikes and the National Road began the process of creating national economic and political unity, but they did not complete it.

In an ironic way, part of what killed the turnpikes was the very same financial device that had built so many of them: the joint-stock company, operating with various kinds of state subsidies. The joint-stock company turned out to be a very flexible instrument, able to finance the construction of many kinds of public works, including some of those that would eventually surpass the turnpikes as modes of transportation: the canals and the railroads.

Canal Building and State Financing

Like road building, canal building was an ancient art: the Romans had built artificial watercourses; so had the Venetians in the Renaissance and the Dutch in the seventeenth century. There had even been something of a canal boom in Great Britain in the last decades of the eighteenth century, which meant that Americans who had traveled to England already knew what canals could accomplish. On paper, canals seemed relatively easy to build. A trough wide enough to allow passage of barges had to be dug; it had to be lined with something that would hold water; dams had to be built to control the flow of water into the canal from natural water sources; locks, which were nothing more than double dams, had to be built wherever natural elevations of the land required that the barges be raised or lowered.

Knowing that something can be built is not, however, quite the same thing as building it. The potential advantages of canals were clear: they could bypass difficult stretches of a river, connect two rivers, connect a lake and a river, or provide a river with a navigable outlet to the ocean. But the risks were as clear as the advantages. Who could be sure what that best route for a canal would be? Who could be sure where locks ought to be constructed? Who could be sure, lacking experience, how much construction would cost? Or even how the necessarily large teams of laborers would be hired, fired, and trained? Not surprisingly, no one was willing, single-handedly or jointly, to take such a series of risks in the turbulent decades that preceded the Revolution.

As soon as both independence and peace seemed assured, however, pent-up interest in canals found some release. George Washington was one of several landowners who invested in the Potomac Company, chartered by the state of Virginia in 1785, for the purpose of building canals wherever they could ease navigation on that river. In 1787, a group of prominent South Carolinians formed the Santee Canal Company for the purpose of uniting the Santee and the Cooper Rivers, thereby providing easy access to the port of Charleston for upland plantations in both Carolinas and in Tennessee. By 1793, the government of Massachusetts had chartered a joint-stock company to build a canal connecting the Charles and the Merrimack Rivers so that the small towns of southern New Hampshire and northeastern Massachusetts would be connected with the great port of Boston. At about the same time, Virginia and North Carolina allowed a group of investors to start building a navigable watercourse through the Dismal Swamp so that small boats could travel between Chesapeake Bay and Albemarle Sound.

But permission to raise money and to build was only the first step down a very arduous path, for in the decade after the Revolution, there was not a single American who had any experience in building a canal. No one in the United States had ever built a watercourse more than several yards long, or designed a dam that would allow an artificial watercourse to be filled from a river. No one had ever lined a trench so that it would not leak—"weep" was the term used in the eighteenth century—or located a mortar that would hold bricks or stones together under water, or built a lock that could be filled without crumbling or a lock door that could be controlled against thousands of pounds of water pressure.

All the early American canal builders were amateurs. The chief engineer of the Potomac Company, James Rumsey, had been a successful millwright; he was in the midst of developing a jet-propelled steamboat (about which more is said below) when he took the engineer's job. John Christian Senf, who was put in charge of the Santee Canal, was a Swedish military engineer who had come to the colonies with the Hessian forces but had chosen to fight on the side of the colonists; he knew a great deal about building fortifications but nothing at all about building watercourses. Laommi Baldwin, supervisor of the Middlesex Canal, was a farmer and a cabinetmaker who had had some experience surveying property boundaries for himself and his neighbors.

Baldwin's predicament exemplifies the problem all of his colleagues faced. Without any previous experience, he had to design a level, twenty-seven-mile trench running through forests and hilly countryside, then supervise the teams of unskilled laborers who would both dig and line that trench. He had to choose the places in which dams would be located and in which the trench would be interrupted by locks. Then he had to design those dams and locks and supervise the workers who would build them. Using traditional hand tools—crowbars, scythes and pitchforks, saws, planes, chisels, anvils, tongs, and hammers—trees had to be felled, stumps pulled, stones removed, rock broken. Tons of hydraulic mortar had to be mixed; miles of stone walls had to be built; massive wooden lock and dam doors had to be constructed; stone foundations for the doors had to be laid.

There were, needless to say, many disappointments and much frustration along the way. Four different routes for the Santee Canal had been proposed, and Senf had to negotiate a route that would satisfy the demands of the designers, several of whom were stockholders in the company. The organizers of the Middlesex Canal, patriots to the core, had hoped to build the canal with the "morals and steadiness of our own people," a polite way of saying "without expert advice from English canal builders," but in the end, they had to hire an experienced British consultant for a short time. Repeatedly over the long years of construction, the original stockholders had to produce more capital to underwrite what would in the twentieth century be called cost overruns. The first lock took three years to build. On the day it opened, Baldwin noted in his diary, "opened—the first lock. Broak and failed."[5]

Despite the difficulties, the Middlesex Canal opened for business roughly a dozen years after construction began. The Santee Canal and the canal of the Potomac Company preceded it by a few years. All these early canals were discussed at some length in Gallatin's report to the Senate in 1808. Construction and maintenance costs were so high, he noted, that the original investors had not yet seen a return on their investment, nor were they likely to see such a return for many years, if ever; the Santee Canal had cost twice what its original projectors had estimated; the Middlesex Canal almost three times. Despite these problems, Gallatin continued, the canals benefited the regions they served, stimulating the economies of areas that had previously been depressed. For this reason, Gallatin argued that the financing of canals ought to be the responsibility of governments, particularly the federal government; since canals provide for the public good, they ought to be publicly financed.

When the economic constraints created by the Embargo of 1807 and the War of 1812 were finally lifted, several state legislatures began to take Gallatin's advice seriously. In Pennsylvania, a canal to connect the Schuylkill with the Delaware had been planned by a joint-stock company but never built; in Maryland, the canal that such a company had built to circumvent the falls of the Susquehannah turned out to be too narrow for the flat-bottomed boats that needed to use it; in New Jersey, the company that had proposed to build a canal between the Raritan and Delaware Rivers had never raised sufficient capital to start the work. State legislatures began to see that they were going to have to borrow money to do what private enterprise had not yet succeeded in doing.

The first, and justly the most famous, of the canals to be built with state funds was in New York. For centuries, the Iroquois had utilized a series of rivers, streams, and brooks to travel by canoe between what the European settlers called the Hudson River in the eastern part of the state and Lake Erie at its western border. In the 1790s, various promoters and legislators had argued that western New York State, which was in those days almost devoid of European settlement, could be opened for commerce if a canal were constructed along that same route; freight could then travel by barge rather than by canoe. The sheer scale of such a project was daunting to private investment; the Erie Canal would have to be 363 miles long, more than ten times longer than the Middlesex. Enormous amounts of capital would be required for construction; who would be willing to take such a risk?

De Witt Clinton, a wealthy real estate owner who was at the time governor of New York, believed that the state itself should be the risk taker. Since a canal between the Hudson River and Lake Erie would benefit the economy of the entire state, Clinton reasoned that such a public good should be financed at public risk. A bill authorizing the issuance of bonds to pay for construction of a state-owned canal was passed in April 1817, and by June, the state had sold enough of the bonds to begin construction. Each bond purchaser was promised that the State of New York would pay back the loan at a fixed rate of interest in a fixed number of years.

In view of the immense size of the undertaking, the legislature had decided to divide the canal route into four sections and to delegate construction of each section to a special commissioner. Of the four men chosen for this task, three were judges and one was a schoolteacher; like Laommi Baldwin before them, all four men were amateur canal builders. Before construction actually began, the commissioners arranged to visit the Middlesex Canal so they could learn how to build stone locks and to locate the right materials with which to mix hydraulic cement. The school-

Virtually the only contemporary picture of some of the equipment used to build the Erie Canal. These are horse-powered cranes (there is a horse inside the base of the crane in the foreground) constructed to lift rubble that was blasted out of solid rock near the western end of the canal. From Cadwallader Colden, *Memoir prepared for the Celebration of the Completion of the New York Canals*, 1825. (Courtesy New York Public Library, Astor, Lenox and Tilden Foundations.)

teacher also had to be taught the basic techniques of surveying by one of the judges. What the experts in Massachusetts could not teach the amateurs from New York, however, was how to hire, train, and supervise the enormous workforce—4,000 men at the peak of construction—that was required if construction was to proceed at the schedule required by the payback period on the bonds. As the work progressed, the commissioners also discovered that they needed to create new pieces of machinery if the work was to stay on schedule: stump pullers and turf cutters, for example.

The construction of the Erie Canal was thus, as historian Elting Morison has put it, "the first—and quite possibly the best—school of general engineering in the country."[6] Amateurs though they were, the commissioners managed to build the canal in record time. The middle section, six locks and seventy-five miles long, opened for business in 1819, just two years after construction began. Within a year, it was humming with traffic, mules and oxen pulling flat-bottomed barges loaded with freight. The revenue from tolls on that middle section was sufficient to finance construction of the remaining sections. A feeder waterway, the Champlain Canal, perpendicular to the main east to west route, connected northern New York and Vermont with the Hudson River by 1823. And the remaining sections of the main route were open for business in 1825. That year, the Erie Canal collected roughly half a million dollars in tolls; the next year, three quarters of a million.

"Clinton's Folly" was a success—one that other states, and some joint-stock companies, quickly tried to imitate. Between 1816 and 1840, some 3,326 miles of artificial waterways were constructed in the United States, enough canal mileage to span the entire continent. In New England, three major canals were built to link interior regions with tidewater. By 1840, several canals connected the mountainous, coal-rich countryside of eastern Pennsylvania with the Delaware River, then marched across New Jersey to connect the Delaware River with New York harbor. In Ohio, the legislature authorized construction of a canal that would connect the Ohio River with Lake Erie, running from Portsmouth in the southern part of the state to Cleveland in the north, a distance of 308 miles. Indiana followed suit several years later with the Wabash and Erie Canal. In Maryland, a mammoth project connected Cumberland, in the western part of the state with the edge of Chesapeake Bay. In Virginia, the James River was opened to barge traffic several hundred miles into the interior of the state.

The total cost of the canals built between 1816 and 1840 was roughly $125 million. Most of that money was either borrowed by or belonged to governmental bodies. The state-owned canals, like the Erie, were financed by public borrowing. But states also used their tax revenues to buy large blocks of shares in private joint stock companies that were chartered to build canals. And governments—local, state, and federal—sometimes assisted the joint-stock companies, either by donating public land as an inducement to build or by selling public land at below-market rates as an inducement for investment.

Some of these later canals were financially successful. Toll revenue on the Erie Canal, for example, was sufficient not only to pay back the bondholders but to augment state coffers for several decades. Some of the later privately owned canals, unlike those before 1810, paid handsome dividends to their shareholders, at least for a time. The vast majority of canals, however, were not profitable. Some routes were badly chosen and were never particularly advantageous for commerce. Other canals, especially the ones that opened in the 1830s, encountered fierce competition from the steamboats that were by then plying the natural watercourses and from the railroads: in some cases, the payback was never more than a fraction of construction costs. Three states (Pennsylvania, Ohio, and Indiana) were brought almost to the verge of bankruptcy by their canal borrowing since tolls never reached projected levels and taxes had to be raised to pay back the bondholders. The canal boom, such as it was, came to an end in the 1840s; by the 1870s, traffic had decreased substantially on even the previously successful canals because of the advent of the railroads. By default, most of the privately owned canals became state and federal property by the end of the nineteenth century, and those that stayed open did so at a net loss to taxpayers.

Like most public works projects, the canals also changed the character of the regions they served. During the canal era, some very small frontier towns—Syracuse, Rochester, Buffalo, Cleveland, Harrisburg, Columbus, Dayton—were transformed into major centers of commerce because they were located on canal routes. Simultaneously, other communities shrank. Gristmills that had once flourished in the rural towns of western New York and Pennsylvania went out of business as farmers discovered that they could get better prices for their wheat at the larger mills in Buffalo and Pittsburgh. Rural distilleries and breweries went out of business also. In one state after another, young men and women from the dying rural towns, and from the farms that surrounded them, began migrating to the canal towns where jobs could be had, accelerating the decline of the older communities, even those communities that had been settled only one or two generations earlier. If the canals did, overall, benefit the economy of the states in which they were built, that did not mean that all the benefits were unalloyed. The transportation revolution was a gain for some people and places and a loss for others.

In their day, however, even the canals that were a disappointment to investors were a boon to freight handlers and nation builders because they drastically lowered the cost of transportation. Moving a ton of grain or salt or ore overland from Buffalo to New York City cost roughly $100 in 1820. Five years later, the same load could have been shipped via canal and river barges to the same destination for $9. Ton mile rates for wagon transport varied between 30 and 70 cents; ton mile rates on the canals varied between 2.70 cents and 4.5 cents. The canals lowered the price that consumers had to pay for goods. Even though they were built in individual states, the canals, because of the nature of the watercourses that they connected, served to bind the states to each other economically. Finally—as Gallatin and Jefferson had hoped—the canals made farming and manufacturing profitable in areas of the country that had previously been underpopulated, making both secession and foreign incursions increasingly unlikely.

Steamboats: Steam Power and State Power

Roads and canals were old technologies, but the steamboat was brand new. And like the canals and the turnpikes, the transportation improvements that followed (literally and figuratively) in its wake had an enormous impact both on the economy and on the politics of the new nation.

In the 1790s, many people believed that a lot of money could be made by someone who could invent a boat powered by something other than wind or human muscle—a boat that could go upstream on America's many rivers as easily as it could go down. The steamboat is a complicated technology consisting of a boiler to supply the steam, an engine that utilizes the steam, a propelling device for the boat, and a design for the hull. Because the steamboat is so complex, there are four different contenders for the honor of being the first to succeed in using the power of steam to propel a vessel: John Fitch, John Rumsey, John Stevens Jr., and Robert Fulton.

Fitch, Rumsey, and Fulton were in some ways very similar. All were born in the colonies (Fitch in Connecticut, Rumsey in Maryland, Fulton in Pennsylvania). Fitch and Rumsey were the sons of farmers; Fulton was the son of a tailor. In their youth, all three were apprenticed to craftsmen. Something of a loner, Fitch had been successful as a brass founder, a silversmith, and a gunsmith, but he had abandoned these trades and had chosen to support himself on the colonial frontiers as an itinerant surveyor and cartographer. Rumsey had been apprenticed to a blacksmith, but he had gone into business for himself as a millwright, eventually settling down on the Potomac River as a mill owner, builder, and, within a few years, proprietor of a boardinghouse. Fulton had been apprenticed to a jeweler; for several years he had worked as a painter of miniatures, but then he had gone to London to study painting and while there had become interested in mechanical inventions. Stevens was very different; son and grandson of wealthy landowners, he had a very large estate and many business interests in Hoboken, New Jersey, on the Hudson River, but he had been classically rather than technically educated.

In the mid-1780s, just after the conclusion of the Revolution, both Fitch and Rumsey were inspired to think of connecting a steam engine to a propulsion source. Fitch's first drawings were of paddle wheels, but he very quickly shifted to water jets—an idea that he took from a paper by Benjamin Franklin—and when that seemed problematic, he settled on a crank and paddle system, rather akin to two sets of automatic oars. Over the course of at least five years, Rumsey worked with variations on the jet propulsion system. Both men needed financial backers in order to build engines and boats. Fitch found his in the Philadelphia region; Rumsey's patron was his neighbor, George Washington. Fitch actually succeeded in building an operating boat; it ran for the entire summer of 1790 on scheduled trips between Philadelphia, Burlington (New Jersey), and Trenton at about eight miles per hour. Rumsey spent several years in Britain and France, consulting with the best machine builders of the day, but he never succeeded in building a boat. Both men filed patent applications on their designs.

Stevens had learned about the possibility of steamboats partly from Fitch (who had applied to the New Jersey legislature for a monopoly when Stevens was a member) and partly from a pamphlet war in which Fitch and Rumsey had engaged when they were both trying to find backers in the 1780s. Stevens thought that on general Newtonian principles the boilers that both Fitch and Rumsey were working with were inefficient, so he designed a new kind of boiler which, when connected to a conventional steam engine, would pump water out the stern of a boat, creating jet propulsion. In 1791, he filed a patent application on his design.

Fitch's boat rotted on a pier in Philadelphia, as the fares it had collected in the summer of 1790 were insufficient to cover the cost of fuel. Rumsey died of a stroke in 1792. Stevens discussed his plans with his brother-in-law, Robert R. Livingston, a politically well-connected New Yorker, but their joint effort to obtain a monopoly on the Hudson River traffic and to build a steamboat according to Stevens's design was derailed by the legal battle that erupted over which of the three patent holders was entitled to royalties from the other two.

In the end, circumstances that Stevens had set in motion meant that none of the original patents would be worth very much. In 1801, Thomas Jefferson appointed Livingston as ambassador to France, and in Paris, Livingston met Robert Fulton. By that time, Fulton was calling himself a civil engineer: he had won an award for a marble-cutting machine and had taken out French patents on devices for digging canals, for building aqueducts, and for raising canal boats on ramps. He had also drawn many designs for steamboats and had even built a few experimental models.

Shortly after Fulton and Livingston met in Paris, they signed a contract: Livingston would put up the money and Fulton would build a trial steamboat. If it were successful, Fulton would return to the United States and build a boat that would ply the Hudson between New York and Albany; the two men would split the profits. Fulton's first boat was completed in France in 1803; unlike the designs patented by Rumsey, Fitch, and Stevens, it had side paddle wheels and used a Watt-type steam engine with a separate condenser.

The French trials were successful. Four years later, in the spring of 1807, a very similar vessel, *The North River Steamboat* (its name was later changed to *The Clermont*), steamed out of New York harbor, heading upriver, against the current, to Albany, at a speed of five miles per hour—thereby inaugurating the steamboat era. Fulton and Livingston had imported a Watt engine for the *North River Steamboat* as well as an English mechanic to install and maintain it, but Fulton had designed the boat, tested it, and supervised its construction. For a number of years, Fulton did not even bother to patent his design; the earlier patentees had exhausted themselves in their contests with each other and Fulton initially did not even believe that he could argue that he had invented something new.

But he had—and entrepreneurs all over the country quickly entered the steamboat business. Through Livingston's political connections, Fulton and Livingston had secured a monopoly on steam transportation in New York. Within a few years, they had added several more boats to their fleet and several more routes to their service: down the Hudson to the Raritan River and New Brunswick, around Manhattan to Long Island Sound, and through the sound to New Haven. Steamboat service in New York harbor proved to be so lucrative that several enterprises began competing illegally with Livingston and Fulton. These newer steamboats had a more capacious hull design and their engines were located above decks (Fulton had placed the engine in the hold in his original design), where they would do less damage if they exploded (which these early steam engines regularly did).

John Stevens, still fascinated by steamboats, began service on the Delaware River in 1809 and the Connecticut River in 1813. Stevens's boats were designed by his son Robert Livingston Stevens and were the first to be wholly made in the United States. By 1820, steamship service had been established on all the tidal rivers of the East Coast and in Chesapeake Bay. In addition, several machine shops that specialized in building engines and boilers for steamboats were flourishing, and the design of the all parts of the boat—the engines, the boilers, the gearing, and the transmission—had been considerably modified from Fulton's original plan so as to be more dependable and efficient.

The steamers of the Great Lakes evolved at about the same time, but in a somewhat different direction from their cousins on the eastern rivers. The Canadian-owned *Frontenac* was providing service on Lake Ontario in 1816 and the *Walk-in-the-Water*, built in Buffalo, was plying Lake Erie two years later. A canal between those two lakes was opened in 1829, and at that point, traffic on the lakes began to expand significantly. During the 1830s and 1840s, the lake steamers grew to be much larger than the eastern river-boats, as they were meant to carry bulky freight (wheat, coal, iron ore) as well as passengers. The *Walk-in-the-Water* weighed in at 330 tons, but by 1860, the average size of a lake steamer was 1,000 tons. Lake steamers built after 1845 were propelled by underwater screw propellers rather than by paddle wheels.

Steam and sail competed for several decades on the east coast and in the Great Lakes, but in the western rivers, steam predominated almost from the beginning. Such rivers as the Mississippi, the Monongahela and the Ohio had numerous shallow places, which had made deep-keeled sailing vessels unusable. They also had some extremely swift currents that made upstream travel not just arduous but also dangerous.

Prior to the advent of steam, river traffic in the west had consisted large-ly of flat-bottomed barges powered by poles, and upstream traffic had been either minimal or nonexistent. Adding a paddle wheel and a steam engine to those flat keelboats and barges was not a difficult undertaking, and many men attempted it.

In 1811, the Fulton-Livingston interests, having first obtained a mo-nopoly on Mississippi traffic from the Territory of Orleans (which later be-came the state of Louisiana), commissioned a steamboat in Pittsburgh and sent it downriver along the Ohio and the Mississippi to New Orleans. Four years later, another boat, aptly named the *Enterprise*, made the return trip upriver. By 1817, there were 17 steamboats operating on western rivers; three years later, there were 69; by 1855, there were 727.

The design of western steamboats diverged significantly from those on the East Coast. Adapted for shallow riverbeds, they were flat-bottomed boats (western rivermen were said to have boasted that they could "set sail" on little more than a heavy dew), which had several above decks and usu-ally a single rear-mounted paddle wheel. In addition, these western river-boats used high pressure steam engines of the sort that had been original-

A classic Mississippi River steamboat. This drawing, "Wooding Up on the Mississippi," by Currier and Ives, emphasizes the smokestacks of the steamboat, rather than the paddlewheel. Because of the shallow draft the boats could not carry much fuel, so they had to stop regularly for refueling. Landowners and entrepreneurs lit fires at riverside (*bottom left*) as signals to riverboat pilots that fuel was for sale; thieves sometimes created decoy fires as lures. (Courtesy New York Public Library.)

ly patented early in the century by the Philadelphia-based mechanic Oliver Evans (see Chapter 4).

The steamboat revolutionized domestic transportation and domestic industry in numerous ways. On the major waterways, particularly those on which upriver traffic had previously been either impossible or very difficult, the cost of transportation declined, in some cases, precipitously. Downstream traffic costs on the Mississippi fell to one quarter of what they had been; upstream traffic, to one tenth. On the Hudson, rates fell 90 percent between 1814 and 1854.

In their day, steamboats were faster than any other mode of transport— even if the speeds they achieved seem tortoiselike by today's standards. Flatboats, for example, had once taken a month to six weeks to float downriver from Pittsburgh to New Orleans, and four months (plus a crew of very strong men) to go back up. In 1819, the *Enterprise* made that trip in twenty-five days, and in the 1850s, boats of considerably greater tonnage were regularly doing it in two weeks or less. On the Hudson, sailing packets had once transported passengers between New York and Albany in anything from two days to two weeks, depending on wind and weather conditions. By steamer in the 1830s, the trip took ten hours.

The steamboats also created new legal and political precedents that would turn out, as aspects of the public-private partnership in transportation, to have immense significance in succeeding years. In English law, governments were allowed to grant monopolies to individuals and to joint-stock companies. In the years immediately following the Revolution, state governments continued the practice, essentially without even thinking twice about it. Livingston had arranged for the company that he had formed with Fulton to have such a monopoly on all steamboat traffic in New York harbor. English law, however, had not been developed to suit a federal system, in which the power of the several federated units had to be balanced. Steamboat traffic in New York harbor was so lucrative that several entrepreneurs were quick to challenge the Livingston-Fulton monopoly in the courts. Why, their attorneys asked, does the legislature of New York have the right to grant a monopoly over waterways that are as much a part of the borders of New Jersey and Connecticut as they are of New York?

The case of *Gibbons v. Ogden*, settled by the United States Supreme Court in 1824, is justly famous in American history for having established the precedent that the federal Congress, and only the federal Congress, could regulate commerce *between* the states. Today, few people recall that the suit was brought for the purpose of ending the Fulton-Livingston monopoly in New York harbor. Ogden had been paying the monopoly a fee for the privilege of operating a steamboat between Manhattan and Elizabeth, New Jersey. Gibbons had been a partner of Ogden's, but after a quarrel, he left the partnership and established his own steamboat line, plying the same route without paying the fee. When both Ogden and Livingston tried to stop him, Gibbons took his case to court. Unsuccessful on the state level, he hired Daniel Webster to plead his case before the Supreme Court. When that court voted in his favor, the Livingston-Fulton monopoly ended, and a precedent was established that would allow the federal government to regulate interstate transportation for years to come.

The steamboat was also the subject of legislation that first established the authority of the federal government to regulate an industry in the interest of public safety. Steamboat travel may have been relatively comfortable, quick, and cheap, but it was still risky. In the first decades of steamboating, many lives were lost because of boiler explosions. In May of 1824, the most disastrous of such accidents—thirteen people killed, several dozen maimed—led to a congressional inquiry, and legislation to ban the use of high-pressure engines was unsuccessfully proposed in that session. But accidents kept happening, and as steamboats grew larger, the death tolls grew too: between 1825 and 1830, forty-two explosions killed 273 people, and a particularly severe explosion near Memphis in 1830 took 50 or 60 lives. Yet another congressional inquiry led to yet another bill—this time to require inspection of steamboat boilers every three months—that also failed to win passage. The Constitution, after all, did not include "insuring the public safety" among the powers reserved to Congress.

In 1831, however, the Franklin Institute in Philadelphia—a voluntary society, recently formed to promote the "mechanical arts and applied sciences"—began a series of experimental studies intended to discover the reasons for steam boiler explosions. That report was presented to Congress in 1836 along with suggestions for appropriate legislation. In his State of the Union address the following winter, President Van Buren urged passage and Congress duly complied—after another bad accident in Charleston had killed 140 people—in July 1838. This law provided that each federal judge would appoint a boiler inspector, who was to examine every steamboat boiler in his jurisdiction twice a year. For this service, the owner of the boiler was to pay the inspector $5, in return for which his license to navigate would be certified. The law also provided that, in suits against boiler owners for damage to persons or property, the fact that a boiler had burst was to be considered prima facie evidence of negligence.

As it happened, the legislation of 1838 did not prevent an increase in the frequency of boiler accidents or in the loss of life (Congress felt impelled to strengthen the law in 1852), but like *Gibbons v. Ogden*, it set an important precedent for the future. Until the problem of bursting boilers, most Americans had believed that the federal government could not, and ought not, interfere with the rights of personal property. Apparently, several hundred boiler-related deaths had convinced many Americans and most congressmen that there was a point at which property rights of some individuals had to give way to the civil rights of others. The boiler legislation thus established the precedent on which all succeeding federal safety-regulating legislation would be based.

The steamboat catalyzed several additional changes. Like the canals and the turnpikes, it lowered the cost and increased the reliability of transportation, thereby helping to create a national market, increase the pace of western settlement, and accelerate the rate of economic change. Like the canals and the turnpikes, it also tied the states and the regions to one another in an increasingly prosperous economic union.

In addition, the steamboat led to the training of the first generation of American machinists. In the 1790s, as we have seen in Chapter 4, there were very few Americans who had the skills necessary to build a steam engine—let alone to repair or maintain one. There were, of course, craftsmen who worked in metal—blacksmiths, brass founders, gunsmiths, and clock makers—but none who were accustomed to boring and smoothing cylinders that had to fit moving pistons and none who knew how to cast and weld iron so that it would hold together under boiler pressures. Both John Rumsey and Robert Fulton had gone abroad not only for their engines but also for the mechanics who were trained to maintain them.

The steamboat boom, the Embargo of 1807, and the War of 1812 led, however, to the creation and then the expansion of American machine shops; the increasing demand for steam engines could not continue to be met by imports from Britain. Oliver Evans, as we have seen, expanded his Philadelphia works and then opened an affiliated operation in Pittsburgh. Nicholas Roosevelt, the owner of a copper mine in Belleville, New Jersey, enlarged the machine shop part of his operations so as to produce engines for Stevens's steamboat lines. Many of Roosevelt's employees were recent immigrants from England and Germany. Several went on to establish successful machine shops of their own; others, once trained, took jobs with steamboat companies, repairing and maintaining engines. By 1820, there were dozens of excellent machine shops in New York, Connecticut, Pennsylvania, and Maryland, all building steam engines for boats. In the years to come, the skills of American mechanics and machinists, first honed in the steamboat trade, would be put to excellent use elsewhere: in the building of power looms, automatic trip-hammers, sewing machines, harvesters, spinning mules—and railroads.

Railroads: Completing a National Transportation System

In the first decade of the nineteenth century, when construction was just starting on the roads and canals and the steamboat was a new idea, the railroad was virtually unimaginable except to a very few technological visionaries. In his report of 1807, Gallatin was already able to guess that steamboats would be traveling on some important rivers and lakes (Fulton, after all, had been asked to write a technical appendix to his report), but the railroad was beyond even his enlightened dreams. Indeed in those days, only a handful of people in the United States seem to have conceived the potential for steam engine–powered transportation on land; one of them was Oliver Evans.

In the latter part of the 1780s, Evans had begun developing an engine that would be small and light enough to be installed in what had previously been a horse-drawn carriage. These experiments bore fruit in some ways, but Evans was never able to attract the money that he needed to pay full-time attention to solving the multiple problems involved in steam transportation on land. In 1813, he wrote a long article exploring the possibility of a railway between New York and Philadelphia "for the transportation of heavy produce, merchandise, and passengers on carriages drawn by steam engines," in which he prophesied that someday "carriages propelled by steam will come into general use, and travel at the rate of 300 miles a day." Such prophecies had struck most of those who had heard them as ridiculous, possibly even insane, and Evans believed that he would not see steam transport on land in his lifetime. "One step in a generation is all we can hope for," he wrote. "If the present generation shall adopt canals, the next may try the railway with horses, and the third generation use the steam carriage."[7]

Evans was right about the potential of the railroad, but wrong in his prediction about when that potential would be fulfilled. Railroad fever began to infect the United States during the 1820s, and by the time another decade had passed, it had become a full-scale epidemic. One generation, rather than two, was required to pass from the canals to the railroads.

Opinions are divided about who should take the credit for being first in American railroading (although all experts agree that the British were about five years ahead of the Americans). If only ideas count, then the honor should probably go to John Stevens, the steamboat entrepreneur, a visionary of Evans's ilk, who received a charter to build a railroad across New Jersey in 1821 (and a similar charter to go across Pennsylvania in 1823). Unfortunately, Stevens was initially unable to attract enough investors or raise enough capital to build on either of the routes he had selected, despite having constructed, with his own hands (he was seventy-six at the time), a functioning steam locomotive.

If only the rails count, then credit should probably go to the Granite Railroad, which opened in Quincy, Massachusetts, in 1826: a set of rails, two miles long, used to haul granite blocks to Boston harbor in open wagons drawn by horses. If only steam locomotion counts, then we ought to pay homage to the managers of the Delaware and Hudson Canal Company, who ordered a locomotive from a manufacturer in Britain in 1828, at the not inconsiderable price of almost $4,000. They intended to install it on a railway that they were building to connect two parts of their canal system so that freight could be loaded off canal boats and onto railroad carriages for transport over land. Unfortunately, that first locomotive to operate in the United States, the *Stourbridge Lion*, had to be put out of service after its very first run because it was too heavy for the tracks; freight on the Delaware and Hudson tracks was hauled by horses for several more years. The Baltimore and Ohio Railroad is sometimes applauded as the nation's first railroad to carry freight and passengers in regular service under steam locomotion because the cornerstone for its track was laid on July 4, 1828, but in truth it operated under horsepower for the first year or two that it was in operation; for a few of those early months, the managers of the B&O experimented with sail-powered cars as well.

Thus the honor of being the first functional railroad system probably belongs to the Charleston and Hamburg Railroad. Its locomotive, *Best Friend of Charleston* (which was also one of the first locomotives built in the United States), pulled a train of cars along six miles of track in December 1830—and began regular service, at twenty-one miles per hour, a month later.

Ten years later, there were 3,326 miles of operating railroad trackage in the United States, almost all of it east of the Appalachian mountains. The Philadelphia and Columbia Railroad connected Philadelphia with the Pennsylvania canal system by 1834. In the same year, the Camden and Amboy created a rail link between Philadelphia and New York harbor. The Mohawk and Hudson ran parallel to the Erie Canal by 1835 (although it had to pay a punitive tax to New York State for taking business away from the Erie Canal), and Boston had three short rail lines connecting it with, respectively, Worcester, Lowell, and Providence by 1836. Coal was being hauled from western Pennsylvania to the port of Philadelphia along the Philadelphia and Reading by 1839. By 1840, a good part of the nation's commerce was beginning to move by rail.

By today's standards, most of these early railroads were very short: when it was completed in 1833 the Charleston and Hamburg, was the longest rail line under single management in the entire world: 136 miles. A few railroads were dedicated lines, built by a particular business for the sole purpose of hauling its own goods to a port or to a terminal. Most, however, were public conveyances (which means that they would carry any passenger who paid the fare and freight from any source) owned by private corporations that had been chartered by state governments. Like many of the turnpike and canal companies, these corporations had the right to sell stock to investors, so as to build and operate rail lines along a specified route. Often the original charters detailed the terms—usually very liberal—by which public land was to be given to the corporation. In addition, as with the canals, many states purchased stock in railroad companies or made loans to them. A few railroads were built, owned, and operated entirely by state governments (the Philadelphia and Columbia, for example, belonged to the Commonwealth of Pennsylvania; Georgia and Virginia also built their own rail lines).

The decade of the 1830s was a period of experimentation in the construction of railbeds, locomotives, and train cars. The railways that had been developed for horse-drawn vehicles had wooden tracks that were capped with iron; under the weight of steam locomotives, those caps often worked loose, impaling cars and, sometimes, people. Robert L. Stevens, the son of John Stevens and the president of the Camden and Amboy Railroad in New Jersey, designed and installed the first fully iron rail in 1831 (a T rail). Choosing a material out of which to build a railbed was not an easy matter either, given the variety in American topography. Some of the early railroad builders tried anchoring rails to granite blocks, while others tried constructing elevated roadbeds. Eventually wooden sleepers embedded in gravel became the norm, but it wasn't until after the Civil War that any effort was made to establish a standard either for the width of tracks or for the construction of cars to run on them.

Early in the 1830s, several American machine shops and foundries began to specialize in the construction of locomotives; by the end of the decade, a distinctly American locomotive style had begun to emerge out of such machine shops as the West Point Foundry in Cold Spring, New York, and the Matthais Baldwin Engine Works in Philadelphia. Like the *Stourbridge Lion* (which weighed eight tons, three tons more than expected), British locomotives had proved to be too heavy for the relatively lightweight tracks that Americans were constructing in such haste. Furthermore, British engines, built to operate across the relatively flat terrain of England, did not generate enough pressure (provide enough horsepower) to pull trains of cars up the steep inclines that American mountainous routes sometimes required. In addition to being both lighter and more powerful, American locomotives also were longer (four sets of double wheels, rather than two) and their forward wheels were mounted on a swivel, or bogie, truck, which allowed the wheels to turn freely with the curves on the track, thus minimizing the likelihood of derailments.

The very earliest railroad passenger cars were constructed by carriage makers and, not surprisingly, were almost identical to stagecoaches, even to the extent of including an elevated seat for a nonexistent coachman. By the end of the 1830s, however, the graceful curves of the carriage had been replaced by a much more practical box shape. These early passenger cars were far from luxurious, however. Cars were linked to each other by chains; seats were wooden planks; the wheel assemblies did not contain springs; brakes were little more than levers that pressed against the wheels: sparks flew out of the boiler constantly. Travelers were likely to end their trips with whiplash, black and blue marks, and scorched clothing.

Railroad travel before the Civil War was, by modern standards, tiresome and dangerous. It was also undependable; trains were put on sidings for long periods of time (as most lines consisted of only one set of tracks) and minor derailments were frequent (on several lines, the passenger contract required that able-bodied men help the conductors put cars back on the tracks). But the railroads were fast, or at least faster than any other mode of transport then available—and they were also cheap. In 1817, fifty days were required to get goods from Cincinnati to New York City by keelboat and wagon. By steamboat down the Mississippi and sailing packet along the coast, the trip was cut to twenty-eight days. By canal boat (the Ohio Canal to Lake Erie; the Erie Canal to the Hudson River; the Hudson River to New York harbor) only eighteen days were required. By railroad in 1850, even including the time that was spent in unloading from one line to another (since each rail line had its own terminals, freight cars and track gauges) only six to eight days were needed. Rates for land transport fell

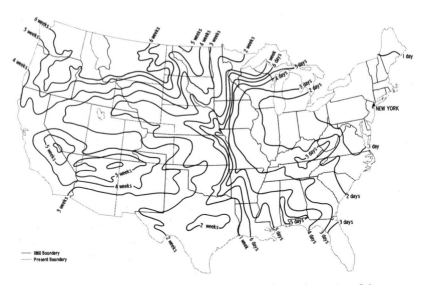

These two maps compare travel times from New York to other parts of the country in 1790 (*top*) and 1860 (*bottom*), after new roads, steamboats, canals, and railroads had transformed the transportation system. (Courtesy Smithsonian Institution.)

from an average of $.17 per ton mile when wagons were hauling over turn-pikes, to $.6 per ton mile at the beginning of the railroad era, to $.1 per ton mile by the outbreak of the Civil War. Passenger fares also collapsed. The trip from Philadelphia to Quebec, which cost $47 (and consumed 100 hours) by steamboat and stagecoach in 1816, cost $19 (and 31 hours) by railroad in 1860.

By 1860, hundreds of American railroad companies were operating 30,600 miles of railroad track in the United States, far more than anywhere else in the world. Turnpikes were beginning to fall into disrepair, and canals were beginning to lose money for their stockholders. Cities at the hubs of rail networks were growing at an even faster pace than they had earlier in the century—and they were beginning to spawn suburbs, connected by rail (the Long Island Railroad, for example, which considers itself the nation's first commuter railroad, began operations in the 1840s). The cost of transporting heavy, bulky items—coal, wheat, cattle, even iron rails—was falling, giving a boost to the economy in general and the economy of the sparsely settled regions in particular. Regions west of the Mississippi River were being settled at an accelerating rate. Some Americans began to consider the possibility of visiting each other over long distances, and others began to think about living in one place but vacationing in another. Perishable foodstuffs such as butter, carrots, and apples could travel farther to market. Armies could travel faster to battlegrounds, as Confederate generals discovered to their dismay.

Once the Civil War had clearly demonstrated the folly of secession, the railroad would complete the economic process of nation building, the process begun with the turnpike, canal, and steamboat. On the afternoon of May 10, 1869, a telegraph operator signaled with the single word, "Done," that a golden spike had been driven into a wooden sleeper in Promontory Point, Utah. A hundred guns were fired in salute in City Hall Park in New York and the bells rang out from Independence Hall in Philadelphia as Americans celebrated the technological and economic unity of their country just four years after its political union had been ratified. The world's first transcontinental railroad was complete, the federal system had been extended from one ocean to another. The ghosts of Jefferson and Gallatin must have been smiling.

Notes

1. *Report of the Secretary of the Treasury on the Subject of Public Roads and Canals* [1808] (New York: Augustus Kelley, 1968), p. 3.

2. *Report of the Secretary of the Treasury,* p. 121.

3. From J. T. Callender, *The American Annual Register for the Year 1796,* pp. 35–36, as quoted in Joseph Austin Durrenberger, *Turnpikes: A Study of the Toll Road Movement in the Middle Atlantic States and Maryland* [1931] (Cos Cob, CT: John E. Edwards, 1968), p. 30.

4. Francis Bailey, *Journal of a Tour in Unsettled Parts of North America, 1796–1797* (London, 1856), p. 107, as quoted in Durrenberger, *Turnpikes,* p. 52.

5. As quoted in Elting E. Morison, *From Know-How to Nowhere: The Development of American Technology* (New York: Basic Books, 1974), p. 27.

6. Morison, *From Know-How to Nowhere,* p. 44.

7. Both quotes are from an article Evans wrote in the magazine *Aurora* in 1813, as quoted in Stewart Holbrook, *The Story of American Railroads* (New York: Crown, 1947), p. 4.

Suggestions for Further Reading

George H. Douglas, *All Aboard! The Railroad in American Life* (New York, 1992).

Joseph Austin Durrenberger, *Turnpikes: A Study of the Toll Road Movement in the Middle Atlantic States and Maryland* [1931] (Cos Cob, CT, 1968).

James Thomas Flexner, *Steamboats Come True: American Inventors in Action* (New York, 1944).

Brooke Hindle, *Emulation and Invention* (New York, 1981).

Stewart H. Holbrook, *The Story of American Railroads* (New York, 1947).

Louis H. Hunter, *Steamboats on the Western Rivers* (Cambridge, MA, 1949).

F. Daniel Larkin, *John B. Jervis: An American Engineering Pioneer* (Ames, IA, 1990).

Albro Martin, *Railroads Triumphant: The Growth, Rejection and Rebirth of a Vital American Force* (New York, 1992).

Elting E. Morison, *From Know-How to Nowhere: The Development of American Technology* (New York, 1974).

George Rogers Taylor, *The Transportation Revolution, 1815–1860* (New York, 1951).

6

Inventors, Entrepreneurs, and Engineers

INDUSTRIALIZATION SOUNDS LIKE AN IMPERSONAL PROCESS, but it was actually one of the great dramas of human history, replete with comedy, tragedy, tenderness, cruelty—the full range of human passions and possibilities. The factories and railroad bridges, the oil wells and steam engines of industrialization were conceived, crafted, and constructed by people, some of whom were motivated by greed, others by altruism. Some were wise, others foolhardy. Some died penniless, others achieved great wealth. Thousands of people were involved in creating industrialized society in the years between 1870 and 1920, but we can simplify the task of describing their contributions by categorizing them into one of three social roles: inventor, entrepreneur, and engineer.

Inventors invent: they create things that have never been created before and sometimes try to earn their living from the funds these new things may generate. Entrepreneurs innovate and diffuse: they figure out how to enter the marketplace to make money with the things that inventors invent. Inventions turn into innovations when they fall into the hands of entrepreneurs who are willing to provide the funds or the hard work or the managerial skill (or all three) that is necessary to turn an invention into an innovation.

Sometimes entrepreneurial skill and inventive skill are combined in the same person. Thomas Edison, for example, is renowned as an inventor-entrepreneur because he invented a feasible electric light, then designed a system that would deliver electricity to lighting customers, then founded a number of companies to manufacture the parts of his system and other companies to provide the electric service—all the while using other people's capital. George Eastman, likewise, invented photographic film on a paper roll, then designed a camera that would use such film, then founded a company that would both manufacture the film and the camera and provide developing services for purchasers.

Some engineers complete the projects that inventors and entrepreneurs initiate. Inventors may have brilliant ideas and entrepreneurs may conceive of the ways in which those ideas can be put into practice, but both inventors and entrepreneurs have to rely on the patience and the skill of engineers to complete any substantial technological project. In the late nineteenth and early twentieth centuries, many engineers built or, to be more precise, designed things that were intended to be built and frequently also directed the efforts of those whose hands did the building: John Augustus and Washington Augustus Roebling, the father and son team that designed and built the Brooklyn Bridge, were engineers of this type. Other engineers discovered; they designed and performed experiments that tested materials and objects or they located resources (such as oil deposits and veins of coal) that nature has hidden from view: the chemists who developed tests to determine whether drinking water was polluted and the geologists who mapped the Mesabi Range were engineers of this variety. Yet other engineers managed, designing and implementing new ways for people to work: efficiency experts, founders of the field now sometimes called industrial engineering, fell into this category. Like inventors and entrepreneurs, some engineers crossed roles: they became inventors when they patented things and entrepreneurs when they founded consulting firms or manufacturing companies.

The best general definition of the term engineer is probably this: engineers do for the sake of production what scientists do for the sake of knowledge. Put a slightly different way, this means that engineers are people who have systematically acquired knowledge about the natural world and who put that knowledge to use in achieving some practical goal.

The Patent System: The Public History of Invention

If there were no such thing as a patent, we would not know very much about inventors. A patent is a temporary monopoly on the economic benefit that can be derived from an invention. As such, a patent turns an idea into a form of property; the person who has a new idea, a patent asserts, can own it in the same way that he or she may own land or money. Once a patent is issued an idea can be monopolized, sold, or leased—in just the same way that land and money can be monopolized, sold, or leased. By issuing a patent, a government pledges to back up the owner's property rights with its policing and judicial powers. A patent makes it profitable for an inventor to "go public" with an invention; in the act of going public, historians get to know something about inventors.

Patents did not exist in the West in ancient and medieval times; in those days if an inventor wanted to maintain exclusive rights to an invention, he or she could only do so maintaining absolute secrecy about the invention. As a result of this secrecy, we who live in later centuries have no way of knowing much about ancient and medieval inventors; they had every good reason to keep themselves anonymous.

Sometime in the fifteenth century, when economic activity was beginning to accelerate in western Europe, the government of Venice, one of the most prominent trading states of the Renaissance, developed the practice of rewarding the inventors of new devices with temporary monopolies. A Venetian law of 1474 was the first to treat an inventor's ideas as property; once a person had been granted a patent, the law prohibited other people from copying the device for ten years. The government of Venice had thus discovered, as many other governments also discovered subsequently, that invention is good for an economy and that invention can be encouraged if the economic interests of inventors are protected.

By the time that the English, French and Dutch had begun establishing colonies in the New World, all three countries were issuing patents. The English patent law, which served as a model for subsequent American laws, was called the Statute of Monopolies; it had been passed by Parliament in 1623—when Jamestown was just getting on its feet and the Puritans had not yet landed on Plymouth Rock. During the seventeenth and eighteenth centuries, very few American colonists actually made much use of their patenting rights under the Statute of Monopolies. The statute was very difficult to enforce across three thousand miles of ocean, and the would-be patentee had to be present in London for several months in order to carry an application from one governmental office to another. The first British patent granted to an American colonist was issued in 1715 to a Pennsylvania woman, Sybilla Masters, who had invented a new mill for "cleaning and curing the Indian corn growing in the several colonies in America," as well as a new technique for "working and staining straw for hats."[1] (Since British law prohibited married women from owning property, the patents were registered in the name of Sybilla's husband, Thomas, although they very clearly stated that she, rather than he, was the inventor.)

In 1787, the Constitutional Convention unanimously passed a motion that authorized Congress "[t]o promote the progress of science and useful arts by securing for limited times to authors and inventors the exclusive right to their respective writings and discoveries." Three years later, in his first presidential message, George Washington asked Congress to write a patent law that would put some teeth into this constitutional clause, a law that would, in Washington's words, "give effectual encouragement . . . to the introduction of new and useful inventions from abroad as well as to the exertions of skill and genius in producing them at home."[2]

Congress responded quickly; the first American patent law was passed in 1790. Although it has been repeatedly amended and reformed, it still expresses the country's fundamental patent policy: "Any useful art, manufacture, engine, machine or device, or any improvement therein not before

The United States.

To all to whom these Presents shall come. Greeting.

Whereas Samuel Hopkins of the City of Philadelphia and State of Pennsylvania hath discovered an Improvement, not known or used before, such Discovery, in the making of Pot ash and Pearl ash by a new Apparatus and Process; that is to say, in the making of Pearl ash 1st by burning the raw Ashes in a Furnace, 2nd by dissolving and boiling them when so burnt in Water, 3rd by drawing off and settling the ley, and 4th by boiling the ley into Salts which then are the true Pearl ash, and also in the making of Pot ash by fluxing the Pearl ash so made as aforesaid; which Operation of burning the raw Ashes in a Furnace, preparatory to their Dissolution and boiling in Water, is new, leaves little Residuum; and produces a much greater Quantity of Salt: These are therefore in pursuance of the Act, entitled "An Act to promote the Progress of useful Arts", to grant to the said Samuel Hopkins, his Heirs, Administrators and Assigns, for the Term of fourteen Years, the sole and exclusive Right and Liberty of using and vending to others the said Discovery, of burning the raw Ashes previous to their being dissolved and boiled in Water, according to the true Intent and Meaning, of the Act aforesaid. In Testimony whereof I have caused these Letters to be made patent, and the Seal of the United States to be hereunto affixed. Given under my hand at the City of New York this thirty first Day of July in the Year of our Lord one thousand seven hundred & Ninety.

G Washington

City of New York July 31st 1790.—
I do hereby certify that the foregoing Letters patent were delivered to me in pursuance of the Act, entitled "An Act to promote the Progress of useful Arts"; that I have examined the same, and find them conformable to the said Act.

Edm: Randolph Attorney General for the United States.

Delivered to the within named Samuel Hopkins this fourth day of August 1790.

Th: Jefferson

The first United States patent, granted to Samuel Hopkins of Philadelphia, for a new process for making potash and pearl ash (see Chapter 1). The patent document was signed by George Washington and Edmond Randolph, Attorney General, and endorsed on the back by Thomas Jefferson, who was Secretary of State, when it was delivered to Samuel Hopkins in August 1790. In the summer of 1790 New York was still functioning as the nation's capitol. (Courtesy Chicago Historical Society.)

known or used" can be rewarded with a patent if some designated person or office (then it was a committee composed of three government officials, now it is the Patent Office) should "deem the invention or discovery sufficiently useful and important."[3] The new law had a provision that was entirely novel in its day. It required inventors to demonstrate not only the usefulness but also the novelty of their devices; they must provide precise, publishable specifications, as well as a miniature working model (this latter provision was dropped late in the nineteenth century because storage of models was becoming impossible).

The specifications feature of American patent law makes the law inherently paradoxical—and has created numerous nightmares for inventors over the years. Several eighteenth century inventors (Franklin and Jefferson in particular) objected to patents. They thought that the public ought to have free access to a new invention because every invention is based on preceding ideas and thus no individual can claim exclusive property rights in an idea. The specifications feature was a compromise intended to preserve the public's right to know in the face of an inventor's right to exploit. In return for the temporary protection they were being offered, inventors had to provide for the public good by disclosing the information that other people would need to reproduce the invention after the patent had expired. Essentially the American patent system tries to maintain trade secrets by discussing them in public.

Especially in the early years of the patent system many inventors ran afoul of this paradox. Having disclosed the details of their inventions to the government (which proceeded to publish them), inventors discovered that other people were exploiting that information without paying either a licensing fee or a royalty. What was the hapless inventor to do but try to press his claims in court and risk expending more in legal fees perhaps than his invention had ever earned? Eli Whitney struggled in this fashion over his cotton gin, as did Oliver Evans over his automatic flour mill (see Chapter 4). Later in the nineteenth century, would-be patentees became more sophisticated: they hired lawyers to help them draft specifications that were not quite specific enough to be infringable. The inherent paradox in the system still remains, however, which is why some inventors do not seek patents and why patent law has become one of the most arcane disciplines—and patent lawyers among the highest paid attorneys.

Once it had become clear, however, that the federal government intended to enforce the law, more and more Americans decided to become inventors. "The act of Congress authorising the issuing patents for new discoveries has given a spring to invention beyond my conception," Thomas Jefferson remarked, just three months after the act became law—and what began as a spring soon developed into a deluge. [4] Although patent statistics are not strictly comparable to each other (the criteria for awarding patents have changed in various ways over the years), they give a general sense of the acceleration of inventive activity in the nineteenth century. In the patent committee's very first year of operation (1790), three patents were granted. Fifty years later, an average of 600 patents were being issued every year—a 20,000 percent increase at a time when the population had only increased about 400 percent. And forty years later, those figures had multiplied fivefold—in 1900, approximately 40,000 applications were filed and about 26,000 were issued, although the population had only doubled.

As more and more people began to try to make their living from invention, and as more and more inventors had recourse to the patent system to protect their intellectual property, inventors lost their anonymity. It may be impossible to find out who was responsible for a seventeenth or eighteenth-century invention, but once the patent law had been passed inventors clamored to have their names associated with their inventions. Indeed as the nineteenth century wore on, some inventors became cultural heroes. Such men as Samuel Morse (the telegraph), Alexander Graham Bell (the telephone), George Eastman (rolled photographic film, the simple box camera), Thomas Edison (the electric light, the phonograph, the motion picture projector), Nicola Tesla (the alternating current motor), and Elmer Sperry (the gyrostabilizer) were celebrated in print and from the pulpit—household names either in their own right or because of the names of the companies they founded. Newspapers quoted their opinions; popular magazines recounted their exploits; huge crowds turned out to hear them lecture; artists clamored for the right to paint their portraits. Part of this cultural notoriety was self-induced. Many nineteenth century inventors were pioneer self-promoters; they needed to call public attention to themselves, sometimes to sell stock in the companies they were trying to form, sometimes to sell their innovative products to an unaccustomed public.

Inventors: Changes between 1820 and 1920

Inventors are creative people, and—like artists and musicians—they tend to be unique individuals, each possessing some idiosyncratic combination of technical creativity, dogged persistence, managerial skill, economic drive, and market insight. Yet because of their status as celebrities, we know a good deal of the biographies of many successful nineteenth- and early twentieth-century inventors, and what we do know suggests some common characteristics in their lives; such commonality, although not perfect, makes possible some rough generalizations about inventors.

None of the earliest nineteenth-century inventors was a brilliant scholar and several lacked formal higher education altogether. Morse attended Yale, but he was an indifferent student and managed to get himself into debt by drinking and partying excessively. Edison never completed school; by the age of thirteen he was working full time. Bell was educated at home by his mother until the age of eleven and then was sent to one of the best high schools in his hometown (Edinburgh, Scotland), but he was a poor student, did not complete the full six-year course of study, and ended his education at the age of fifteen. Sperry attended an excellent primary school in Cortland, New York, but he chose the business track rather than the academic track in high school and completed only two years of the three-year course, having failed almost all his courses in his very first term.

As the nineteenth century drew to a close, however, and as the scientific basis for the electrical and chemical industries became both more complex and more obvious, this pattern changed. Gifted amateurs could no longer cope; inventors were drawn, with increased frequency, from the ranks of those who had acquired specialized, advanced training. Charles Martin Hall, who figured out how to separate aluminum from its ores by electrolysis, was a graduate of Oberlin College. Nicola Tesla, whose patent on an alternating current (a.c.) motor was purchased by the Westinghouse

Company in 1888, had attended of one of Austria's finest engineering colleges for several years. Charles Proteus Steinmetz (first director of research and development for the General Electric Company, holder of 195 patents on various devices, many of them related to high-voltage a.c. transmission) had graduated from the University of Breslau and had studied mechanical engineering at a technical college in Zurich. And William Burton, Robert Humphreys, and F. M. Rogers, inventors, in 1909, of the petroleum-refining technique called thermal cracking, all had doctorates in chemistry. After 1900, self-taught men and women could still become successful inventors (the Wright brothers are a famous example; see Chapter 11) but only if they concentrated on technical fields, such as aeronautics, that were just starting to develop.

A remarkable number of inventors had poor verbal skills. They found reading and writing disagreeable and frustrating; no doubt this is part of the reason why, in the early decades of the nineteenth century, when science and mathematics were largely absent from the curriculum, so many were poor students. Sperry reported, for example, that he had a "terrible fear of being a total failure owing to my difficulty with spelling, grammar and languages."[5]

On the other hand, these inventors were notable for their facility in other, nonverbal realms of activity. They may have struggled with words and sentences, but they did not struggle with spaces, objects, forms, and notes: they were nonverbal thinkers—and unusually creative ones at that. Morse was earning his living as an artist when he first got the idea that led to the telegraph; he was a painter, a portraitist, and a sculptor; a founding member of the National Academy of Design; and the first to hold the position of Professor of Painting and Sculpture at New York University. Bell was a superb musician: as a child he could read music by sight; throughout his life he could recall melodies at will, sometimes lying sleepless at night with musical passages running through his head; his mother hoped he would pursue a career as a concert pianist. Sperry was gifted with his hands; even as a small child he built model waterwheels and windmills. Tesla did the same thing; as an adolescent, he tried to construct a windmill powered by the beating wings of June bugs.

Many of the nineteenth-century inventors for whom biographical information exists were the sons of middle-class families who had for various reasons experienced periods of poverty. This fact may help us to understand why they were driven to seek commercial rather than scientific success and why they had the incredible levels of persistence that the development of an invention requires. Charles Goodyear is perhaps the most extreme example of this pattern. The man who discovered the process of vulcanizing rubber (which prevents rubber items from melting when they get hot) was the son of a hardware manufacturer. The family was sufficiently comfortable to allow Charles to enter and finish high school (this was in the 1820s, when such a high level of education was not common), but a few years later—the young man having joined his father's business—both father and son were seriously in debt. In those days, one of the penalties for indebtedness was a jail term. Charles Goodyear went to jail several times before his belief in the potential of the rubber industry finally paid off; indeed, his first experiments on vulcanization were conducted in a jail cell.

Edison went to work at thirteen because his family needed his income, his father's once prosperous business having met reverses. After the age of fifteen (when he left home to become a telegraph operator), Edison was entirely self-supporting, living in squalid quarters, consuming meager meals at restaurants. Morse's family was economically comfortable. They supported him for many years when he was a student, but eventually their funds dried up, and for many years Morse lived on a more or less hand-to-mouth basis, traveling up and down the eastern seaboard hoping to garner commissions for portraits. Even his appointment at New York University carried no salary other than the fees his students (who at first were not numerous) were willing to pay.

The bright idea that produces a successful invention may be the unique creation of one solitary mind, but the process of development which turns that bright idea into a workable reality, requires collaborators, and each successful inventor eventually recognized this necessity. The story of Morse's telegraph is exemplary: when Morse first tried his hand at inventing (in the 1820s), he collaborated with one of his brothers on a design for a fire-engine pump (jokingly, they called it "Morse's Patent Metallic Double-Headed Ocean Drinker and Deluge Spouter") and, following that, he worked on a marble-cutting machine with an acquaintance who was an amateur sculptor.

In 1832, while on board ship (he was returning to the United States after studying art in London for several years), Morse conceived the fundamental idea for a telegraph: a device that would send messages over long distances by means of an electric current; a battery to create an electric circuit; a dot and dash code, which could be transmitted by opening and closing a circuit; a receiver, which would use an electromagnet to move a roll of paper (through a clockwork mechanism) and mark the paper (through a lever mechanism) in response to the opening and closing of the circuit.

The idea excited him, but it was just an idea, and—lacking funds, a thorough knowledge of electric phenomena, and mechanical skill—he was forced to work on it by himself and only in the time that he could create between his teaching obligations at NYU. Four years passed, unsuccessfully. In those years, Morse managed to learn a lot about electricity, to complete his code, and to build several working models of his telegraph, but he never managed to send a message more than forty feet. Still without funds, but recognizing that he needed help, he turned to a fellow NYU faculty member, Leonard D. Gale, professor of geology and mineralogy, who, because of his scientific researches, had experience in working with electric circuits. Over the course of the next year, in return for the promise of a partnership in the invention, Gale helped Morse develop a larger battery and an electromagnetic device with more windings. In 1837, the two men took on a third partner, Alfred Vail, a recent graduate of NYU; Vail was more skilled with his hands than either Gale or Morse and, perhaps most crucially, he had access through his father to both funds and an iron foundry.

Still later on, after Morse had patented the device and its code (Gale and Vail received, by contract, shares in the future royalties) and after the three partners had convinced Congress to allocate $30,000 so that they might build an experimental telegraph line between Baltimore and Washington, a fourth collaborator was added, Ezra Cornell, an itinerant plow salesman. Cornell, at Morse's request, had designed a special plow that could open the ground, lay lead pipe (which was to carry the copper wires), and close the trench in one operation. Later, after the team had discovered that wire in pipes shorted out uncontrollably, Cornell suggested stringing the wires on poles, which turned out to be the successful solution.

The initial idea for an invention, Morse had discovered, is just the beginning of an arduous process that often requires more skills, more patience, and more time than a single individual is likely to possess. Every successful inventor of the nineteenth century learned this crucial lesson; some learned it through bitter experience; others by studying the lives of the inventors who had preceded them. Alexander Graham Bell traveled the experience route, perhaps because he did not intend to earn his living as an inventor but rather as a teacher of the deaf. When Bell first conceived the idea for a multiple telegraph (a device that would allow several messages to be sent simultaneously along one electric wire), he feared that he would not be able to get an American patent on his idea (because he was not yet an American citizen), and so he worked on it entirely alone, concerned to keep his fundamental idea secret. But he needed to purchase equipment for his experiments and to have some pieces of his equipment tailor-made; this was expensive, and Bell had little money. When he finally realized that he was in a race against other inventors (both for the multiple telegraph and for a second idea, a telegraph that could carry sounds), he borrowed some money from two wealthy men and hired an assistant, a mechanically clever young man by the name of Thomas Watson. Watson had been employed by the instrument maker that Bell patronized. In recognition of his crucial role in the developmental work on the telephone, Watson was made the first full-time employee of the Bell Telephone Company; he also received a one-tenth interest in all of Bell's patents.

Late in the nineteenth century, Edison, Sperry, and Eastman, each of them consciously aware that inventive activity was collaborative, established special laboratories to encourage that kind of collaboration. In 1878, when he began the search for a successful electric lighting system, Edison had already established his laboratory in Menlo Park, New Jersey (with funds that he had acquired from previous successful inventions): there was a scientific and technical library, a chemical laboratory, an electric testing facility, and a machine shop (later a glassblowing operation was added). Several dozen people were employed at Menlo Park—among them Francis Upton, a physicist who had taken advanced training in the Berlin laboratory of Hermann von Helmholtz; Charles Batchelor, an expert mechanic; and John Kreusi, a superb draftsman and machinist, who directed the construction of whatever model equipment Edison required.

Sperry had the idea for his first invention—an electric generator—in 1880, but he knew that he could not afford to hire a draftsman, pay a lawyer, and rent a machine shop, so he entered into an agreement with a local manufacturer of wagons (who happened to be his father's employer). In return for a half interest in the eventual returns, Sperry was put on a salary and provided with tools and supplies, assistance from skilled workers (including a superb machinist) as well as the advice of the company's executives (one of whom, Hugh Duffey, was a mechanical engineer and inventor) and a patent lawyer. In all of his subsequent enterprises, Sperry following this initial model, made certain that he was surrounded by skilled collaborators (including, in later years, his own sons) and that the manufacturing and financial aspects of his companies were delegated to others so that he and his collaborators could concentrate on invention and development.

Eastman started out in the photography industry by inventing a machine that would coat glass plates with the photosensitive gelatin emulsions that were then being used and by raising capital to become a manufacturer of the plates. In the early 1880s, he hired a local camera maker, William H. Walker, to work with him in developing a new system of photography based on a paper roll film rather than on a glass plate. Competition in the photographic industry was keen, however, and by the 1890s Eastman had decided that, as a business strategy, his only chance of staying ahead of his competitors was to improve his products every year. Accordingly he established an Experimental Department in his factory and hired chemists and physicists to lead it.

Early on in this process, Eastman discovered that since he lacked scientific training, he could not understand the work that was being done by the people he had hired in his "invention factory." That did not lessen his enthusiasm for the strategy. In 1912, he transformed the department into a separate facility—a research laboratory with a well-trained physicist as its director, a staff of twenty, and an operating budget of $50,000; by 1920, both the staff and the budget had quadrupled. Also by 1920, other companies—General Electric, Bell Telephone, Westinghouse, to name just a few—had established industrial laboratories, following the path first trod by Sperry and Eastman. The heyday of the independent inventor was at an end and the era of the corporate inventor—someone who is on salary, and who assigns patents to an employer—had begun. The need for inventors to collaborate had been enshrined, by the independent inventors who first recognized the need, in an institution: the industrial laboratory.

Many people (including some inventors themselves) would like to paint the process of invention in romantic terms, carried out by unusual men and women, very different from the ordinary run of mortals: more inclined to

Thomas Edison was both a brilliant inventor and an expert entrepreneur, as this sketch and note from one of his laboratory notebooks suggests. The handwriting seems to be that of Edison's employee and collaborator Francis Upton. (Courtesy U.S. Department of the Interior, National Park Service, Edison National Historic Site).

take risks, more persistent, more devoted to their work, certainly more creative. Without necessarily detracting from that romantic vision (for there can be no doubt that people like Edison, Sperry, Morse, and Bell were cut from a different, and very special, cloth), the history of invention quite properly reveals that it is only a partial truth. Few inventive undertakings were brought to a successful conclusion in the absence of some very mundane prerequisites. One of them was an economic incentive to invent; in the United States, this was provided by a strong, well-supported social institution: the patent system. Another mundane prerequisite, as we have just seen, was the cooperation of people on whom the creative inventor relied in order to achieve success: the collaborators who provided the goods and the services that sustained inventive development. Some of those collaborators provided technical and scientific skills that the inventor did not have, others provided well-machined models or excellently scaled drawings. Some collaborators provided legal advice, others, financial counsel; some (about whom less is known) must have provided meals and emotional support after long days and nights in the laboratory. Yet others provided—and it is to these others that we must now turn—the two mundane requirements without which all the others would have been for naught: money and management.

Entrepreneurs: Innovation and Diffusion

The French word *entrepreneur* means "to undertake," not in the sense of preparing bodies for burial but in the sense of attempting to shoulder a great burden, to take a great risk. Entrepreneurs are innovators and diffusers rather than inventors; they are the people who have the money, the skill, and the determination necessary to bring an inventor's idea into the marketplace. Sometimes inventors have also been entrepreneurs, but entrepreneurship is not the same thing as inventiveness. Inventors dream impossible dreams; entrepreneurs have the resources and the skills to try making those dreams come true.

In the nineteenth century there were many such innovative entrepreneurs, all convinced that there were many pots of gold at the end of technology's rainbow. Some entrepreneurs managed to find those pots of gold; they are the ones about whom biographies are written. We know less about those who failed or who were only moderately successful, which is why no one really knows what makes entrepreneurs different from the rest of us and the successful entrepreneurs different from the unsuccessful.

The years between 1870 and 1930 were the golden age of entrepreneurship, the best of times to be an economic pioneer, because of the cooperative assistance of local, state, and federal governments. There probably has been no time in American history in which American governments were quite so willing to be quite so generous with entrepreneurs of all kinds. Between 1870 and 1920, governmental assistance to economic pioneers was the watchword of the day, and governmental regulation of their actions was either minimal or completely nonexistent. The geographical size of the country was growing with every passing year and so was the size of its population. Economic development of the areas they lived in was important to many, probably most, Americans and therefore to the representatives they elected: farmers on the frontiers wanted railroads that would bring their crops to market quickly just as much as city residents wanted factories that would provide jobs and dredged harbors that would lower the price of imports.

Before the Civil War, as we learned in Chapter 5, governments had assisted in building some of the earliest technological systems by investing in them, particularly those that were concerned, as were the canals and the railroads, with transportation. Monopolies created by legislatures were another tool that pioneering entrepreneurs used in their attempts to protect risky investments by controlling the market: many of the earliest steamboat lines and telegraph companies were granted such monopolies. The Army Corps of Engineers helped dredge harbors and clear riverbeds so that private companies could open new frontiers to riverboat traffic; state governments regularly lent their surveyors to lay out railroad tracks and their construction crews to assist in the building.

After the Civil War, governmental assistance to private industry became both more extensive and more sophisticated. The most famous of the forms in which that assistance was rendered was the land grant to the railroads: huge tracts of government-owned land were delivered to the railroads virtually free of charge. Some of this land was to become a railroad right-of-way, but some (indeed the vast majority of it) was intended to be sold off by the railroads to underwrite the huge costs of construction, a governmentally sanctioned subsidy to entrepreneurship.

The land grant was not, however, the only mechanism by which governments in this period were willing to provide assistance to entrepreneurs. State governments were willing to provide legal shields behind which monopolies could hide. When the company that John D. Rockefeller had created to vertically integrate his oil interests was declared illegal in Ohio, he moved his operation to New Jersey, where the state legislature was more amenable. The federal government was willing to provide tariff assistance. When American steel manufacturers discovered that the English monopoly on the manufacture of railroad tracks was difficult to break, they went to Congress, asked for, and promptly got a protective tariff to help them dominate the market. Generally speaking, the needs and desires of entrepreneurs tended to have very high priority on the agendas of American politicians between the end of the Civil War and the beginning of the Great Depression.

Regulation was minimal; environmental protection barely recognized as necessary; the interests of workers understood to be almost completely beyond the purview of the government. Even when Congress, for example, became sufficiently irritated by the manipulative practices of the railroad magnates to create the first major regulatory commission (the Interstate

Commerce Commission, which began operating in 1887 with a mandate to control freight charges on the railroads), the enabling legislation gave the commission very few enforcement powers and the courts repeatedly whittled away those it did possess. The net result was that in subsequent years the railroads continued not only to increase their monopolistic tendencies but also their profits. There was no minimum wage, few requirements that those who operated dangerous equipment be licensed, no necessity for protecting the safety of workers—until 1906, there was no requirement that there be any kind of truth in advertising. The entrepreneur had almost free rein and, most of the time, did not hesitate to exercise it.

The iron and steel industry was one of the fundamental industries on which American industrial power was based. We can learn something about the multifaceted character of entrepreneurship in the period of industrial growth by examining that industry and the biographies of some of the men who attempted to make their fortunes in it.

At the beginning of the nineteenth century, there were three forms of iron used in manufacturing processes: wrought iron, cast iron, and steel (see Chapter 3). Steel was the rarest, most desirable, and most expensive form of iron: desirable because it is easy to shape when red hot and very hard when cold; rare and expensive because its manufacture required more refining processes and very skilled workers. Steel contains 0.25 to 1.25 percent carbon; until the middle of the nineteenth century, all of the several different ways of making it were both difficult and time-consuming. The expansion of railroading, however, created a demand for cheaper ways to produce steel since steel turned out to be the best material for railroad tracks, train axles, and wheels. In 1847, the owner of an iron furnace in Kentucky began experimenting with what he hoped would be a new, and lucrative, production process.

William Kelly had been born in Pittsburgh in 1811, the son of a well-to-do landowner, and educated in the public schools there. His early career was unremarkable; he became the junior partner in a dry-goods firm in Pittsburgh. After meeting and marrying a young woman from Eddyville, Kentucky, he moved to that community and, with his brother, purchased iron-ore lands and an adjacent furnace. William Kelly was going to manage the furnace; his brother would handle the finances. One of the continuing concerns for the Kellys, as well as for other makers of wrought iron, was the high cost of fuel; finding ways to limit fuel bills was a high priority.

One day, Kelly noticed something that surprised him in his blast furnace; molten pig iron became white hot when air was blown on it. Perhaps, Kelly thought, this phenomenon could be used to develop a "fuel-less" form of pneumatic refining. Experimenting further, he came to the conclusion that the carbon that is mixed in molten pig iron can be burned out by an air blast. He also concluded that, among its other virtues, pneumatic refining would be an exceptionally cheap way to make steel—if he could control it. Kelly continued to experiment secretly with this technique throughout the 1850s in a special converter (essentially a large kettle in which pig iron can be heated) that he built deep in the woods. In 1856, however, he learned that Henry Bessemer, an Englishman, had been granted a patent in the United States on the same process. Kelly immediately filed for a patent himself and managed to convince the patent examiners of his priority. In 1857, he was declared, within the confines of the United States, the original inventor of the process.

Unfortunately, Kelly was never able to reap the full financial benefits from his patent; successful as an inventor, he was not a skillful entrepreneur. Within months of winning his patent, he had gone bankrupt (partly as a result of the financial panic of 1857, partly because he had let his business slide while attending to his experiments) and had sold his patent for $1,000 to his father. The elder Kelly, thinking his son inept, bequeathed the patent to his daughters. In later years, Kelly's sisters, agreeing with their father, refused to return the patent to their brother (although they did bequeath it to his children).

Kelly had no choice, then, but to become financially dependent on other people in order to exploit his own patent; he had neither sufficient financial resources to build a suitable refining furnace nor full control of licensing under his patent. Seven more years passed before his experiments were completed and a furnace that fit the requirements of his process was built. During that time, three other men had invested their considerable resources in his enterprise. By 1864, they had succeeded in blowing steel successfully at the specially built Wyandotte Iron works near Detroit, Michigan. At the very same time, however, an eminent mechanical engineer, Alexander Lyman Holley, was successfully raising money to build a steel plant designed under the American rights to the Bessemer patents (although Kelly had the patent on the process, Bessemer had patents on some of the equipment that made the process economical).

Holley's father had been a successful cutlery manufacturer and governor of the state of Connecticut. Refusing to go to Yale, as his family wished, he had taken a degree in engineering from Brown University and had subsequently worked as a locomotive operator, a draftsman, and an engineering journalist. In undertaking to bring the Bessemer process to the United States, he was acting simultaneously as an entrepreneur, inventor, and engineer: entrepreneur when he raised the money to pay the Bessemer licensing fees and build the plant, engineer when he designed it, inventor when he patented various modifications on the original equipment. But, like Kelly, he never had complete control over the enterprise that he had created.

Within a few years, the men who had invested in Kelly and those who had invested in Holley decided that it was a better plan to merge their operations rather than to fight interminable legal battles. The Kelly interests received three tenths of the stock in the new company; the Holley interests, seven tenths—a decision over which neither man had much control and that Kelly bitterly resented. Kelly did not die a poor man; he received a goodly sum from his patents and, after the merger of the two companies, formed a company of his own to manufacture axes. Nonetheless, he did not reap anything like the benefit that Bessemer did (about $10 million in royalties and a knighthood); neither did he receive the public recognition that, to this day, still attaches to the name of his rival. Holley did not die impoverished either; he went on to a more than moderately successful career as a consultant engineer and became one of the founders of the American Society of Mechanical Engineers.

The person who reaped the largest financial benefit from Kelly-Bessemer steel was not Kelly, Bessemer, or Holley but rather a brilliant financier who knew precious little about making steel but a precious lot about making money: Andrew Carnegie. Carnegie had been born in Scotland in 1835; his family had emigrated in 1848 because his father, a handloom weaver, had been ruined (economically and emotionally) by the introduction of the steam-powered factory loom. Opportunities for handloom weaving were not much more abundant in Pittsburgh, where the Carnegies settled, than they had been in Scotland. At the age of fourteen, Andrew Carnegie went to work, first as a menial employee in a textile factory, then as a messenger boy for a telegraph company. Always aggressively looking to get ahead, Carnegie taught himself to receive and send Morse messages and was quickly promoted from messenger boy to operator.

In 1852, the Pennsylvania Railroad extended its lines to Pittsburgh and opened an office there; the head of this western division of the railroad, Tom Scott, made daily trips to the telegraph office to send messages up and down his line. Soon he realized that Carnegie was an excellent operator and that he ought to have a telegraph operation in his own office. At that point, Carnegie became an employee of the Pennsylvania Railroad. Under Scott's tutelage, he also began investing in stocks (occasionally borrowing the necessary funds either from Scott or from local bankers): first in an express freight company and then in a company that manufactured sleeping cars and leased them to railroads. This second investment turned into a gold mine: within five years, Carnegie's initial $217 investment was bringing in $5,000 a year in dividends—vastly more than his salary with the railroad.

Carnegie spent thirteen years as an employee of the Pennsylvania Railroad, all the while learning how to operate a complex business. By the time he left, he had succeeded his former boss as the superintendent of the western division, had made several more brilliant investments—some oil wells in western Pennsylvania—and had become a partner in a company that built durable iron bridges for railroads. In 1868, three years after he left the railroad, Carnegie was a wealthy man, earning over $50,000 a year; he was then thirty-three years old.

The year 1868 was also the year in which the Kelly interests and the Holley interests merged their patent rights. Carnegie knew the potential uses of steel in railroading. Having spent most of his life in western Pennsylvania, he also knew that new techniques for turning the bituminous coals of western Pennsylvania into sulfur-free coke had been developed. Sulfur-free coke was an inexpensive replacement for charcoal, the expensive fuel that had so troubled Kelly decades earlier (sulfur is an unwanted contaminant in smelting). He thought the time was ripe to build a huge steel producing plant using the Kelly-Bessemer process, and he knew that he could raise the money to do it. By 1872, the plant was in operation. Designed by Holley and named after Edgar Thomson (then the president of the Pennsylvania Railroad, Carnegie's best customer), the plant was located in Braddock, Pennsylvania (where it could be serviced by both the Pennsylvania and Baltimore and Ohio Railroads), and was a colossal success from the very beginning. Its ten converters produced high-grade steel, which could be sold profitably at $65 a ton, well below the going market rate of $70 a ton.

Having instituted the centralized management style that he had first learned while working for the Pennsylvania Railroad, Carnegie was able, over the course of the years, to press the Thomson plant (and others that he subsequently built) to achieve greater and greater efficiency: proportionally less capital going in and more steel coming out. Profits soared. By 1900, Carnegie Steel was vertically integrated; Andrew Carnegie (and a small number of partners) owned and controlled the iron-ore fields of northern Minnesota, the great fleet of steamboats that brought ore over the Great Lakes, the rail lines that made deliveries to company mills, and the coal mines and coke ovens that supplied the fuel. Carnegie had even stopped selling his finished products to wholesalers and had instead developed his own staff of sales personnel. When the Thomson plant opened in 1872, Britain was far and away the world's leading supplier of steel. By 1900, twenty-eight years later, Carnegie's company alone produced more than all the plants in Britain combined. A year later Carnegie, now aging, sold out to the Morgan interests, and the King of Steel became the World's Richest Man, retiring from his last and largest undertaking with a mere half a billion dollars in bonds.

Kelley, Holley, and Carnegie were quite obviously very different from each other, but each was, in his own way, an entrepreneur. Indeed, all the great technological systems of the industrial age were built by men like Kelly and Holley, men who had the economic vision to see the potential in an invention, and also by men like Carnegie, who were neither engineers nor inventors but rather financiers, possessed of the business skills (to raise money, to manage subordinates, to curry favor with legislators, to harass competitors, to manipulate the courts) necessary to build vast systems on the basis of those inventions.

Another way of putting the point about technological change and entrepreneurship is to say that inventions have to be widely diffused before they begin to have social and economic impact. Entrepreneurs, successful ones, take inventions and turn them into systems; entrepreneurs are, therefore, the people responsible for diffusion.

We can see this clearly in the case of the telegraph system. Samuel F. B. Morse may have invented the telegraph and the code that bears his name,

but he was not the person responsible for building the vast network of telegraph wires that crisscrossed North America and western Europe by 1875.
Indeed, not one but several dozen men (all, in their own way, entrepreneurs) were responsible for that accomplishment, and a historian or a prize
giver who went about trying to settle the mantle on just one head would
have a hard time deciding on whom it should lie. Alfred Vail, son of the
wealthy proprietor of an iron and brass works, actually built Morse's first
working instruments. Along with his father and brother, Vail financed the
inventor's initial experiments. Amos Kendall was a lawyer who had served
as postmaster general under Andrew Jackson. He did the legal and financial work for Morse's Magnetic Telegraph Company, which means that he
was in charge of licensing Morse's patents to the companies that would
build and operate telegraph lines. Cyrus Field, who was a wealthy New
York paper merchant, unified dozens of competing telegraph companies
into one large conglomerate. By the 1860s the American Telegraph Company had an effective monopoly on telegraph service for the entire Atlantic
seaboard; it also connected the seaboard to England by laying the first Atlantic cable. Hiram Sibley, a successful businessman from Rochester, New
York, organized another conglomerate for the purpose of creating a unified telegraph system from the Great Lakes to the Pacific Ocean; he called
it Western Union. Western Union later absorbed American Telegraph, so
that it controlled telegraph service across the entire nation. Each of these
men had some entrepreneurial skill. Together they built the massive telegraphic system that revolutionized (see Chapter 7) communications in the
United States in the middle years of the nineteenth century.

In like manner, the credit (or the blame, depending on your point of
view) for any of the great technological systems lies as much with entrepreneurs as it does with inventors. One can reasonably ask, for example,
whether Alexander Graham Bell would have ever filed his patents on the
telephone if his prospective father-in-law, Gardiner Hubbard (a wealthy
Massachusetts attorney, who had made—and partially lost—a fortune by
investing in gas and water utilities), had not withheld his daughter's hand
until the telephone experiments were finished and a patent application was
actually filed. Bell had little business sense and few entrepreneurial skills,
and he knew it. He held stock in the various businesses that were built
around his patent, but took no active role in the management of them,
thereby leaving the creation of the telephone network to others. One of
those others was Theodore M. Vail (no relation to Alfred Vail), who started out in 1878 as the general manager of the Bell Telephone Company
(which was licensing franchises to other entrepreneurs who wanted to set
up local telephone companies under the Bell patents). Vail ended up creating the long distance telephone association that was, until recently, one
of the nations most extensive monopolies: the American Telephone and
Telegraph Company. Bell became an exceedingly rich man on the basis of
his patents (the telephone patent is thought to have been the single most
valuable patent in American history), but he was not responsible for build-

This painting by Daniel Huntington (c. 1880) depicts the entrepreneurs and inventors who were responsible for the Atlantic Cable project as adventurous and sagacious men. It also reminds us that successful entrepreneurship requires collaboration, just as successful invention does. Cyrus Field is second from right and Samuel F. B. Morse is fifth from the left. (Courtesy New York City Partnership and Chamber of Commerce.)

ing the system that made him rich and that turned the telephone into an integral part of American life within the space of fifty years.

Entrepreneurs undertake great risks in the hope of reaping great profits. Those who succeeded in the period between 1870 and 1920 had the money and the skills to build large, profitable companies. These companies were technological; they diffused the products of inventive activity, building the great systems (telephone, telegraph, railroad, iron and steel, petroleum, manufacturing) on which the economic health of the country came to depend. Successful entrepreneurs, at least in the period between 1870 and 1920, were people who had the vision to understand the potential uses of new technologies coupled with the political and economic skills to create and sustain large, complex organizations. But they could not build those organizations and those systems single-handedly. They needed to hire large numbers of technically competent people, especially people who had sufficient training in scientific theory and research methods to make improvements in the technologies as the systems continued to develop: engineers.

Engineers: Changes between 1820 and 1920

The roots of engineering go far back in time to the Roman Empire, when the people who could design and maintain bombardment machines (called *ingenium* in Latin, "battle engines" in English) were members of the military. In the nineteenth century, however, as a result of industrialization, technically trained people began to play a role in civilian life—and that was when a significant number of people began identifying themselves as members of a new profession, engineering. The 1850 census was the first one to include "engineer" as an occupation and just over two thousand people (all men, probably) were included in that category. Some of these two thousand engineers were probably drivers of steam locomotives. Others may have been tending and feeding stationary steam engines. The majority, however, were probably doing the work that we would now designate by the term *civil engineering*: building canals, surveying for railroad lines, supervising construction crews. The vast majority of those people had acquired their skills through some form of on-the-job training.

John B. Jervis (1795–1885) was widely acknowledged to be the leader of this first generation of American engineers. Jervis was the son of a farmer and lumberman in upstate New York. When he was sixteen years old, he took a job helping to cut a path, sixty feet wide, through a cedar swamp along the proposed route of the Erie Canal. Within a few months, he had learned surveying; during the next seven years, he made his way up the career ladder that extraordinary construction job had created, learning each job before he progressed to the next one: axeman, target man, rodman, tally man, surveyor, resident engineer for a subdivision, superintendent of an operating division. By the time he was thirty-five, Jervis was probably the most experienced civil engineer on the American continent: he had designed the country's first operating railroad (part of the Delaware and Hudson canal system in 1828), had written the specifications for the first steam locomotive (the Stourbridge Lion), and had invented the mechanism (the swiveling four-wheeled bogie truck) that allowed locomotives to take curves at relatively high speeds without derailing—all without any additional formal education. What he needed to know, he had learned from watching and listening to others, from experimenting and attempting, from reading.

Few of the engineers of those pioneering days—the men who surveyed the routes for the country's first railroads, built the bridges that first spanned its interior rivers, or designed the power trains for its first steam driven factories—had had any formal education either in the sciences or in what were then called the technical arts. The founders of civil engineering learned their craft working on the canals and railroads. The founders of what would later be called mechanical engineering learned theirs as apprentices and laborers in machine shops. James B. Francis, who was the chief engineer in charge of the locks and canals that served the industrial city of Lowell, Massachusetts (and who designed a cheap and efficient turbine to replace the upright waterwheels then in use), learned his trade by being apprenticed to his father, an English engineer. William Sellers, who designed what turned out to be the standard screw thread as well as all of the structural steel used on the Brooklyn Bridge, apprenticed in his uncle's machine shop in Philadelphia before opening his own business as a maker of machine tools. Ellis Sylvester Chesbrough, who planned and built the sewage system that made the growth of Chicago possible, started his career surveying for railroad lines. James Buchanan Eads, who—among many other accomplishments—designed and built ironclad gunboats that patrolled the Mississippi during the Civil War, had only a year or two of formal education and began his career at age nineteen as a clerk on a Mississippi riverboat.

Indeed in the early years of the Republic, only on-the-job training could have taught a young person how to practice engineering because most schools did not teach anything even vaguely resembling science, let alone engineering. Lower schools, based on the model of the church schools of England and Europe, taught children to read, to write, and to do simple mathematical computations. Upper schools (which were attended by only a minute fraction of the population) taught more gentlemanly subjects, in the style of the traditional classical education: Latin, Greek, rhetoric, literature, modern languages, geometry. In the early colleges (such as William and Mary, Harvard, Yale, Columbia), science was barely mentioned (except for an occasional bit of astronomy) and technology not at all.

The only exception to this general lack of training opportunities either in scientific or in technical subjects was West Point, established in 1811 as the nation's first military academy. There—in the tradition established by the Romans centuries earlier—some practical subjects were taught to the men who might someday be called on to build bridges and fortifications or to see to it that cannons were properly installed and roads properly maintained. In addition—in the tradition established by the French a decade or two earlier—theoretical subjects, such as Newtonian mechanics, were also taught at West Point. Some West Point graduates became staff members of the Army Corps of Engineers, which oversaw some of the most important civil engineering projects of the nineteenth century. By the second half of the nineteenth century, however, the American educational landscape had markedly changed and school culture was beginning to replace shop culture as the most likely precursor to an engineering career. New schools—polytechnic institutes they were sometimes called—were being created, encouraged, and endowed by people who believed that industrialization could best be furthered by training scientifically educated and technically skilled young men.

In 1815, the American inventor Benjamin Thomson (best known by his British name, Count Rumford) willed $1,000 a year (which was then an enormous sum) to Harvard for a course of lectures to teach "the utility of the physical and mathematical sciences for the improvement of the useful arts, and for the extension of the industry, prosperity, happiness and well-being of society." In 1823, Amos Eaton, a young graduate of Yale, convinced Stephen van Rensselaer, a wealthy landowner and entrepreneur, to help establish a polytechnic academy in Troy, New York: Rensselaer Polytechnic Institute, the first American institution that gave a bachelor's degree based on a fundamentally scientific and technical curriculum (including such subjects as analytic geometry, determinative mineralogy, rational mechanics, machine design, and spherical projections). In 1845, nearby Union College (in Schenectady, New York) created a special curriculum in civil engineering, and two years later, the wealthy mill owner Abbott Lawrence donated $50,000 to start the Lawrence Scientific School at Harvard. Brown, Dartmouth, and Yale added engineering to their course offerings in the 1850s. The Naval Academy, founded in Annapolis in 1845, began offering courses in steam engineering (the theory, design, and maintenance of the steam engines on ships) early in the 1860s, thus helping to train the second and third generations of mechanical engineers. The Morrill Act, passed in 1862, provided substantial grants of federal land to states that would establish colleges of "agriculture and mechanic arts," and this accelerated the trend toward technical education. There were seventy academic engineering colleges in the United States in 1872, eighty-five by 1880. By the turn of the century, 10,000 young men a year (and a handful of young women) were entering engineering colleges as freshmen, five times as many future engineers as had been counted in the entire workforce fifty years earlier.

Indeed, in the years between 1870 and 1920, as the economy became increasingly industrialized, engineering became an increasingly popular career. The census recorded 7,000 engineers in 1880 and 136,000 in 1920, almost a twentyfold rise. More and more of these fourth- and fifth-generation engineers had been produced by school culture rather than shop culture. Washington Roebling, who built the Brooklyn Bridge, was a graduate of Rensselaer Polytechnic Institute (which he hated). Horace Vaughn Winchell who mapped the Mesabi range and made a systematic study of its iron deposits, had studied economic geology at the University of Michigan. Arthur D. Little, who developed the crucial concept of unit operations in chemically based industries, was a graduate of the Massachusetts Institute of Technology. Franklin Sprague, who built the first electric intraurban railroad in Richmond, Virginia, had received his technical education at the Naval Academy. In those decades, a person might still become an engineer without benefit of a degree (George Herman Babcock, son and grandson of mechanics and cofounder of the famous machinery-building firm Wilcox and Babcock, for example, or Kate Gleason, who inherited her father's machine shop and then pioneered the use of concrete blocks in construction), but such exceptions became fewer and fewer.

Unlike his predecessors, the typical early twentieth-century engineer was a man with a bachelor of science degree, the product of school, not shop training. By 1920, the possession of a degree had become a necessary requirement for employment as an engineer. Among other things, this edu-

The distinguishing mark of the late nineteenth century engineer was scientific training, and the distinguishing mark of the late nineteenth century scientific and technical colleges was the scientific laboratory. This is a chemistry lab at MIT in the 1880s. (Courtesy MIT Museum.)

cational requirement meant that women were virtually excluded from the profession. Most of the private engineering schools were all male; the public schools, nominally coeducational by law, had managed to arrange matters so that only a handful of women had actually succeeded in obtaining degrees. The typical early twentieth-century engineer was also a man who had been born into a middle-class family: the child of a merchant, a small businessman, a manufacturer, or a prosperous farmer. Many of the engineering colleges charged tuition and even those that were free still required full-time attendance, not something a poor boy was likely to find feasible. Most engineers were also native born, white, and Protestant—very different from many of the workers, whether skilled or unskilled, whom they had to supervise. The engineer wore a jacket and a white shirt to work; artisans and laborers wore blue or gray. The engineer went to the same church, lived in the same neighborhoods, sent his children to the same schools, and came from the same ethnic stock as his employer. By the end of the nineteenth century, engineering had become, in short, a profession. The engineer had become completely differentiated both from the artisan class of his predecessors and from the laboring classes that he supervised.

As engineering became a profession, various engineers felt the need to form organizations that would protect their group interests. They wanted such professional associations to be somehow different both from the guilds that had once protected artisans and from the unions that were beginning to protect laborers. The first such enterprise was the Franklin Institute, organized in Philadelphia in 1824, and although it was later supplanted by such organizations as the American Society of Civil Engineers and the American Institute of Mining Engineers, its history precisely reflects the social transformation of nineteenth-century engineering.

The founders of the Franklin Institute were a group of prosperous machine-shop owners in Philadelphia. They had two intentions: first, to provide a library and public lectures on the sciences and the crafts so that working people could learn and, by learning, improve themselves—in short, a way to formalize on-the-job training. Their second intention was more self-serving. They wanted to learn from each other, to break the bonds of secrecy that had once shrouded the crafts in mystery, to experiment, to publish the results of their experiments—in short, to behave like scientists, to make the "arts" (their term) more scientific and also more profitable.

As the years wore on, however, it became increasingly clear that the Franklin Institute (and other enterprises like it that were founded in industrial cities up and down the eastern seaboard) was going to fail at its first objective and succeed at its second. Apprentices, journeymen, and masters just weren't terribly interested in attending lectures on the principles of dynamics or kinematics, assuming that they could have found the energy to stay awake after a long day's employment. But machine-shop owners, inventors, canal builders, and surveyors were, on the other hand, very interested in keeping abreast of the latest developments in their fields, in finding out what had just been patented—in learning what chemists had just discovered about the bleaching properties of chlorine, for example, or what tests physicists had devised to estimate the efficiency of engines. Science, or better yet, the scientific method, was becoming part of the growing social distinction between the engineer and the artisan. The professional engineer, unlike the artisan, was thought to be someone who was interested in advancing his career through experimentation, inquiry, and the flexibility that theoretical skills can provide.

The growth of national professional associations for engineers was both evidence for and an encouragement to that growing distinction. The Franklin Institute was fine in its way, progressive engineers argued, but it wasn't a membership organization; it could neither withhold membership nor confer it and therefore could not act to certify and uphold the engineer's status. In order to achieve that, groups of engineers in various localities began forming clubs, organizations that could decide whom to admit and whom to turn away: in Baltimore in 1839, in Boston in 1848, New York in 1852, in St. Louis in 1868, in Chicago a year later. These local clubs were the foundations on which national professional engineering associations came to be built.

The first such "national" was ASCE, the American Society of Civil Engineers, which grew out of the local engineers club that was founded in

New York in 1852. More or less inactive for the first fifteen years of its existence, ASCE was revived by a group of wealthy, college-educated engineers in 1867. The work of these men kept them traveling. Builders of bridges and tunnels, railway lines and aqueducts, they moved around from place: sometimes on the frontiers, sometimes in the cities, one year in Winnetka, Illinois, the next, perhaps in Denver, Colorado. A national organization suited the pattern of their careers better than a local one ever could. In its revived form, ASCE also had very different membership requirements. As a local, it had admitted "all persons professionally interested in the advancement of engineering," but as a national, it had ranks of membership. These rankings were the mechanism by which the society hoped to act as a certifier of professional status. Anyone could join as an associate, but to be a full member (which meant to have voting privileges), a man (membership was explicitly restricted to men) had to have been in professional practice for at least five years and to be "in substantial charge" of an engineering work. Civil engineers had to be managers, not laborers; mechanics, masons, machinists, and women need not apply.

During the last decades of the nineteenth century, as more and more industries were created on the basis of new inventions and new sciences, new categories of engineering education and practice developed. By the 1870s, one professional association could no longer accommodate the needs of the various engineering subspecialties. The papers presented at meetings of the civil engineers (and published in their journal) simply did not interest those engineers who were searching for minerals in the far West or designing printing presses in Boston. Thus the American Institute of Mining Engineers (AIME) was organized in 1871, the American Society of Mechanical Engineers (ASME) followed in 1880, and the American Institute of Electrical Engineers (AIEE) four years later, in 1884. As the nineteenth century turned into the twentieth, additional professional associations for additional subspecialties were created, testimony to the continuing diversification of engineering: the Society of Naval Architects and Marine Engineers (1893), the American Society of Heating and Ventilating Engineers (1894), the Society of Automotive Engineers (1904), the American Institute of Chemical Engineers (1908), and the Institute of Radio Engineers (1912).

In structure and intentions these organizations were relatively similar, despite the varying professional activities of their members. Formed by economically comfortable and well-educated engineers, the engineering elite, they all had membership requirements, which necessitated some criteria for belonging and some procedures for elections. All met annually to hear scientific papers; all annual meetings were accompanied by numerous social gatherings that permitted members to meet, greet, and "network" with each other. All the societies published professional journals, which contained news items as well as printed versions of scientific lectures. In addition, all the societies became involved in the twin processes of setting standards for education (what courses must an electrical engineer take? do civil engineers need to know German?) and for production (should there be standardized grades for lubricating oil? should all screws manufactured in the United States have the same pitch? should there be a uniform code governing the materials used and the strength of the seals on steam engines?).

Within each of these subprofessional groupings, engineers were diversifying socially as well. The founders of the engineering associations, the elite engineers, had been men of a very particular stripe: consultants (if they were civil or mining engineers) or the owners of their own businesses (if they were mechanical, electric, or chemical engineers). Arthur D. Little's career is close to a paradigm case, almost precisely mirroring Jervis's half a century earlier. After Little graduated from MIT (in the early 1880s with a degree in chemistry), he took a job as the superintendent of a paper mill. On the job, he learned how to manage the equipment and the raw materials (some of them chemicals) needed to turn wood into pulp and pulp into paper. For a short period after that, Little became something of an entrepreneur: with various partners, he built start-up paper mills and then sold out to his partners. By 1886, however, he had gone into business for himself as a consulting chemist, advising manufacturers in the chemical industries about how to upgrade their facilities and improve their products—just as Jervis had contracted with various railroads and municipal administrations. Little was one of the founding members of the American Institute of Chemical Engineers; when he died in 1935, his consulting company, Arthur D. Little, Inc., was the largest commercial industrial research laboratory in the world.

Like Little, the elite engineers who had founded the professional associations were financially successful, entrepreneurial men. They had fashioned for themselves a work world in which they were in substantial charge of engineering works and they were accustomed to a good deal of both hard work and personal autonomy. Unfortunately, more and more technical graduates came on the job market around the turn of the century, proportionately fewer and fewer could expect to rise to the top in the same fashion. In addition, in that same period of time American small businesses were increasingly being converted into (or being engulfed by) large corporations, which meant that proportionately fewer and fewer technical graduates could become proprietors of their own business. The fields then burgeoning, electrical and chemical engineering, were capital intensive; enormous sums of money were needed to get started in business, and only very large businesses were likely to be successful.

The confluence of all these trends meant that, by 1920, the average engineer was neither a proprietor nor a consultant, but rather an employee. According to one informed estimate of the late 1920s, no more than 25 percent of all engineers who had graduated before 1909 were consultants or proprietors. Indeed, in large organizations such as the Bell Telephone Company or the Carnegie Steel Corporation or the General Electric Company, engineers were most likely to be found in the ranks of middle management. They were still in substantial charge of some tiny portion of the company's work, but unlike their predecessors in the profession they had very little autonomy; they were expected to be team players—implementers of decisions, not makers of them.

In such large corporations (and in the public works departments of large municipalities), beginning engineers, even graduates of the most prestigious engineering schools, started at the bottom, doing such routinized chores as mechanical drawing or calculations of the flow rates of fluids in pipes, for years on end, at relatively low pay. These newly minted engineers were also discovering that their only chances for advancement involved leaving their technical training behind and becoming corporate executives; "it became necessary," the historian Monte Calvert has concluded, "for an engineer to leave the engineering of materials and enter the engineering of men"—becoming pushers of paper, rather than designers of machines.[6]

Thus by 1920, the profession of engineering, barely a century old, was rent by divisions. Some of these divisions were created by the development of new engineering disciplines. Engineering educators, the people who taught at engineering colleges, worried about whether there were core courses—central bodies of information and theory that every engineer ought to know, whatever subspecialty he would choose to pursue. Was there any central body of either knowledge or practice that might unify engineers, whether they were working with electromagnetic waves or pharmaceuticals or building materials? Was there anything that made the profession of engineering unique, different from science and different from the crafts? Some engineers who were not elite members of their professions and who worked in subordinate positions in large corporations fought for the creation of a single professional association (the American Association of Engineers, founded in 1915) that would, they hoped, unite all engineers, whatever their subspecialty, in the fight to improve the poor working conditions that all engineers faced day to day on the job.

Yet other divisions were created by the unique history of engineering, the way in which engineering sciences had emerged from the technical arts over the course of the nineteenth century. How much practical, as opposed to theoretical, education should an engineer have? Did schools need to develop what were then called cooperative programs, so that every engineering student spent some time working in industry, recreating the on-the-job training that had characterized the early days of the profession? Or should schools bend more to the scientific model of education, providing courses in scientific theory (as opposed to engineering practice) and giving each student ample opportunity to work in teaching laboratories, learning the methods and nuances of research?

Finally, there were ethical divisions between engineers, focused on questions that were created by the social situation in which engineers found themselves as employees of large organizations. In the late nineteenth and early twentieth century, engineers were the only professional people who were likely to be employed by large, profit-oriented organizations or by large tax-funded organizations, both ultimately concerned with keeping costs down. Physicians and lawyers generally were in private practice; they made their own financial decisions about finances and about how best to spend their time. Many clergymen were employed by their individual congregations, and congregations were usually more concerned with their religious and social functions than with their financial needs. Teachers and professors worked for small school districts or for small colleges and universities, all not-for-profit organizations. Only the engineer had to worry all the time about the bottom line or had to be constrained by supervisors who worried about the bottom line.

To whom or what, then, did the engineer owe his loyalty? To the standards of his profession? To his employer, without whom neither his job nor his profession would exist? To the public, which puts its trust in his professional judgment, and ultimately pays the bill? If professionals are supposed to be able to make autonomous judgments based on their special competence, how can an engineer exercise that autonomy when he is just one cog in a very big wheel and when the bottom line must always be considered? If professional engineers are supposed to adhere to the scientific method—which demands open divulgence of information—how can they continue to work for organizations that, for economic advantage, require complete secrecy about new products and processes?

In the early decades of the nineteenth century, men like John B. Jervis thought that the answers would be simple. "A true engineer," Jervis wrote, "first of all, considers his duties as a trust and directs his whole energies to discharge the trust with all the solemnity of a judge on the bench. He is so immersed in his profession that he has no occasion to seek other sources of amusement, and is therefore always at his post. He has no ambition to be rich, and therefore eschews all commissions that blind the eyes and impair fidelity to his trust."[7] Unfortunately, as the American economy expanded, as the foundations for the great technological systems were laid, as engineers became professionals, and then as they became both more numerous and more specialized, many found it increasingly difficult to live up to, or even give credence to, Jervis's ideal. Twentieth-century engineers found themselves caught in the middle, squeezed between the selflessness, the altruism (thought to be one of the hallmarks of professional work), and the reality of their social situation as employees of large organizations.

Despite these divisions, engineers were crucial players in the drama of industrialization. Between 1870 and 1920, the American economy became fully industrialized. This meant, as we will learn in the next chapter, that the economy of the nation became enmeshed in and dependent on several enormous technological systems. Technological systems are networks of artifacts and organizations. The artifacts were invented by inventors; the organizations were created by entrepreneurs. But the engineers built the systems, doing the day-to-day skilled work that made it possible for messages to be transmitted nationwide by wires, for petroleum to flow nationwide through pipes, for steam engines to carry freight and people nationwide along a complex transportation network, for electricity to be generated at considerable distances from where it would be used. The discovering, designing, calculating, managing, and implementing that constitute engineering work are neither as dramatic nor as romantic as the more heralded work of inventors and entrepreneurs. But without engineers, neither the great technological systems nor industrialized society could ever have been built.

Notes

1. As quoted in Anne MacDonald, *Feminine Ingenuity: A History of American Women Inventors* (New York: Ballantine, 1992), p. 3. See also Frederick B. Tolles, "Masters, Sybilla," *Notable American Women, 1607–1950*, Vol. II (Cambridge, MA: Belknap Press of Harvard University Press, 1971), p. 508.

2. U.S., *Annals of Congress,* 1st Cong., 2nd Sess., I, 932, as quoted in Bruce W. Bugbee, *Genesis of American Patent and Copyright Law* (Washington, DC: Public Affairs Press, 1967), p. 137.

3. "Patent Act, 1790," Henry L. Ellsworth, *Digest of Patents Issued by the United States, 1790 to January 1, 1839* (Washington, DC: Patent Office, 1840), p. 3.

4. Jefferson to Benjamin Vaughan, 27 June 1790, as quoted in Bugbee, *Genesis,* p. 37.

5. Sperry, in a letter to a cousin, May 26, 1920, as quoted in Thomas Parke Hughes, *Elmer Sperry, Inventor and Engineer* (Baltimore: Johns Hopkins Press, 1971), p. 13.

6. Monte Calvert, *The Mechanical Engineer in America, 1830–1910* (Baltimore: Johns Hopkins Press, 1967), p. 231.

7. As quoted in Elting Morison, *From Know-How to Nowhere: The Development of American Technology* (New York: Basic Books, 1974), p. 68.

Suggestions for Further Reading

Robert V. Bruce, *Bell: Alexander Graham Bell and the Conquest of Solitude* (Boston, 1973).

Bruce W. Bugbee, *Genesis of American Patent and Copyright Law* (Washington, DC, 1967).

Daniel H. Calhoun, *The American Civil Engineers: Origins and Conflict* (Cambridge, 1960).

Monte A. Calvert, *The Mechanical Engineer in America: 1830–1910* (Baltimore, 1967).

W. Bernard Carlson, *Innovation as a Social Process: Elihu Thomson and the Rise of G. E., 1870–1900* (New York, 1991).

Alfred D. Chandler, *The Visible Hand: The Managerial Revolution in American Business* (Cambridge, 1977).

Margaret Cheney, *Tesla: Man Out of Time* (Englewood Cliffs, NJ, 1981).

Vary T. Coates and Bernard S. Finn, *A Retrospective Technology Assessment: Submarine Telegraphy: The Transatlantic Cable of 1866* (San Francisco, 1979).

Carolyn Cooper, *Patents and Invention* (Chicago, 1991).

Robert Friedel, *Zipper: An Exploration in Novelty* (New York: 1994).

Robert Friedel and Paul Israel, *Edison's Electric Light: Biography of an Invention* (New Brunswick, NJ, 1986).

Brooke Hindle, *Emulation and Invention* (New York, 1981).

Paul Israel, *From the Machine Shop to the Industrial Laboratory: Telegraphy and the Changing Context of American Invention, 1830–1920* (Baltimore, 1992).

Reese Jenkins, *Images and Enterprises: Technology and the American Photographic Industry, 1839 to 1925* (Baltimore, 1979).

Ronald R. Kline, *Steinmetz: Engineer and Socialist* (Baltimore, 1992).

Edwin T. Layton, Jr., *The Revolt of the Engineers: Social Responsibility and the American Engineering Profession* (Cleveland, 1971).

Stuart R. Leslie, *Boss Kettering* (New York, 1983).

A. Michal McMahon, *The Making of a Profession: A Century of Electrical Engineering in America* (New York: 1984).

Andre Millard, *Edison and the Business of Innovation* (Baltimore, 1990).

Daniel Nelson, *Frederick W. Taylor and the Rise of Scientific Management* (Madison, WI, 1980).

David F. Noble, *America by Design: Science, Technology and the Rise of Corporate Capitalism* (New York, 1977).

Robert C. Post, *Physics, Patents and Politics: A Biography of Charles Grafton Page* (New York, 1976).

Leonard Reich, *The Making of American Industrial Research: Science and Business at GE and Bell* (New York: 1985).

Terry S. Reynolds, *75 Years of Progress: A History of the American Institute of Chemical Engineers* (New York, 1983).

Terry S. Reynolds, ed., *The Engineer in America* (Chicago: 1991).

Bruce Sinclair, *A Centennial History of the American Society of Mechanical Engineers, 1880–1980* (Toronto, 1980).

Bruce Sinclair, *Philadelphia's Philosopher Mechanics: A History of the Franklin Institute, 1824–1865* (Baltimore, 1974).

Robert Sobel, *The Entrepreneurs: Explorations within the American Business Tradition* (New York, 1974).

Robert L. Thompson, *Wiring a Continent* (Princeton, 1947).

Joseph Frazier Wall, *Andrew Carnegie* (New York, 1970).

George Wise, *Willis R. Whitney, General Electric and the Origins of U.S. Industrial Research* (New York, 1985).

7

Industrial Society and Technological Systems

BETWEEN 1870 AND 1920, the United States changed in ways that its founders could never have dreamed possible. Although American industrialization began in the 1780s, the nation did not become an industrialized society until after the Civil War had ended. The armistice agreed to at Appomattox signaled, although the participants probably did not realize it, the beginning of the take-off phase of American industrialization. Having begun as a nation of farmers, the United States became a nation of industrial workers. Having begun as a financial weakling among the nations, by 1920 the United States had become the world's largest industrial economy.

What did this transformation mean to the people who lived through it? When a society passes from preindustrial to industrial conditions, which is what happened in the United States in the years between 1870 and 1920, people become less dependent on nature and more dependent on each other. This is one of history's little ironies. In a preindustrial society, when life is unstable, the whims of the weather and the perils of natural cycles are most often to blame. In an industrial society, when life is unstable, the whims of the market and the perils of social forces are most often to blame. Put another way, this means that in the process of industrialization individuals become more dependent on one another because they are linked together in large, complex networks that are, at one and the same time, both physical and social: technological systems.

Industrialization, Dependency, and Technological Systems

Many Americans learned what it means to become embedded in a set of technological systems in the years between 1870 and 1920. Today we have become so accustomed to these systems that we hardly ever stop to think about them; although they sustain our lives, they nonetheless remain mysterious. In the late twentieth century, people have tended to think that, if anything, industrialization has liberated them from dependency, not encased them in it, but that is not the case. We can see this clearly by imagining how a woman might provide food for a two-year-old child in a nonindustrialized society.

In a hunter-gatherer economy, she might simply go into the woods and collect nuts or walk to the waterside and dig for shellfish. In a premodern agricultural community (such as the one that some of the native peoples of the eastern seaboard had created), she might work with a small group of other people to plant corn, tend it, harvest it, and shuck it. Then she herself might dry it, grind it into meal, mix it with water, and bake it into a bread for the child to eat. In such a community, a woman would be dependent on the cooperation of several other people in order to provide enough food for her child, but all of those people would be known to her and none of them would be involved in an activity in which she could not have participated if necessity had demanded.

In an industrialized economy (our own, for example), an average woman's situation is wholly different. In order to get bread for a child, an average American woman is dependent on thousands of other people, virtually all of them totally unknown to her, many of them living and working at a considerable distance, employing equipment that she could not begin to operate, even if her life (quite literally) depended on it and even if she had the money (which isn't likely) to purchase it. A farmer grew the wheat using internal combustion engines and petroleum-derivative fertilizers. Then the wheat was harvested and transported to an organization that stored it under stable conditions, perhaps for several years. Then a milling company may have purchased it and transported it (over thousands of miles of roads or even ocean) to a mill, where it was ground by huge rollers powered by electricity (which itself may have been generated thousands of miles away). Then more transportation (all of this transportation required petroleum, which itself had to be processed and transported) was required: to a baking factory, where dozens of people (and millions of dollars of machinery) were used to turn the flour into bread. Then transportation again: to a market, where the woman could purchase it (having gotten herself there in an automobile, which itself had to be manufactured somewhere else, purchased at considerable expense, and supplied with fuel)—all of this before a slice of it could be spread with peanut butter to the delight of a two year old.

The point should, by now, be clear. People who live in agricultural societies are dependent on natural processes: they worry, with good reason, about whether and when there will be a drought or a flood, a plague of insects or of fungi, good weather or bad. People who live in industrial societies are not completely independent of such natural processes, but are more so than their predecessors (many floodplains have been controlled; some droughts can be offset by irrigation). At the same time, they are much more dependent on other people and on the technological systems that other people have designed and constructed. The physical parts of these systems are networks of connected objects: tractors, freight cars, pipelines, automobiles, display cases. The social parts are networks of people and organizations that make the connections between objects possible: farmers, bakers, and truck drivers; grain elevators, refineries, and supermarkets.

Preindustrialized societies had such networks of course (some of them are described in Chapter 2), but in industrialized societies, the networks are more complex and much denser—all of which makes it much harder for individuals to extricate themselves. A small change very far away can have enormous effects very quickly. Daily life can be easily disrupted for reasons that ordinary people can find hard to understand, and even experts can have difficulty comprehending.

People live longer and at a higher standard of living in industrial societies than in preindustrial ones, but they are not thereby rendered more independent (although advertising writers and politicians would like them to think they are) because, in the process of industrialization, one kind of dependency is traded for another: nature for technology. Americans learned what it meant to make that trade in the years between 1870 and 1920. We can begin understanding what they experienced if we look at some of the technological systems that were created or enlarged during those years.

The Telegraph System

The very first network that Americans experienced really looked like a network: the elongated spider's web of electric wires that carried telegraph signals. The fact that electricity could be transmitted long distances through wires had been discovered in the middle of the eighteenth century. Once a simple way to generate electric currents had been developed (a battery, or voltaic pile, named after the man who invented it, Alessandro Volta) many people began experimenting with various ways to send messages along the wires. An American portrait painter, Samuel F. B. Morse, came up with a practicable solution (see Chapter 6). Morse developed a transmitter that emitted a burst of electric current of either short or long duration (dots and dashes). His receiver, at the other end of the wire, was an electromagnet, which, when it moved, pushed a pencil against a moving paper tape (thus recording the pattern of dots and dashes). The most creative aspect of Morse's invention was his code, which enabled trained operators to make sense out of the patterns of dots and dashes.

In 1843, after Morse had obtained a government subvention, he and his partners built the nation's first telegraph line between Baltimore and Washington. By 1845, Morse had organized his own company to build additional lines and to licence other telegraph companies so that they could build even more lines, using the instruments he had patented. In a very short time, however, dozens of competing companies had entered the telegraph business, and Morse had all he could do to try to collect the licensing fees to which he was entitled. By 1849, almost every state east of the Mississippi had telegraph service, much of it provided by companies that were exploiting Morse's patents without compensating him.

Beginning around 1850, one of these companies, the New York and Mississippi Valley Printing Telegraph Company, began buying up or merging with all the others; in 1866, it changed its name to the Western Union Telegraph Company. In the decades after the Civil War, Western Union had an almost complete monopoly on telegraph service in the United States; a message brought to one of its offices could be transmitted to any of its other offices in almost all fairly large communities in the United States. Once the message was delivered, recipients could pick it up at a Western Union office. During these decades, only one company of any note succeeded in challenging Western Union's almost complete monopoly on telegraph service. The Postal Telegraph Company specialized in providing pick-up and delivery services for telegrams; yet even at the height of its success, it never managed to corner more than 25 percent of the country's telegraph business.

In 1866, when Western Union was incorporated, it already controlled almost 22,000 telegraph offices around the country. These were connected by 827,000 miles of wire (all of it strung from a virtual forest of telegraph poles, many of them running along railroad rights of way), and its operators were handling something on the order of 58 million messages annually. By 1920, the two companies (Western and Postal) between them were managing more than a million miles of wire and 155 million messages. Yet other companies (many of the railroads, for example, several investment banking houses, several wire news services) were using Western Union and Postal Telegraph lines on a contractual basis to provide in-house communication services (the famous Wall Street stock ticker was one of them).

As a result, as early as 1860, and certainly by 1880, the telegraph had become crucial to the political and economic life of the nation. Newspapers had become dependent on the telegraph for quick transmission of important information. The 1847 war with Mexico was the first war to have rapid news coverage, and the Civil War was the first in which military strategy depended on the quick flow of battle information over telegraph lines. During the Gilded Age (1880–1900), the nation's burgeoning financial markets were dependent on the telegraph for quick transmission of prices and orders. Railroad companies used the telegraph for scheduling and signaling purposes since information about deviations in train times could be quickly transmitted along the lines. The central offices of the railroads utilized telegraph communication to control the financial affairs of their widely dispersed branches. When the Atlantic cable was completed in 1866, the speed and frequency of communication between nations increased, thereby permanently changing the character of diplomatic negotiations. The cable also laid the groundwork for the growth of international trade (particularly the growth of multinational corporations) in the later decades of the century.

In short, by 1880, if by some weird accident all the batteries that generated electricity for telegraph lines had suddenly run out, the economic and social life of the nation would have faltered. Trains would have stopped running; businesses with branch offices would have stopped functioning; newspapers could not have covered distant events; the president could not have communicated with his European ambassadors; the stock market would have had to close; family members separated by long distances could not have relayed important news—births, deaths, illnesses—to each other. By the turn of the century, the telegraph system was both literally and figuratively a network, linking together various aspects of national life—making people increasingly dependent on it and on one another.

The Railroad System

Another system that linked geographic regions, diverse businesses, and millions of individuals was the railroad. We have already learned (in Chapter 5) about the technical developments (the high-pressure steam engine, the swivel truck, the T-rail) that were crucial to the development of the first operating rail lines in the United States in the 1830s. Once the technical feasibility of the railroad became obvious, its commercial potential also became clear. The railroad, unlike canals and steamboats, was not dependent on proximity to waterways and was not (as boats were) disabled when rivers flooded or canals froze.

During the 1840s, American entrepreneurs had began to realize the financial benefits that railroading might produce and railroad-building schemes were being concocted in parlors and banks, state houses, and farm houses all across the country. By the 1850s, a good many of those schemes had come to fruition. With 9,000 miles of railroad track in operation, the United States had more railroad mileage than all other western nations combined; by 1860, mileage had more than trebled, to 30,000 miles.

The pre–Civil War railroad system was not yet quite a technological system because, large as it was, it still was not integrated as a network. Most of the existing roads were short-haul lines, connecting such major cities as New York, Chicago, and Baltimore with their immediate hinterlands. Each road was owned by a different company, each company owned its own cars, and each built its tracks at the gauge (width) that seemed best for the cars it was going to attempt to run and the terrain over which the running had to be done. This lack of integration created numerous delays and additional expenses. In 1849, it took nine transshipments between nine unconnected railroads (and nine weeks of travel) to get freight from Philadelphia to Chicago. In 1861, the trip between Charleston and Philadelphia required eight car changes because of different gauges. During and immediately after the Civil War, not a single rail line entering either Philadelphia or Richmond made a direct connection with any other, much to the delight of the local teamsters, porters, and tavern keepers.

The multifaceted processes summed up under the word "integration" began in the years just after the Civil War and accelerated in the decades that followed. The rail system grew ever larger, stretching from coast to coast (with the completion of the Union Pacific Railroad in 1869), penetrating into parts of the country where settlement did not yet even exist. There were roughly 53,000 miles of track in 1870, but there were 93,000 miles by the time the next decade turned, and 254,000—the all-time high—by 1920. In that half century, the nation's population tripled, but its rail system grew sevenfold; the forty-eight states of the mainland United States became physically integrated, one with the other.

The form of the rail system was just as significant as its size. By 1920, what had once been a disjointed collection of short (usually north–south) lines had been transformed into a network of much longer trunk lines (running from coast to coast, east–west), each served by a network of shorter roads that connected localities (the limbs) with the trunks. Passengers could now travel from New York to San Francisco with only an occasional change of train and freight traveled without the necessity of transshipments. What had made this kind of integration possible was not a technological change, but a change in the pattern of railroad ownership and management.

From the very beginning of railroading, railroad companies had been joint-stock ventures (see Chapter 5). Huge amounts of capital had been required to build a railroad: rights of way had to be purchased, land cleared, bridges built, locomotives ordered, passenger cars constructed, freight cars bought. Once built, railroads were very expensive to run and to maintain: engines had to be repaired, passengers serviced, freight loaded, tickets sold, stations cleaned. Such a venture could not be financed by individuals, or even by partnerships. Money had to be raised both by selling shares of ownership in the company to large numbers of people and by borrowing large sums of money by issuing bonds.

As a result, both American stockbroking and American investment banking were twin products of the railroad age. Some of America's largest nineteenth-century fortunes were made by people who knew not how to build railroads, but how to finance them: J. P. Morgan, Leland Stanford, Jay Gould, Cornelius Vanderbilt, and George Crocker. These businessmen consolidated the railroads. They bought up competing feeder lines; they sought control of the boards of directors of trunk lines; they invested heavily in the stock of feeder roads until the feeders were forced to merge with the trunks. When they were finished, the railroads had become an integrated network, a technological system. In 1870, there had been several hundred railroads, many of which were in direct competition with each other. By 1900, virtually all the railroad mileage in the United States was either owned or controlled by just seven (often mutually cooperative) railroad combinations, all of which owed their existence to the machinations of a few very wealthy investment bankers.

As railroad ownership became consolidated, the railroad system became physically integrated. The most obvious indicator of this integration was the adoption of a standard gauge, which made it unnecessary to run different cars on different sets of tracks. By the end of the 1880s, virtually every railroad in the country had voluntarily converted to a gauge of 4 feet, 8½ inches in order to minimize both the expense and the delays of long distance travel. On this new integrated system, the need for freight and passengers to make repeated transfers was eliminated; as a result, costs fell while transportation speed increased.

The railroad system had a profound impact on the way in which Americans lived. By 1900, the sound of the train whistle could be heard in almost every corner of the land. Virtually everything Americans needed to maintain and sustain their lives was being transported by train. As much as they may have grumbled about freight rates on the railroads (and there was much injustice, particularly to farmers, to grumble about) and as much as they may have abhorred the techniques that the railroad barons had used to achieve integration, most Americans benefited from the increased operational efficiency that resulted.

In the years in which population tripled and rail mileage increased seven times, freight tonnage on the railroads went up elevenfold. Cattle were going by train from the ranches of Texas to the slaughterhouses of Chicago; butchered beef was leaving Chicago in refrigerated railroad cars destined for urban and suburban kitchens. Lumber traveled from forests to sawmills by train; two-by-four beams to build houses on the treeless plains left the sawmills of the Pacific Northwest on flatcars. Some petroleum went from the well to the refinery by train; most kerosene and gasoline went from the refinery to the retailer by train. Virtually all the country's mail traveled by train, including cotton cloth and saddles, frying pans and furniture ordered from the mail-order companies that had begun to flourish in the 1880s.

Even as fundamental and apparently untransportable a commodity as time was affected by the integration of the rail system, for scheduling was an important facet of integration. People who were going to travel by train had to know what time their trains would leave, and if connections had to be made, trains had to be scheduled so as to make the connections possible. Schedules also had to be constructed, especially on heavily trafficked lines, to ensure that trains did not collide. But scheduling was exceedingly difficult across the long distances of the United States because communities each established their own time on the basis of the position of the sun. When it was noon in Chicago, it was 12:30 in Pittsburgh (which is to the east of Chicago) and 11:30 in Omaha (to the west). The train schedules printed in Pittsburgh in the early 1880s listed six different times for the arrival and departure of each train. The station in Buffalo had three different clocks.

Sometime in the early 1880s, some professional railroad managers and the editors of several railroad publications agreed to the idea, first proposed by some astronomers, that the nation should be divided into four uniform time zones. By common agreement among the managers of the country's railroads, at noon (in New York) on Sunday, November 18, 1883, railroad signalmen across the country reset their watches. The zones were demarcated by the 75th, 90th, 105th, and 120th meridians. People living in the eastern sections of each zone experienced, on that otherwise uneventful Sunday, two noons, and people living in the western sections, skipped time. Virtually everyone in the country accepted the new time that had been established by the railroads, although Congress did not actually confirm the arrangement by legislation for another thirty-five years. Such was the pervasive impact of the integrated rail network.

The Petroleum System

In 1859, a group of prospectors dug a well in a farmyard in Titusville, Pennsylvania. Although they appeared to be looking for water, the prospectors were in fact searching for an underground reservoir of a peculiar oily substance that had been bubbling to the surface of nearby land and streams. Native Americans had used this combustible substance as a lubricant for centuries. The prospectors were hoping that if they could find a way to tap into an underground reservoir of this material, they could go into the business of selling it to machine shops and factories (as a machine lubricant, an alternative to animal fat) and to households and businesses (as an illuminant, an alternative to whale oil and candles).

The prospectors struck oil—and the American petroleum industry was born. Within weeks the news had spread, and hundreds of eager profiteers rushed into western Pennsylvania, hoping to purchase land, drill for oil, or find work around the wells. The Pennsylvania oil rush was as massive a phenomenon as the California gold rush a decade earlier.

The drillers soon discovered that crude petroleum is a mixture of oils of varying weights and characteristics. These oils, they learned, could be easily separated from one another by distillation, an ancient and fairly well-known craft. All that was need was a fairly large closed vat with a long outlet tube (called a still) and a fire. The oil was heated in the still and the volatile gases produced would condense in the outlet tube. A clever distiller (later called a refiner) could distinguish different portions (fractions) of the distillate from each other, and then only the economically useful ones needed to be bottled and sent to market.

The market for petroleum products boomed during the Civil War: northern factories were expanding to meet government contracts; the whaling industry was seriously hampered by naval operations; railroads were working overtime to transport men and materiel to battlefronts. By 1862, some 3 million barrels of crude oil were being processed every year. Under peacetime conditions the industry continued to expand; by 1872, the number of processed barrels had trebled.

Transportation of petroleum remained a problem, however. The wells were located in the rural, underpopulated Appalachian highlands of Pennsylvania, not only many miles away from the cities in which the ultimate consumers lived, but also many miles away from railroad lines that served those cities. Initially crude oil had been collected in barrels and had been moved (by horse and cart or by river barges) to railroad-loading points. There the barrels were loaded into freight cars for the trip to the cities (such as Cleveland and Pittsburgh) in which the crude was being refined and sold. The transportation process was cumbersome, time-consuming, and wasteful; the barrels leaked, the barges sometimes capsized, the wagons—operating on dirt roads—sometimes sank to their axles in mud.

Pipelines were an obvious solution, but a difficult one to put into practice given that no one had ever before contemplated building and then maintaining a continuous pipeline over the mountainous terrain and the long distances that had to be traversed. The first pipeline to operate successfully was built in 1865. Made of lap-welded cast-iron pipes, two inches in diameter, it ran for six miles from an oil field to a railroad loading point and had three pumping stations along the way. This first pipeline carried eighty barrels of oil an hour and had demonstrated its economic benefits within a year. Pipeline mileage continued to increase during the 1870s and 1880s (putting thousands of teamsters out of business), but virtually all of the lines were relatively short hauls, taking oil from the fields to the railroads. Throughout the nineteenth century and well into the twentieth, the railroads were still the principal long-distance transporters of both crude and refined oil. After the 1870s, the drillers, refiners, and railroads gradually dispensed with barrels (thus putting thousands of coopers out of business) and replaced them with specially built tank cars, which could be emptied into and loaded from specially built holding tanks. As it was being constructed, the network of petroleum pipelines was thus integrated into the network of railroad lines. It was also integrated into the telegraph network. Oil refineries used the telegraph system partly to keep tabs on prices for oil in various localities and partly to report on the flow of oil through the lines.

The most successful petroleum entrepreneurs were the ones who realized that control of petroleum transportation was the key ingredient in control of the entire industry. The major actor in this particular economic drama was John D. Rockefeller. Rockefeller had been born in upstate New York, the son of a talented patent medicine salesman, but he had grown up in Cleveland, Ohio, a growing commercial center (it was a Great Lake port and both a canal and railroad terminus), and had learned accountancy in a local commercial college. His first job was as a bookkeeper for what was then called a commission agent, a business that collected commissions for arranging the shipment of bulk orders of farm products. A commission agent's success depended on getting preferential treatment from railroads and shipping companies. Rockefeller carried this insight with him, first when he went into a partnership as his own commission agent and then, in 1865, when he became the co-owner of an oil refinery in Cleveland.

Rockefeller and his associates were determined to control the then chaotic business of oil refining. They began by arranging for a secret rebate on oil shipments from one of the two railroads then serving Cleveland. Then in the space of less than a month, using the rebate as an incentive, they managed to coerce other Cleveland refiners into selling out and obtained control of the city's refining. Within a year or two, Rockefeller was buying up refineries in other cities as well. He had also convinced the railroads that he was using that they should stop carrying oil to refineries owned by others, so that he was in almost complete control of the price offered to drillers. In the early 1870s, a group of drillers banded together to build pipelines that would take their oil to railroads with which Rockefeller wasn't allied. Rockefeller responded to this challenge by assembling a monopoly on the ownership of tank cars (since the pipelines did not go all the way to the refineries and railroad tank cars were still necessary), and by 1879, he had been so successful in squeezing the finances of the pipeline companies that their stockholders were forced to sell out to him. In that year, as a result of their control both of refineries and pipelines, Rockefeller and his associates controlled 90 percent of the refined oil in the United States.

Having bought up the competing pipelines (having let other people take the risks involved in developing new technologies for building and maintaining those lines), Rockefeller was quick to see their economic value. In 1881, one of his companies completed a six-inch line from the Pennsylvania oil fields to his refinery in Bayonne, New Jersey—the first pipeline that functioned independently of the railroads. By 1900, Rockefeller had built pipelines to Cleveland, Philadelphia, and Baltimore, and Standard Oil (Rockefeller's firm) was moving 24,000 barrels of crude a day (he still used the railroads to move the oil after it had been refined).

By that point, hundreds of civil and mechanical engineers were working for Rockefeller's pipeline companies (which held several patents on pipeline improvements), and several dozen chemists and chemical engineers were working in his refineries (and developing new techniques, such as the Frasch process for taking excess sulfur out of petroleum). In addition, Standard Oil was pioneering financial, management, and legal techniques for operating a business that had to control a huge physical network, spread out over several states. Since the laws dealing with corporations differed in each state and since some of them prevented a corporation in one state from owning property in another, one of Rockefeller's attorneys worked out a corporate arrangement so that Standard Oil had a different corporation in each state in which it operated (Standard Oil of New Jersey, Standard Oil of Ohio, and so forth). The stockholders in each corporation turned their stock over to a group of trustees, who managed the whole enterprise from New York—the famous Standard Oil Trust, of which Rockefeller himself was the single largest stockholder and therefore the major trustee. (The trust, as a way to organized a complex business, was soon picked up in tobacco and sugar refining and other industries involved in large-scale chemical processing, leading Congress, worried about the monopolistic possibilities, to pass the Sherman Anti-Trust Act in 1890.)

By 1900, the Standard Oil Trust (which had successfully battled antitrust proceedings in court) controlled most of the oil produced in Pennsylvania, and it owned most of the new oil fields that had been discovered in Ohio and Indiana. Rockefeller's almost complete stranglehold on the industry wasn't broken until oil was discovered early in the twentieth century in Texas, Oklahoma, Louisiana, and California, outside the reach of the pipelines he controlled and the railroads with which he was associated. Increased competition was accompanied by the continued growth not only of the pipeline network, but also of the industry as a whole: 26 million barrels of petroleum were processed in 1880, 45 million in 1890, 63 million in 1900, 209 million in 1910 (as gasoline was just beginning to edge out kerosene as the most important petroleum product), and 442 million in 1920 (when the Model T had been in production for almost eight years).

Like the telegraph and the railroad (and in combination with the telegraph and the railroad), the oil pipeline network had become a pervasive influence on the American economy and on the daily life of Americans. In the last decades of the nineteenth century, a very large number of Americans, especially those living outside of the major cities, used one of its products, kerosene, for heating and lighting their homes and for cooking. During the same decades, American industry became dependent on other fractions of petroleum to lubricate the machinery with which it was producing everything from luxurious cloth to common nails. Finally, in the early decades of the twentieth century, with the advent first of the internal combustion engine fueled by gasoline and then of automobiles and trucks powered by that engine, Americans discovered that access to petroleum was becoming a necessary condition not only of their working lives but also of their leisure time.

The Telephone System

Technologically the telephone was similar to the telegraph, but socially it was very different. The device patented by Alexander Graham Bell in 1876 was rather like a telegraph line: voices rather than signals could be transmitted by electric current because the transmitter lever and the receiving pencil had been replaced by very sensitive diaphragms. Aware of the difficulties that Morse had encountered in reaping profits from his patents—and aware that he had no head for business—Bell decided to turn over the financial and administrative details of creating a telephone network to someone else.

The businessmen and the attorneys who managed the Bell Telephone Company did their work well. While the railroad, telegraph, and petroleum networks had been integrated by corporate takeovers, the telephone system was integrated, from the very beginning, by corporate design. A crucial decision had been made early on: Bell Telephone would manufacture all the telephone instruments, then lease the instruments to local companies, which would operate telephone exchanges under license to Bell. This meant that for the first sixteen years of telephone network development (sixteen years was then the length of monopoly rights under a patent), the Bell Telephone Company could dictate, under the licensing agreements, common technologies for all the local telephone systems. Bell could also control the costs of telephone services to local consumers.

Because of this close supervision by one company, the telephone system was integrated from the very beginning. Between 1877 and 1893, the Bell Telephone Company, through its affiliated local operating companies, controlled and standardized virtually every telephone, every telephone line, and every telephone exchange in the nation. Indeed in the 1880s, the officers of Bell were confident that they could profitably begin long-distance service (that is, service that would connect one local operating company with another) precisely because all of the operating companies were using its standardized technology. Bell needed to hire physicists and electrical engineers to solve the technical problems involved in maintaining voice clarity over very long wires, but the organizational problems involved in connecting New York with Chicago and Chicago with Cleveland turned out to be minimal.

On the assumption that the telephone system would end up being used very similarly to the telegraph network, the officers of Bell had decided that their most important customers would be other businesses, particularly those in urban areas. They decided, as a marketing strategy, to keep rates fairly high, in return for which they would work to provide the clearest and most reliable service possible. By the end of the company's first year of operation, 3,000 telephones had been leased, 1 for every 10,000 people. By 1880, there were 60,000 (1 per 1,000), and when the Bell patents expired in 1893, there were 260,000 (1 per 250). About two thirds of these phones were located in businesses. Most of the country's business information was still traveling by mail and by telegraph (because businessmen wanted a written record of their transactions), but certain kinds of businesses were starting to find the telephone very handy: in 1891, the New York and New Jersey Telephone Company served 937 physicians and hospitals, 401 pharmacies, 363 liquor stores, 315 stables, 162 metalworking plants, 146 lawyers, 126 contractors, and 100 printing shops.

After the Bell patents expired, independent telephone companies entered the business despite Bell's concerted effort to keep them out. By 1902, there were almost 9,000 such independent companies, companies not part of the Bell system. When the organizers of the Bell system had analogized the telephone to the telegraph, they had made a crucial socio-logical mistake. They understood that in technological terms the telephone was similar to the telegraph, but they failed to understand that in social terms it was quite different. The telephone provided user-to-user commu-nication (with the telegraph there were always intermediaries). In addition, the telephone was a form of voice communication; it facilitated emotional communication, something that was impossible with a telegraph. In short, what the organizers of the Bell system had failed to understand was that people would use the telephone to socialize with each other.

The independent companies took advantage of Bell's mistake. Some of them offered services that Bell hadn't thought to provide. Dial telephones were one such service, allowing customers to contact each other without having to rely on an operator (who sat at a switchboard, manually con-necting telephone lines, one to another, with plugs). Operators were no-torious for relieving the boredom of their jobs by listening in on conver-sations, something many customers wanted to avoid. Party lines were another such service. Anywhere from two to ten residences could share the same telephone line and telephone number, which drastically lowered the costs of residential services. Many lower-income people turned out to be willing to put up with the inconvenience of having to endure the ringing of telephones on calls meant for other parties in exchange for having tele-phone service at affordable rates.

Yet other independent companies served geographic locales that the Bell companies had ignored. This was particularly the case in rural areas where there were farm households. Bell managers apparently hadn't thought that farmers would want telephones, but it turned out that they were wrong. Farm managers used telephones to get prompt reports on prices and weathers. Farm households used telephones to summon doctors in emer-gencies and to alleviate the loneliness of lives lived far from neighbors and relatives. In 1902, relatively few farm households had telephones, but as the independent companies grew, so did the number of farm-based cus-tomers; by 1920, just under 39 percent of all farm households in the Unit-ed States had telephone service (while only 34 percent of nonfarm house-holds did).

All this competition in telephone service had the net effect that any economist could have predicted: prices for telephone service fell, even in the Bell system. In order to keep the system companies competitive, the central Bell company had to cut the rates that it charged its affiliates for the rental of phones, and these savings were passed on to consumers. In New York City, as just one example, rates fell from $150 for 1,000 calls in 1880 to $51 in 1915 (figures adjusted for inflation).

As a result, in the period between 1894 and 1920, the telephone net-work expanded profoundly. Middle-class people began to pay for tele-phone service to their homes. Farm households became part of the tele-phone network (in record numbers). Retail businesses began to rely on telephones in their relations with their customers. By 1920, there were 13 million telephones in use in the country, 123 for every 1,000 people. Eight million of those 13 million phones belonged to Bell and 4 million to independent companies that connected to Bell lines. In just forty years, the telephone network, which provided point-to-point voice communication, had joined the telegraph, railroad, and petroleum networks as part of the economic and social foundation of industrial society.

The Electric System

Like the telegraph and telephone systems, the electric system was (and still is) quite literally a network of wires. Physicists, who had been experimenting with electricity since the middle of the eighteenth century, knew that under certain conditions electricity could produce light. Unfortunately, the first devices invented for generating a continuous flow of electricity—batteries—did not create a current strong enough for illumination. However, in 1831 the British experimenter Michael Faraday perfected a device that was based on a set of observations that scientists had made a decade earlier: an electric current will make a magnet move and a moving magnet will create an electric current. Faraday built an electric generator (a rotating magnet with a conducting wire round around it)—a device that could, unlike the battery, create a continuous flow of current strong enough to be used for lighting.

Within a short time, the generator was being used to power arc lamps in which the light (and a lot of heat) was produced by sparking across a gap in the conducting wires. Arc lamps were first used in British and French lighthouses in the 1860s; the generator that created the electricity was powered by a steam engine. A few years later, arc lamps were also being used for street lighting in some American cities. Unfortunately, arc lamps were dangerous; they had to be placed very far away from people and from anything that might be ignited by the sparks. By the mid-1870s, several people in several different countries were racing with each other to find a safer form of electrical lighting, the incandescent lamp. In such a lamp, light would be derived from a glowing, highly resistant filament and not a spark; but the filament had to be kept in a vacuum so that it wouldn't oxidize (and disappear) too fast.

Thomas Alva Edison won the race. In 1878, when Edison started working on electrical lighting, he already had amassed a considerable reputation (and a moderate fortune) as an inventor. His first profitable invention had been the quadruplex telegraph, which could carry four messages at once, and he had also made successful modifications to the stock ticker, the telegraph system for relaying stock prices from the floor of the stock exchange to the offices of investors and brokers. These inventions had enhanced his reputation with Wall Street financiers and attorneys. In 1876, when he decided to become an independent inventor, building and staffing his own laboratory in Menlo Park, New Jersey, and again in 1878, when he decided that he wanted his laboratory to crack the riddle of electric lighting, he had no trouble borrowing money to invest in the enterprise.

Actually, they were enterpris*es*. From the beginning, Edison understood that he wanted to build a technological system *and* a series of businesses to manage that system. The first of these businesses was the Edison Electric Light Company, incorporated for the purpose of financing research and development of electric lighting. Most of the stock was purchased by a group of New York financiers; Edison received stock in return for the rights to whatever lighting patents he might develop. Once Edison had actually invented a workable lightbulb (it had a carbonized thread as its filament), he proceeded to design other devices, and create other companies, that would all be parts of the system. The Edison Electric Illuminating Company of New York, founded in 1880, was created to build and maintain the very first central generating station providing electric service to customers. When this station opened its doors in 1882 (as its site Edison chose the part of Manhattan with the highest concentration of office buildings), it contained several steam-driven generators (built to Edison's design by the Edison Machine Company) and special cables to carry the electricity underground (made by the Edison Electric Tube Company). Customers who signed up for electric service had their usage measured by meters that Edison had invented; their offices were outfitted with lamp sockets that Edison had designed into which they were to place lightbulbs that another Edison company manufactured.

Information about this new system spread very fast (thanks to publicity generated by the Edison Electric Light Company), and within a few months (not even years), entrepreneurs were applying to Edison for licenses to build electric generating plants all over the country, indeed all over the world. Having been designed as a system, the electrical network grew very fast. There was only one generating plant in the country in 1882, but by 1902, there were 2,250, and by 1920, almost 4,000. These plants had a total generating capacity of 19 million kilowatts. Just over a third of the nation's homes were wired for electricity by 1920, by which time electricity was being used not only for lighting but also for cooling (electric fans), ironing (the electric iron replaced the so-called sad iron quickly), and vacuuming (the vacuum cleaner was being mass-produced by 1915).

The Edison companies (some of which eventually merged with other companies to become the General Electric Company) were not, however, able to remain in control of the electric system for as long (or as completely) as the Bell companies were able to dominate the telephone business or Standard Oil the petroleum business. Part of the reason for this lay in the principles of electromagnetic induction, which can be used to create electric motors as well as electric generators. The same experimenters who were developing electric generators in the middle years of the nineteenth century were also developing electric motors, and one of the first applications of those motors was in a business very different from the lighting business: electric traction for electric intraurban streetcars, often known as trolley cars. The first of these transportation systems was installed in Richmond, Virginia, in 1888 by a company owned by Frank Sprague, an electrical engineer who had briefly worked for Edison.

Sprague had invented an electric motor that, he thought, would be rugged enough to power carriages running day in and day out on city streets. As it turned out, the motor had to be redesigned, and redesigned again, before it worked very well, and Sprague also had to design trolley poles (for conducting the electricity from the overhead wires to the carriage) and a controlling system (so that the speed of the motor could be varied by the person driving the carriage). In the end, however, the electric streetcar was successful, and the days of the horse-pulled carriage were clearly numbered. Fourteen years after Sprague's first system began operating, the nation had 22,576 miles of track devoted to street railways.

Electric motors were also being used in industry. The earliest motors, like the streetcar motors, had been direct current (d.c.) motors, which needed a special and often fragile device (called a commutator) to transform the alternating current (a.c.) produced by generators. In 1888, an a.c. motor was invented by Nikola Tesla, a Serbian physicist who had emigrated to the United States. Tesla's patents were assigned to the Westinghouse Company, which began both to manufacture and to market them. At that point, the use of electric motors in industry accelerated. The very first factory to be completely electrified was a cotton mill, built in 1894. As electric motors replaced steam engines, factory design and location changed; it was no longer necessary to build factories that were several stories high (to facilitate power transmission from a central engine) or to locate them near water sources (to feed the steam boilers). The first decade of the twentieth century was a turning point in the use of electric power in industry as more and more factories converted; by 1901, almost 400,000 motors had been installed in factories, with a total capacity of almost 5 million horsepower.

In short, the electrical system was more complex than the telephone and petroleum systems because it consisted of several different subsystems (lighting, traction, industrial power) with very different social goals and economic strategies; because of its complexity, no single company could dominate it. By 1895, when the first generating plant intended to transmit electricity over a long distance became operational (it was a hydroelectric plant built to take advantage of Niagara Falls, transmitting electricity twenty miles to the city of Buffalo), there were several hundred companies involved in the electric industry: enormous companies such as Westinghouse and General Electric that made everything from generators to lightbulbs; medium-sized companies, such as the ones that ran streetcar systems or that provided electric service to relatively small geographic areas; and small companies, which made specialized electric motors or parts for electric motors. Despite this diversity, the electric system was unified by the fact that its product, electric energy, had been standardized. By 1910, virtually all the generating companies (which, by now, had come to be called utility companies) were generating alternating current at sixty cycles per second. This meant that all electric appliances were made to uniform specifications and all transmission facilities could potentially be connected to one another. By 1920, electricity had supplanted gas, kerosene, and oils for lighting. In addition, it was being used to power sewing machines in ready-made clothing factories, to separate aluminum from the contaminants in its ores, to run projectors through which motion pictures could be viewed, to carry many thousands of commuters back and forth, and to do dozens of other chores in workplaces and residences. As transmission towers marched across the countryside and yet another set of wire-carrying poles were constructed on every city street, few Americans demonstrated any inclination to decline the conveniences that the youngest technical system—electricity—was carrying in its wake.

The Character of Industrialized Society

As inventors, entrepreneurs, and engineers were building all these multifarious technological systems, Americans were becoming increasingly dependent on them. Each time a person made a choice—to buy a kerosene lamp or continue to use candles, to take a job in an electric lamp factory or continue to be a farmer, to send a telegraph message instead of relying on the mails, to put a telephone in a shop so that customers could order without visiting—that person, whether knowingly or not, was becoming increasingly enmeshed in a technological system. The net effect of all that construction activity and all those choices was that a wholly new social order, and wholly different set of social and economic relationships between people, emerged: industrial society.

In industrial societies, manufactured products play a more important economic role than agricultural products. More money is invested in factories than in farms; more bolts of cloth are produced than bales of hay; more people work on assembly lines than as farm laborers. Just over half (53 percent) of what was produced in the United States was agricultural in 1869 and only a third (33 percent) was manufactured. In 1899 (just thirty years later), those figures were reversed: half the nation's output was in manufactured goods and only a third was agricultural, despite the fact that the nation's total farm acreage had increased rapidly as a result of westward migration. Manufacturing facilities were turning out products that were becoming increasingly important aspects of everyday life: canned corn and lightbulbs, cigarettes and underwear.

In a preindustrial society, the countryside is the base for economic and political power. In such societies, most people live in rural districts. Most goods that are traded are agricultural products; the price of fertile land is relatively high; and wealth is accumulated by those who are able to control that land. Industrialized societies are dominated by their cities. More people live and work in cities than on farms; most goods are manufactured in cities; most trade is accomplished there; wealth is measured in money and not in land. Furthermore, the institutions that control money—banks—are urban institutions.

As the nineteenth century progressed, more and more Americans began living either in the rural towns in which factories were located (which, as a result, started to become small cities) or in the older cities that had traditionally been the center of artisanal production and of commerce. Native-born Americans began moving from the countryside to the city; many newly arrived Americans (and there were millions of newcomers to America in the nineteenth century) settled in cities. Just over half of all Americans (54 percent) were farmers or farm laborers in 1870, but only one in three was by 1910. Some American families underwent the rural–urban transition slowly: a daughter might move off the farm to a rural town when she married, and then a granddaughter might make her fortune in a big city. Others had less time: a man might be tending olive groves in Italy one day and working in a shoe factory in Philadelphia two months later.

During the 1840s, the population of the eastern cities nearly doubled, and several midwestern cities (St. Louis, Chicago, Pittsburgh, Cincinnati) began to grow. In 1860, there were nine port cities that had populations over 100,000 (Boston, New York, Brooklyn, Philadelphia, Baltimore, New Orleans, Chicago, Cincinnati, and St. Louis)—by 1910, there were fifty. Just as significantly, the country's largest cities were no longer confined to the eastern seaboard or to the Midwest. There were several large cities in the plains states, and half the population of the far west was living not in its fertile valleys or at the feet of its glorious mountains, but in its cities: Los Angeles, Denver, San Francisco, Portland, and Seattle. By 1920, for the first time in the nation's history, just slightly over half of all Americans lived in communities that had more than 10,000 residents.

Money was flowing in the same direction that people were; by 1900, the nation's wealth was located in its cities, not in its countryside. The nation's largest businesses and its wealthiest individuals were in its cities. J. P. Morgan and Cornelius Vanderbilt controlled their railroad empires from New York; Leland Stanford and Charles Crocker ran theirs from San Francisco; John D. Rockefeller operated from Cleveland and New York; Andrew Carnegie, at least initially, from Pittsburgh. Probably by 1880, and certainly by 1890, stock exchanges and investment bankers had become more important to the nation's economic health than cotton wharves and landed gentry.

This transition to an urban society had political consequences because political power tends to follow the trail marked out by wealth (and, in a democracy, to some extent by population). In the early years of the nineteenth century, when the independent political character of the nation was being formed, most Americans still lived on farms and American politics was largely controlled by people who earned their living directly from the land. After the Civil War, city residents (being both more numerous and more wealthy) began to flex their political muscles and to express their political interests more successfully. The first twelve presidents of the United

States had all been born into farming communities, but from 1865 until 1912, the Republican party, then the party that most clearly represented the interests of big business and of cities, controlled the White House for all but eight years, and those eight years were the two terms served by Grover Cleveland, who before becoming president had been the mayor of Buffalo, New York.

The transition to an urban society also had economic and technological consequences. In a kind of historical feedback loop, industrialization caused cities to grow and the growth of cities stimulated more industrialization. Nineteenth-century cities were, to use the term favored by urban historians, walking cities. Since most residents could not afford either the cost or the space required to keep a horse and carriage, they had to be able to walk to work or to work in their own homes. Since businesses also had to be within walking distance of each other, this meant that as cities grew they became congested; more and more people had both to live and to work within the same relatively limited space. With congestion came disease; all nineteenth-century American cities were periodically struck by devastating epidemics: cholera, dysentery, typhoid fever.

Even before they understood the causes of these epidemics, city governments became convinced that they had to do something both to relieve the congestion and to control the diseases. Streets had to be paved, running water provided, sewers constructed, new housing encouraged. This meant that reservoirs had to be built, aqueducts and pumping stations constructed, trenches dug, pipes purchased, brickwork laid, new construction techniques explored. All of this municipal activity not only stimulated American industry but also served as a spur to the growth of civil engineering.

In addition, in the years between 1870 and 1920, many American cities actively stimulated industrialization by seeking out manufacturing interests and offering operating incentives to them. Many of the nation's older cities found themselves in economic trouble as railroad depots become more important than ports as nodes in the country's transportation system. In their distress, these cities decided that their futures lay not in commerce but in manufacturing, and they began to seek out manufacturing entrepreneurs to encourage industrial growth. By that time, the steam engine having been perfected and its manufacture made relatively inexpensive, manufacturers had ceased to depend on waterwheels as a power source, which meant that they could easily (and profitably) establish their enterprises in cities rather than in the countryside; the development of the electric motor only served to increase this potential.

Minneapolis became a center of flour milling, Kansas City of meatpacking, Memphis of cotton seed oil production, Rochester of shoe manufacture, Schenectady of electric equipment, New York of ready-made clothing, Pittsburgh of steel and glass manufacture. Local banks helped manufacturers start up in business and local politicians helped recruit a docile labor force, all in the interests of stabilizing or augmenting a city's economy. Nationwide the net result was a positive impetus to the growth of industry; the processes of industrialization and urbanization are mutually reinforcing.

If American cities grew prodigiously during the second half of the nineteenth century, so, too, did the American population as a whole: between 1860 and 1920, the population of the United States more than tripled (from 31 million to 106 million). Some of the increase was the result of a high natural birthrate; in general, American families were larger than what is needed to keep a population at a stable size from one generation to the next. In addition, as the result of improvements in public health and improvements in the food supply, the death rate was declining and life expectancy was rising. People were living longer and that meant that in any given year a declining proportion of the total population was dying. On top of this, immigrants were arriving in record numbers. The figures are astounding; the total, between the end of the Civil War and the passage of the Immigration Restriction Acts (1924), came to over 30 million people. Like their native-born contemporaries, immigrants had a high birthrate and a declining death rate and more of their children lived past infancy and then enjoyed a longer life expectancy, all of which further contributed to the mushrooming size of the American population. This startling population increase—almost 20 percent per decade—reflects another crucial difference between societies that have become industrialized and those that have not. In a preindustrialized society the size of the population changes in a more or less cyclical fashion. If the weather cooperates and the crops are bounteous and peace prevails, people remain reasonably healthy and many children live past infancy; over the course of time the population will grow. But eventually the population will grow too large to be supported by the available land or the land itself will become infertile. Droughts may come or heavy rains; locusts may infest the fields or diseases may strike the cattle. Men will be drawn off to battle just when it is time to plow the fields or soldiers engaged in battles will trample the wheat and burn the barns. Then starvation will ensue. People will succumb to disease; fewer children will be born, and more of them will die in infancy. The population will shrink.

Under preindustrial conditions, such population cycles have been inexorable. Sometimes the cycle will take two generations to recur, sometimes two centuries, but it has recurred as long as there have been agricultural peoples who have been keeping records of themselves. Industrialization breaks this cyclical population pattern. Once a country has industrialized, natural disasters and wars do not seem to have a long-term effect on the size of its population; the rate of increase may slow for a few years or so, but there is still an increase. And the standard of living keeps rising as well. People stay relatively healthy; they live longer lives. Generally speaking, they can have as many (or as few) children as they want, knowing that, also generally speaking, most of their children will live past infancy. This is the salient characteristic that makes underdeveloped countries long for development: industrialized countries seem able to support extraordinarily large populations without any long-term collapse either in the size of the population or in the standard of living.

Industrialized countries can do this because agriculture industrializes at the same time that manufacturing does. In the transition to industrialization, what is happening on the farm is just as important as what is happening in the factories since, to put it bluntly, people cannot work if they cannot eat. These social processes—sustained growth of the population and the industrialization of agriculture—are interlocked. Both were proceeding rapidly in the United States between the years 1870 and 1920 as American farmers simultaneously pushed west and industrialized, settling new territory and developing more productive farming techniques. As the frontier moved westward roughly 400 million new acres were put under cultivation: virgin prairie became farms, fertile mountain valleys were planted in orchards, grassy hills became grazing land for sheep and cattle. The total quantity of improved acreage (meaning land that had been cleared or fenced or otherwise made suitable for agricultural use) in the United States multiplied two and a half times between 1860 and 1900.

This alone would have considerably expanded the nation's agricultural output, but newly introduced agricultural implements profoundly altered the work process of farming (particularly grain growing) and increased its productivity. The first of these was the reaper (patented by Cyrus McCormick in 1834 and in limited use even before the Civil War). The reaper, which was pulled by horses, replaced hand labor. Once a reaper had been purchased, a farm owner could quadruple the amount of acreage cut in one day or fire three day laborers who had previously been employed for the harvest or greatly increase the acreage put to plow (since the number of acres planted had always been limited by what could be reaped in the two prime weeks of harvest).

The reaper was followed by the harvester (which made binding the grain easier), followed by the self-binder (which automatically bound the grain into shocks), and—in the far west—followed by the combine, a steam-driven tractor (which cut a swath of over forty feet, then threshed and bagged the grain automatically, sometimes at the rate of three 150-pound bags a minute). In those same years, haymaking was altered by the introduction of automatic cutting and baling machinery, and plowing was made considerably easier by the invention of the steel plow (John Deere, 1837) and the chilled-iron plow (James Oliver, 1868), both of which had the advantage of being nonstick surfaces for the heavy, wet soils of the prairies.

The net result, by 1900, was that American farmers were vastly more productive than they had been in 1860. Productivity has two facets: it is a measure both of the commodities being produced and of the labor being used to produce them. Statistics on wheat production indicate how radically American agriculture was changing in the second half of the nineteenth century. In 1866, there were roughly 15.5 million acres devoted to wheat production in the United States; farmers achieved average yields of

9.9 bushels per acre, resulting in a total national production of about 152 million bushels. By 1898, acreage had roughly trebled (to 44 million), yields had almost doubled (to 15.3 bushels per acre), and the total production was 675 million bushels.

All this was accomplished with a marked saving of labor. By the hand method, 400 people and 200 oxen had to work ten hours a day to produce 20,000 bushels of wheat; by the machine method, only 6 people (and 36 horses) were required. Farms were getting larger, ownership was being restricted to a smaller and smaller number of people and more machinery was required for profitable farming (between 1860 and 1900, the annual value of farm implements manufactured in the United States went from $21 million to $101 million)—at the same time, the farms were becoming more productive.

What this means, put another way, was that a smaller proportion of the nation's people were needed to produce the food required by its ever larger population. Some people left their farms because they hated the farming life, some because they could not afford to buy land as prices began to rise, some because they were forced off the land by the declining profitability of small farms. The farming population (this includes both owners and laborers) began to shrink in relation to the rest of the population.

New transportation facilities and new food-based industries made it easier and cheaper for the residents of cities and towns to eat a more varied diet. The fledgling canning industry was spurred by the need to supply food for troops during the Civil War. After the war, the canners turned to the civilian market, and by the 1880s, urban Americans had become accustomed to eating canned meat, condensed milk (invented by Gail Borden in 1856), canned peas, and canned corn. The Heinz company was already supplying bottled ketchup and factory pickles to a vast population, and the Campbell's company was just about to start marketing soups. By 1900, cheese and butter making had become largely a factory operation, made easier and cheaper by the invention of the centrifugal cream separator in 1879.

After the Civil War, the railroads replaced steamboats and canal barges as the principal carriers of farm products (from wheat to hogs, from apples to tobacco), thus both shortening the time required to bring goods to market and sharply lowering the cost of transportation. After the 1880s, when refrigerated transport of various kinds was introduced, this trend accelerated: even more products could be brought to market (butchered meat, for example, or fresh fish) in an even shorter time. New refrigeration techniques transformed beer making from a home to a factory operation; by 1873, there were some 4,000 breweries in the United States with an output of 10 million barrels a year. Commercial baking had also expanded and Americans were becoming fond of factory-made crackers and cookies. In the end, then, another historical feedback loop had been established, a loop connecting industrialization with agricultural change. Industrialization made farming more productive, which made it possible for the population to increase, which created a larger market for manufactured goods, which increased the rate of industrialization.

Conclusion: Industrialization and Technological Systems

By 1920, a majority of Americans had crossed the great divide between preindustrial and industrial societies. The foods they ate, the conditions under which they worked, the places in which they lived—all had been transformed. The majority of Americans were no longer living on farms. They were eating food that had been carried to them by one technological system (the railroad) after having been processed by machines that were powered by a second (electricity) and lubricated by a third (petroleum). If they wanted to light their domiciles at night or heat their dwelling places during cold weather, they could not avoid interacting with one or another technological system for distributing energy—unless they were willing to manufacture their own candles (even then, they might have ended up buying paraffin from Standard Oil). The social ties that bound individuals and communities together—someone has been elected, someone else has died, young men are about to be drafted, a young woman has given birth—were being carried over, communicated through, and to some extent controlled by technological networks that were owned by large, monopolistically inclined corporations. More people were living longer lives; fewer babies were dying in infancy; the standard of living for many Americans (albeit not for all) was rising. And at the very same time, because of the very same processes, people were becoming more dependent on each other.

Early in the nineteenth century the process of industrialization had appeared (to those who were paying attention) as a rather discrete undertaking: a spinning factory in a neighboring town, a merchant miller up the river, a railroad station a few miles distant. By the end of the century, virtually all Americans must have been aware that it had become something vastly different: a systematic undertaking that had created interlocking physical and social networks in which all Americans—rich or poor, young or old, urban or rural—were increasingly enmeshed.

Suggestions for Further Reading

Alfred D. Chandler, *Strategy and Structure: Chapters in the History of American Industrial Enterprise* (Garden City, NY, 1966).

Howard P. Chudacoff and Judith E. Smith, *The Evolution of American Urban Society* (Englewood Cliffs, NJ, 1988).

Robert W. Garnet, *The Telephone Enterprise* (Baltimore, 1985).

Richard F. Hirsh, *Technology and Transformation in the American Electric Utility Industry* (Cambridge, 1989).

August Giebelhaus, *Business and Government in the Oil Industry: A Case Study of Sun Oil* (Greenwich, CT, 1980).

Thomas Parke Hughes, *American Genesis: A Century of Invention and Technological Enthusiasm, 1870–1970* (New York, 1989).

Thomas Parke Hughes, *Networks of Power: The Electrification of Society, 1880–1920* (Baltimore, 1983).

Malcolm MacLaren, *The Rise of the Electrical Industry During the 19th Century* (Princeton, 1943).

Glen Porter, *The Rise of Big Business: 1860–1910* (New York, 1973).

Mark H. Rose, *Cities of Light and Heat: Domesticating Gas and Electricity in American Homes* (University Park, PA, 1995).

Nathan Rosenberg, *Exploring the Black Box: Technology, Economics and History* (Cambridge, 1982).

Nathan Rosenberg, *Technology and American Economic Growth* (New York, 1972).

George David Smith, *The Anatomy of a Business Strategy: Bell, Western Electric and the Origins of the American Telephone Industry* (Baltimore, 1985).

Carlene Stephens, *Inventing Standard Time* (Washington, DC, 1983).

John F. Stover, *American Railroads* (Chicago, 1961).

Neil H. Wasserman, *From Invention to Innovation: Long Distance Telephone Transmission at the Turn of the Century* (Baltimore, 1985).

Harold Williamson and Arnold Daum, *The American Petroleum Industry.* 2 vols. (Evanston, IL, 1959).

8

Daily Life and Mundane Work

INDUSTRIALIZATION HAD A PROFOUND IMPACT on the working lives of most ordinary Americans. Unfortunately, we cannot easily generalize about precisely what that impact was because working people are very different from each other. The introduction of the open-hearth process, for example, had one kind of impact on the manager of a steel mill and a very different kind on the men who tended the hearth; similarly, the introduction of indoor plumbing had one kind of impact on a housewife who did her work alone and a very different kind on the housewife who had many servants. Any generalization about the impact of industrialization on Americans that does not distinguish Americans, at the very least, by occupation, gender, race, and class is bound to be wrong.

Farmers and Unexpected Outcomes

Even within one occupational category, there can be a great deal of diversity. In the years between 1870 and 1920, the word "farmer" was used to designate very different kinds of people. Some were landowners; others were tenants, sharecroppers, or wage laborers. Some planted tobacco in Virginia; others gathered apples in Oregon, harvested wheat in Kansas, tended dairy cows in Wisconsin, or grew corn and beans in fertile valleys in what was not yet called Arizona. Some had once been slaves; others were the children of slaves. Some were recent immigrants; others were the grandchildren and great grandchildren of eighteenth-century colonists. Some tended only one crop (by 1880, monocropping was already common in some parts of the country), most had several. Some were very rich, but the majority were not, and some were close to starvation occasionally. Some worked on two-acre plots; others owned, and worked, thousands of acres.

Some American farmers were not directly affected by industrialization in the years between 1870 and 1920. Tobacco growers provide one example. Tobacco was (indeed still is) one of the most labor-intensive crops; the cultivation method that was commonly used in the American South in the waning decades of the nineteenth century is estimated to have required 370 hours of labor *per acre* to bring the crop to market. Tobacco families planted the tiny seeds (by hand) in the winter; while waiting for the seeds to germinate, they cut the wood that would be used to cure the tobacco later in the summer. Once germinated, the seedlings had to be transplanted (also by hand) into fields that had been broken by the plow and smoothed by the harrow (mules provided the heavy labor for this work). During the growing season, the fields had to be repeatedly plowed and hoed in order to keep the soil loose and to control weeds; the rows also had to be regularly plied in order to prune suckers off the plants and to remove worms (also done by hand). Harvesting was also painstaking work because, once cut, the tobacco leaves had to be bundled and tied securely to a stick so that they could be hung in a barn for curing. Even the curing process was arduous; a wood fire had to be carefully watched for thirty-six straight hours so that the temperature in the curing barn would vary in just the proper way: the cured tobacco had to be just dry enough to burn easily, but not so dry that it crumbled on being touched. In some locales, the work processes of curing were organized not as familial but as communal labor; tobacco families shared in the construction and maintenance of curing barns, and they shared in the work of maintaining and watching the curing fires.

Although many people tried, no method was found for mechanizing these processes or in any other way reducing the drudgery until the 1960s (except for a mechanical transplanter, which was introduced around 1910 and changed the worker's posture from stooping to sitting). As a result, while the tobacco farmers of 1920 might be riding to town in automobiles or sending their produce to market in railroad cars, when they were out in the fields, they were working in precisely the same way—with precisely the same tools—that their predecessors had used decades earlier.

The same was true in cotton farming (for the mechanical picker had not yet arrived) and, to some extent, in dairying (since the automatic milker was still far in the future). Mechanization and population growth had increased the demand for what these farmers could produce: mechanization lowered the price of the goods (cigarettes, for example, and cotton cloth) that were made from farm products, and population growth increased the number of people who could afford to buy the goods. However, neither mechanization nor population change had altered the way in which these farmers grew their crops.

Such farmers were, however, in the minority. The majority of American farmers in the Midwest, the plains, the Prairies, and the far west were wheat growers, and wheat culture changed profoundly, as we learned in Chapter 7, in the second half of the nineteenth century. Each of the steps in wheat culture (plowing, planting, cultivating, harvesting, threshing, winnowing, and hauling) had been successfully mechanized by 1880 as the result of the introduction of such new tools as the iron plow, the gang plow (which could plow more than one row at a time), the spring-toothed harrow (which would trip, instead of sticking, when the harrow encountered roots and rocks), the harvester (which could be pulled by an animal), and the binder (which automatically tied up bunches of grain). Many of these new pieces of machinery were expensive, but rural banks were often willing to provide credit, and the expense declined during the latter decades of the nineteenth century as manufacturers introduced mass-production techniques. As a result, many farmers invested in these new, laborsaving technologies.

When used together, these new tools saved enormous amounts of time (from sixty-one to three hours of labor for every bushel of wheat harvested, according to the estimate of Leo Rogin, a labor economist). Labor saved inevitably meant costs reduced (from nineteen to ten cents a bushel, also according to Rogin's estimate). But time and money saved were not the only significant changes wrought by mechanized grain culture. With the new equipment, farms could grow larger. By hand methods, a farmer could harvest only about 7.5 acres per farm worker during the peak weeks of the autumn; by machine methods, 135 acres could be reaped. Fewer workers were needed per acre even at the peak times of the agricultural

This was one of the expensive pieces of capital equipment that transformed the work processes of grain farming: an early 1880s combine, pulled by twenty-one mules. It could cut, thresh and bind wheat in one operation. (Courtesy New York Public Library.)

year; purchase of a binder or a harvester meant that a farmer did not need to hire as many seasonal laborers.

In addition, much of the new equipment was single-purpose equipment, and as a result, farmers who had borrowed money to purchase it tended to shift from subsistence farming to monocropping, farming for profit. Prior to the nineteenth century, most farming had been conducted not for profit, but for subsistence; farmers worked as hard as they had to work in order to keep themselves and their families living comfortably. They also tended to communalize their labor (in such activities as barn raisings, shucking bees, and quilting bees) as much as geography, weather, and transportation facilities would permit. As subsistence as a goal gradually gave way to profitability as a goal, farmers began choosing their crops and purchasing their equipment not with an eye to what their families needed, but with an eye to what the markets needed—and what bankers were willing to consider as suitable collateral for loans. In addition, communal work processes faded as farmers began to see themselves in competition with their neighbors and as they began to understand that the new equipment provided the means to lighten the labor without the need to share it. Profit-oriented, market-oriented farming required new skills, and these required new forms of education. Young men and women began learning what they needed to know not at their parent's knees, but at agricultural colleges or in classes provided by agricultural extension agents. Thus the new plows and the reapers, the threshers and the combines, disrupted grain farmers' lives, changing the relations between rural borrowers and creditors, between owners and employees, and between parents and children in numerous subtle ways.

Many of the farmers who were able to purchase farm machinery and expand the size of their holdings in the last decades of the nineteenth century prospered, but not all did. When some farms get larger, other farms have been sold, which means that some family farms have gone out of business. And when some farmers no longer need to hire as many seasonal laborers, that means unemployment for some people at the very same time that it means less drudgery for others. If a new binder allows a farmer to gather up twice the number of sheaves with half the number of workers or a plow allows the farmer to break twice the number of acres in half the amount of time, then someone other than the farmer is being put out of work.

Between 1860 and 1900, the population of American cities and rural towns expanded while the population on its farms declined. This means that large numbers of people were leaving the land, emigrating, as it were, from the rural life to the urban life. Clearly some wanted to go and were pleased to have left, but just as clearly, others never meant to be dispossessed. That latter group—people who had been forced to sell their farms, people who could no longer find seasonal labor in rural areas—did not prosper; for them, mechanization of farms was clearly not beneficial.

Neither was it beneficial for the growing number of families who were becoming farm tenants. In the Midwest (where the bulk of the new farm machinery was being used), there was a remarkable growth in tenancy in the waning years of the nineteenth century. In Kansas, for example, the percentage of farm tenants to all farmers grew from 16.3 to 35.2 in the years between 1880 and 1900. Many of these new tenants were victims of the growth in farm size and thus, indirectly, victims of mechanization; unable, for whatever reason, to sustain themselves on their small farms, they eventually sold out to neighbors or to investors who wanted larger parcels. There may be some cultures in which tenancy carries no social stigma; American culture, for good or ill, is not one of them. Farm owners who became farm tenants believed that they were declassed, and they were right; tenants were rarely as prosperous as owners, and even when they were prosperous, they still did not have the sense of independence that ownership confers and that Americans value.

Even the farmers who prospered under the new regime of industrialized agriculture may have experienced some negative effects of mechanization. Between 1870 and 1900, farm output increased spectacularly; the production of wheat increased almost four times; of corn, three and a half times; of barley, some six and a half times. Such production increases meant that more and more farm produce had to be sold on foreign markets, markets that varied in ways that farmers (and most everyone else) had difficulty understanding. Higher production also led, as it usually will in an unregulated capitalist market, to lower prices. The price of wheat declined from $2.06 per bushel in 1866 to $.95 in 1874 to $.45 in the bad years following the depression of 1893; corn followed a similar pattern. Thus the years in which production was increasing were also, not surprisingly, the years in which farmers, especially in the grain-belt, were growing increasingly dissatisfied.

Part of the production increases were attributable to increases in the amount of land under cultivation (this was the period, remember, in which the plains, the prairies, and the far west were being settled), but part of it was also due to mechanization; more acres were being farmed at less expense and with higher yields. We now know that every technological change brings with it both expected and unexpected outcomes. Farmers who purchased farm machinery in the second half of the nineteenth century certainly expected their labor and their costs (after the machinery had been paid for) to decrease, and some of them may also have expected their yields to increase; surely most were glad when this happened. But they did not also expect, at the very same time, that market prices for farm products would plummet.

Falling prices was the unexpected outcome of farm mechanization, and falling prices led to a great deal of human misery and political unrest in rural areas. Thus the years of increased farm output in the United States were also the years of the so-called agrarian revolt: membership in the Grange (a farmer's organization founded in 1867) increased; the Farmers Alliance movement claimed over a million members in 1890; the Populist political party emerged. Farmers tried to get federal and state governments to control the economy in a fashion that would benefit them directly: to stay on the gold standard, to regulate railroad freight rates, to introduce protective tariffs.

There is a corollary to what might be called the principle of unexpected outcomes: if the immediate benefits of a new technology are clear (and the immediate benefits of a new harvester or a new plow were unquestionably clear to many nineteenth-century farm-owning families), then people will be unlikely to blame the technology for what turn out to be distant losses—and this is especially true if there are agents other than technology that can be blamed. Thus nineteenth-century farm families who struggled against falling prices blamed their plight on the railroads or the mill owners or big city merchants or corrupt politicians, but not on their harrows, drills, and threshers. Grain-farming families carefully repaired their machinery when it was broken and traded it in for new (and improved) models when they could. Agricultural historians may now know that mechanization was part of the economic problem in the agricultural sector in the waning decades of the nineteenth century (an especially large part if the families had borrowed money to purchase the machinery), but few nineteenth-century farmers made the same connection.

For city dwellers, farm mechanization was an unmitigated benefit; in those days before government regulation, lower farm prices meant lower food prices, and lower food prices are almost always a good thing for consumers. For farmers, however, we must conclude that mechanization was a mixed blessing. Some farming families liked it and profited from it; some were touched by it not at all; some were touched in ways that they did not at all appreciate; and some, when the long-term as well as the short-term effects were evaluated, experienced losses as well as gains.

Skilled and Deskilled Workers

What was true for farmers in this period was also true for industrial workers: sewing-machine operators and plumbers, miners and teamsters, potters and cigar makers, hat finishers and shipwrights. By 1920, industrial workers constituted 40 percent of the total labor force. Like farmers, industrial workers were very diverse, differing by region and by ethnic origin, by gender and by level of skill, by wage rates and by age, by marital status and by race, by union membership or nonmembership. In order to assess the impact of industrialization on such a diverse group, we will need once again to make a few social distinctions; the most useful distinction is between skilled and unskilled workers.

The adjective "skilled" when applied to the noun "workers" actually denotes a combination of several different variables, none of which has any simple connection to what one might think "skill" is: the amount of time that it takes for a person to be trained; the level of wages that the person is paid; the amount of supervisory work the person is expected to do on the job; the extent to which either the quality or the quantity of production would be affected if the person were not there. Generally speaking, skilled workers receive more training than unskilled ones, and they earn more; unskilled workers are rarely placed in a position to supervise other workers, and they are easily replaced without much loss either to the quality or the quantity of what is being produced. In the years between 1870 and 1920, most skilled workers in the United States were white men of British or Germanic stock. There were a few notable exceptions to these generalizations (such as the Spanish Catholic cigar makers of Tampa, Florida, or the female paper finishers in Lee, Massachusetts, or the black steelworkers in Birmingham, Alabama), but they were exceptions of the sort that prove the rule.

Most (probably the majority) of skilled workers were literate and many were politically active; these were the workers who formed the first craft unions, local associations that represented the interests of people who had particular skills (for example, cigar makers, miners, carpenters, masons). For skilled workers, the period between 1870 and 1920 was not a happy time. Mechanization was as much of a bane to shop workers as it was a boon to shop owners. For owners, mechanization meant increases both in productivity and (often) in profits; for workers, it all too frequently meant deskilling—a process in which people who are proud of their skills, and who have invested time and energy in acquiring them, cannot possibly be expected to appreciate.

"Deskilling" is a deceptively simple word; in fact, the process that it denotes is both complex and multifaceted, as the history of carpentry in the nineteenth century illustrates. At the beginning of the century, carpenters—people who constructed wooden buildings and building components—were one of the oldest, and best organized, craft groups in the country. Americans had been building their homes, barns, churches, and schools out of wood for generations. Not surprisingly, carpenters had prospered. In the eighteenth century, in the burgeoning colonial cities, carpenters began to create the kind of social system defined (in Chapter 3) as a guild of master craftsmen.

The colonial master carpenter maintained a shop, which was often part of, or attached to, his home. Tools were stored in this shop, and especially during the winter months (when outdoor work was difficult), milling, joining, dressing, and finishing of construction components (such as door frames, window sashes, wall paneling, moldings, and floors) were done there. The master indentured apprentices and hired journeymen whose work he supervised both in the shop and at the construction site. Apprentices typically lived with the master's family, and journeymen (who were paid wages) found housing in other ways. The master also purchased all of the raw materials (principally the lumber) needed to complete a job. Master carpenters might build a house for a set fee for a client who owned a piece of property or they might purchase a piece of property and build on it on speculation, hoping eventually to sell the building and the lot at a good price. In several cities, carpenters were successful in creating guilds (the Carpenters Company of Philadelphia, founded in 1724, is one exam-

This early nineteenth century print depicts journeymen carpenters using hand tools to create and finish the component parts needed in the construction of buildings. (Wood engraving by Alexander Anderson, n.d., Courtesy Print Collection, Miriam and Ira D. Wallach Division of Art, Prints and Photographs, The New York Public Library, Astor, Lenox and Tilden Foundations.)

ple), which tried to regulate such matters as the terms of apprenticeships and the prices for various kinds of work. Following the classic European pattern, these guilds principally served the interests of their senior members, the masters.

In the early decades of the nineteenth century, however, journeymen carpenters—skilled men who worked for wages—began organizing in opposition to their masters. The journeymen had multiple complaints. They were angry, for example, about the irregularity of their employment: since they worked by the day they could be completely laid off in the winter months or partially laid off during spells of rainy weather; to make up for down time, good weather days could mean twelve to fourteen hours of work, and then the employed man had no time to enjoy the good weather with his friends or his family. The journeymen wished that, at the very least, masters would organize their work better, so that journeymen could be guaranteed ten-hour days between March and October.

Perhaps emboldened by the egalitarian rhetoric of the revolutionary period or later by the ascent of Andrew Jackson (Champion of the Common Man) to the presidency, journeymen carpenters began to band together and search for ways in which to achieve their goals through collective action. Up and down the eastern seaboard in the period between 1825 and 1837, journeymen carpenters went out on strike thirty-five times, often engaging other artisans in sympathy strikes. Virtually all of those strikes failed, and the reasons for failure are instructive.

Before the Civil War, these nascent trade unions (for this is precisely what the journeymen's associations were) had very little trouble attracting members (by 1837, some 300,000 journeymen carpenters had joined a local association), but they had a great deal of trouble achieving solidarity. For generations upon generations, journeymen had been trained to think of themselves as allies of their masters' precisely because someday the journeymen expected to become masters themselves. In 1825, when the journeymen went out on strike in Boston, the master carpenters expressed considerable surprise at their behavior. The masters could not afford to cut the working day to ten hours, they argued, and, in any event, "[j]ourneymen of good character and of skill, may expect very soon to become masters, and like us the employers of others; and by the measure which they are now inclined to adopt, they will entail upon themselves the inconvenience to which they seem desirous that we should now be exposed."[1] Thus, it proved easy for the masters to break the strikes because with very little inducement some of the journeymen could easily be convinced to come back to work.

And it was—again, for telling reasons—very easy for the masters to produce inducements; in Boston in 1832, for example, the master carpenters had accumulated $20,000 in a fund to break the strike of their journeymen, and with this fund they were able to offer double wages to anyone who would agree to come back to work. The money had been donated not by the master carpenters, but by merchants and capitalists, people we would now describe as real estate developers. In the early decades of the nineteenth century, urban construction had become a very profitable enterprise. Investors did not want their projects delayed by strikes (for time is money, especially when you are building with borrowed money), and they did not want labor strife to spread into the other building trades, such as bricklaying and plumbing. Hence the donations.

Hence also the basic reason for the journeymen's discontent. Even before technological change affected carpentry, economic change was already altering the terms on which artisanal work could be conducted. Men with capital, men with no allegiance to, or familiarity with, the traditions of the guild system, were becoming builders. Such men cared not a whit about whether an apprentice was properly trained or where the journeyman had learned his craft, but only about getting a building built in the least possible time at the least possible expense. If subdividng the work made economic sense, they subdivided the work, hiring one master carpenter to do the frame, while another did the floors, leaving the masters to worry about how to pay their journeymen while still making a (now subdivided) profit. If there was a sufficient supply of newly arrived immigrant carpenters or of semiskilled farm boys who had recently drifted into town, then the builders might avoid the masters and the journeymen altogether and just hire their workers directly—at piece rates or hourly wages.

In short, in the period before the Civil War, master carpenters were finding themselves in a terrible economic squeeze, and this was part of the reason why their journeymen were so discontented. The social ladder that had once looked so sturdy—the ladder that took a young man from apprenticeship to mastery—was beginning to shake. Deskilling had not yet begun, but the traditions of carpentry, and the traditional expectations of both masters and journeymen, were already destabilized.

After the Civil War, the onslaught of the machines began, and the shaky ladder collapsed completely. Carpenters, of course, had always used tools—handsaws and augurs, hammers and planes, sanders and chisels—but the tools that began to come on the market in the middle decades of the nineteenth century were different: most were powered by steam engines rather than muscles; many were automatic rather than manual. In the 1850s, clever investors began installing these machines—borers, compound carvers, power sanders, cut-off saws—in factories and started hiring unskilled workers to tend them. New York State, for example, had fifty-eight planing mills with capital goods and inventories worth $131,000 in 1850; fifty years later, factory planing was a $22 million industry. Factory-made doors and ceiling boards, flooring, moldings, and window sashes were replacing the handmade products, which had once been laboriously cut, carved, joined, and finished by masters, apprentices, and journeymen.

At first, master carpenters welcomed factory-made products, for they realized that such items could save untold hours of hard labor and thereby lower production costs. Journeymen, of course, were not nearly as pleased since the labor being saved was their own; the master who purchased window sashes from a factory was a master who could lay off his journeymen in the winter. Eventually, as builders began using factory-made products as well, the masters echoed the discontent of the journeymen, for with factory-made products, subcontracting masters could be dispensed with also, and builders could employ the unskilled or semiskilled laborers rather than the (considerably more expensive) skilled artisans. Between 1848 and 1897, machine-made construction components, according to one estimate, reduced production costs by 80 to 90 percent—a savings that real estate developers could not possibly be expected to forgo, even if, in the forgoing, they would be saving an endangered social species.[2]

If from the builder's point of view the last decades of the nineteenth century witnessed a transition from hand to machine methods of production, those same decades from the worker's point of view witnessed the transition from being an artisan to being a worker. That transition was not sudden and it was not uniform; in one place the transition may have been completed two decades before it began in another; in some places, hand and machine methods existed side-by-side for a time. But by the end of the nineteenth century, and certainly by 1920, the artisanal system in carpentry was clearly dead.

And many workers mourned its passing. With considerable bitterness, skilled carpenters came to realize that they could no longer expect to be both self-employed and prosperous. Master carpenters no longer wished, or even expected, to have their sons take over their work. Young boys no

This mid-nineteenth-century advertisement is for some of the machines that replaced both the hand tools and the work of journeymen carpenters. (Courtesy Smithsonian Institution.)

longer bothered with seven-year apprenticeships; all the essential skills of carpentry could be learned in much less time, and builders were willing to train boys on the job. In 1820, a journeyman carpenter could have reasonably expected to become a master someday. By 1860, that hope was dashed, but a journeyman could at least have reasonably expected to work for the same employer (a subcontracting master) for many years. By 1900—with all the masters replaced by builders and all the work minutely subdivided—a skilled carpenter would no longer be called either a journeyman or a master (since those terms had lost their meaning), and he might have had as many as twenty or thirty employers in the course of any single year.

Such were the multiple meanings of deskilling. What had once taken days of painstaking, careful work for a master to produce could now be run up in a few minutes by a thirteen-year-old operating a machine. If the average carpenter had once known how to carve a complex molding out of a length of wood, he now only needed to know how to miter factory-carved pieces together at the corners—and what was there to be proud of in that? If an apprentice carpenter had once had to serve for seven years, the basics of the (now subdivided) trade could be taught to a newcomer in a little over seven days—and what loyalty could be engendered by that? If such an apprentice had once reasonably expected that he might someday become a master himself, he could now only hope that if he joined a union and the union was sufficiently strong (and sufficiently wealthy), he might see some improvement in his wages or his working conditions. Upward social mobility was, however, clearly out of the question. In the machine age, once a worker, always a worker—and not even a particularly well-paid one at that.

What subdivision and mechanization were doing to the carpenters, they were also doing to the typesetters and the coopers, the cigar makers and the machinists, the iron puddlers and the hatters, the potters and the tailors. In all these crafts, deskilling meant a long-term economic loss to the individual worker; over the course of a lifetime, the factory hand who was paid wages could not hope to achieve even the moderate prosperity that had once been the master artisan's lot. In addition, deskilling was accompanied by a deeply felt emotional loss, the disappearance of a special on-the-job culture that artisans particularly valued: a culture that stressed independence, autonomy, manliness, humor, loyalty, adventure, and group solidarity. Such a culture, born and bred in small workshops over the course of centuries, could not easily be sustained in enormous, impersonal factories.

Thus for many skilled artisans, industrialization meant profoundly negative changes in almost every aspect of their working lives. Where workshops had once been small and intimate, they were now large and impersonal. Where owners had once been workers and teachers, they were now bankers and stockholders—far distant in social class and physical space. Where prosperity had once been the reward of patience, thrift, and hard work, it was now only a faint, and embittering, memory. The earliest of the national labor unions (the Knights of Labor, formed in 1869; the Federation of Organized Trades and Labor Unions, formed in 1881 and succeeded in 1886 by the American Federation of Labor) grew in soil that had been watered by that bitterness.

There were, however, some skilled workers for whom mechanization was not a bane but a benefit. In some industries new jobs were created *as the result of mechanization*—the signalmen on railroads, for example, or the linotype operators for newspapers—these jobs required new kinds of skills, but skills nonetheless. In yet other industries, even the most mechanized workplaces still required some individuals to act as shop floor supervisors, and these men and women were regarded, by themselves and by others, as skilled laborers. For all these reasons, in the half century after the Civil War—as the industrial sector of the economy was expanding—the absolute number of skilled laborers in the country was actually rising. Despite mechanization and despite deskilling, skilled laborers were still a substantial segment of the labor force. The nature of their work had changed; the character of their relations with their employers had changed; their long-term prospects for both prosperity and autonomy had altered—but they were still an important factor, both in local and national politics as well as in the productive process.

The history of skilled work in the iron and steel industry provides insights into both the changing nature and the changing significance of skilled labor as industrialization proceeded. In the early decades of the nineteenth century, pig iron (the product of blast furnaces; see Chapter 3) was made into steel in establishments called puddling furnaces, and steel ingots were made into steel products (such as train tracks) in other establishments called rolling mills. Both puddling furnaces and rolling mills employed highly skilled workers who worked in teams. Puddlers, for example, were responsible for stirring the huge containers of molten iron, adding various chemicals to the brew, and deciding when to pour out the contents of the containers. Roughers (who worked in the rolling mills) were responsible for arranging hot steel ingots, feeding them into the rolling machinery, and deciding how much pressure was to be applied on how many passes through the machines so as to best weld the ingots together and best determine the shape, and the strength, of the final product.

Puddlers and roughers both had to know a lot about the variations in the materials they were processing, and they had to exercise considerable manipulative skill as they handled very hot, very heavy, very awkwardly shaped, and very expensive batches of iron and steel; several years of work as an apprentice member of a team were required before a junior puddler or rougher was regarded as able to take charge. The senior men in the puddling and rolling crews were the highest paid workers in the industry, supervising and training the junior members of their crews. In some furnaces and mills, these senior men operated as internal sub-contractors; they were paid a lump sum for every unit of iron or steel that they produced and then divided that sum up between themselves and the junior members of their crews. Although they weren't exactly owner-workers, as the master carpenters were, those skilled ironworkers who could subcontract in this way retained certain crucial features of the master-apprentice system.

The advent of the Bessemer-Kelly process changed all these conditions of labor. Purification of metals and ores was accomplished not by stirring but by applying very high heat over the course of several hours. The character of the molten material was determined by taking samples at various times during the melting process and sending these samples to an in-plant laboratory to be analyzed; decisions about additives and temperatures were made in those laboratories by engineers and chemists. In the new mills, several furnaces were placed sequentially within one very large shed, and materials were transported between those furnaces by steam-powered cranes. Bessemer-Kelly steelmaking produced vastly larger, and heavier, ingots than puddling had produced, thereby transforming the rolling process. The employees of the new rolling mills had less physical contact with the steel ingots; instead they operated machines (cranes, shears, rollers) that manipulated the ingots.

Clearly the skills that had been valued in puddlers and roughers were not valued in the new mills. Standardization of the product also meant that less discretion needed to be exercised in production, and what discretion there was was exercised in laboratories by engineers and chemists, not on the plant floor by workers. Furthermore, the fact that skilled workers operated machines meant that they were less likely to work in teams, less likely to train other workers, and less likely to arrange for internal subcontracting. The use of machines for handling and transporting materials meant that sheer brute strength and muscular coordination were no longer essential prerequisites for success on the job.

Despite all these changes, a fairly substantial proportion of ironworkers and steelworkers were still classified by themselves and by others as skilled and semiskilled workers. A comprehensive report done in 1910 by the office of the U.S. Commissioner of Labor estimated that, on the basis of wage rates, in the open-hearth plants of Pittsburgh and the Great Lakes regions, 29 percent of all the workers were skilled and 22.6 percent were semiskilled. The very highest paid of the skilled workers earned almost six times as much as the very lowest paid unskilled workers. Successful operation of the machinery in these plants did not require the long periods of on-the-job training that had previously characterized the work, say, of puddlers and roughers, but it did require *some training* (unskilled workers, many of whom were assigned janitorial tasks, received no training at all), a good deal of manual dexterity, and attention to detail. The charging machine operator (whose machine loaded the furnaces with pig iron) required, according to the report, "considerable experience and much agility of mind and hand." The pit crane operator, who moved the enormous steel ingots into the rolling and shearing machinery, needed "considerable skill to charge and draw ingots with the rapidity and accuracy required."[3] For maximum efficiency, all the furnaces and processing machines were arranged sequentially; this meant that under optimum conditions a continuous flow of goods could be achieved. High levels of productivity thus depended on maintaining those optimum conditions, a difficult thing to do when each machine is tended by a different person, the environment is very dirty, the loads are very heavy, and each step in the process cannot begin until the one before it has been completed successfully. Thus productivity (and profits) could be achieved only when skilled and semiskilled machine operators were willing to cooperate with each other (with regard to timing) and were able to diagnose and remedy bottlenecks in the production process when they occurred. As much as plant owners and designers may have wished it otherwise, the productivity of these new plants thus still depended on both the skills and the morale of the highest paid workers, which is, of course, part of the reason why they were not only the highest paid, but also frequently the most militant.

When compared with workers who had been highly paid decades before, these skilled and semiskilled workers were employed in jobs that required vastly less training time, less brute strength, less discretionary activity, less supervisory work, and less autonomy on the job. On the other hand, they were still the aristocrats of the workforce. They were still being paid relatively high wages, their work was not mind-numbingly repetitive, and they retained some control over their daily activities and some control (when they were able and willing to unite to exercise it) over both the quantity and the quality of what they produced.

The cranemen and rollers of the open-hearth steel mills were similar to the engineers, brakemen, and signalmen on the railroads; the lathe operators and machinists in the metal goods factories; the linotypers and pressmen in newspaper plants; the conductors on the streetcar lines—skilled workers whose jobs had been created, not destroyed, by mechanization. Wise employers tried to keep the turnover rates among such workers to a minimum, and wise workers were able to use their various forms of control to extract concessions from employers. Unlike artisans in the older crafts, they had no golden age to yearn for nostalgically since their crafts had not existed in times past. They may not have regarded mechanization entirely as a blessing—the possibility that their work, like the work of the carpenters, could also be subdivided and automated was an omnipresent threat—but they didn't regard it entirely as a curse either.

Unskilled Workers

Of the 10 million people who were nonfarm manual workers in the United States in 1900, a third were called "operatives and kindred workers" and yet another third were relegated to the category "laborers." The chances are that all of the people in the first category, and many of the people in the second, were unskilled workers, the people who tended the relatively automatic machinery in factories, the people who carried unfinished goods and raw materials from one work position to the next, the people who cleaned the factory floor when the workday was done. These were the people whose low wages had replaced the high wages of craftsmen and apprentices made redundant by mechanization.

The term operative had been coined by the British (who were pioneers of mechanization) to describe unskilled workers in factories. With the exception of the bureaucrats at the Census Bureau, Americans were never quite comfortable with the condescension built into the word "operative." When feeling optimistic and progressive, they liked to speak of "factory hands" which had the virtue of sounding a bit traditional (since it was derived from farmhand) and a bit more humane (since it gave primacy to the person and not to the machine). When feeling pessimistic, however, some Americans replaced factory hand by "wage slave," which was, if not more humane, at least somewhat closer to the truth. By 1920, several millions of Americans—the ancestors of many of the Americans who are reading this book—were mired in repetitive, dangerous, oppressive, dead-end, low-paying, insecure jobs, the jobs that had been created when the work of traditional craftsmen had been subdivided, automated, and mechanized.

In the first phase of industrialization, as we saw in Chapter 4, factory hands had come principally from the native-born population. Factories powered by water were located in the countryside and tended to employ people who lived in the immediate vicinity—an impoverished widow and her children, a small farmer trying to work off some of the debt on his land, young people who were not needed on their parents' farms or who were trying to build a dowry for themselves or lay aside some money before going farther west. Few of these early operatives regarded their employment in the mills as anything more than a temporary expedient or a seasonal variation, a way to earn some cash and then move on to something else, with hope, farm ownership.

As the nineteenth century wore on, however, the progress of industrialization and mechanization meant that the number of factory jobs was increasing faster than the native-born white population. In addition, the increase in competition between a growing number of firms meant that wages fell and working conditions in the factories became more onerous. Furthermore, with the advent of steam engines and later electric motors, businessmen started building factories not in the countryside, but in the cities. The net result of these three processes was that, in every region of the country except the South, fewer and fewer native-born, rural, white Americans could be induced to work in the mills. As a result, factory work increasingly became the domain either of recent immigrants, or of non-white native-born Americans: the French Canadians and the Irish dominated the textile mills of Massachusetts; African American men and women sweated in the steam laundries of almost every major city; Jewish women were employed as cigar molders in Pennsylvania; Polish men were butchering cows and Polish women were stuffing sausages for the Chicago meat-packers; Italian men were making cans in New York; Mexican women were packing tomatoes in California.

In factories from one coast to another, the skilled and unskilled workers could thus be distinguished from one another not just by the wages they were paid but also by the color of their skin and the country of their birth. Sex and age were also important factors. In some industries (women and children's clothing, for example) the vast majority of workers who were designated as unskilled were women. In other industries (textile spinning and weaving, for example) large numbers of children were employed as unskilled workers; nationwide in 1910, some 1,637,000 workers were girls and boys between the ages of ten and fifteen.

Unskilled factory labor was the kind of work that various kinds of people were initially grateful to have. An immigrant, if he or she had the right connections, could get off the boat on Monday and begin working by Wednesday—and a few years later, that immigrant might be promoted to a skilled or semiskilled job or might have acquired enough cash to return home and buy land. A sharecropping white family could go together into a textile mill in the hope that, in a few months or years of employment, enough cash could be laid by to pay off the loans that had accumulated on their land. A young black woman might take a job in a spinning mill to avoid the even more onerous work of domestic service. A woman with an injured husband could take a job stitching the soles onto shoes, assuming that she would quit once her husband was well enough to return to work; or she might send her children into the factory in the hope that in a few months they could return to school. Very, very few people entered into this particular sector of the labor force in the hope of making a lifelong career in an unskilled position—but for many reasons, many people unfortunately ended up doing precisely that.

The work that operatives did was both boring and repetitive: the same task, repeated over and over again, hundreds, even thousands of times a day. A worker on one of Henry Ford's automotive assembly lines described this kind of employment graphically:

> Henry [Ford] has reduced the complexity of life to a definite number of jerks, twists and turns. Once a Ford employee has learned the special spasm expected from him he can go through life without a single thought or emotion. When the whistle blows he starts to jerk and when the whistle blows again he stops jerking, and if that isn't the simple life, what is?[4]

Ten or twelve hours might separate the whistle that blew in the morning from the whistle that blew at night; for unskilled workers, this meant ten or twelve hours of mindless, frequently exhausting labor, interrupted only by breaks for meals.

Many of the people who entered the unskilled labor force had been accustomed to the time-use patterns of agricultural labor; some days they worked hard, other days not so hard; chores varied from week to week and from season to season. Although the weather and other natural patterns could be hard taskmasters and although labor in the fields could be as mindless as labor in the factories, to some extent agricultural workers could set their daily schedules themselves. Factory work was, of course, very different. The daily schedule was set by the factory manager, not by the worker—and for the manager the best use of the machinery was the overriding concern, not the best use of the worker. Every day in the mill tended to be the same as every other day in the mill; machinery could be an even more difficult taskmaster than the weather because its demands were both constant and unremitting. The managers of the International Harvester Corporation recognized the difficulty that their largely Polish, formerly peasant, employees faced in trying to accustom themselves to industrial time discipline. A booklet that the company prepared in order to help their workers learn English read, in part:

> I hear the whistle. I must hurry.
> I hear the five minute whistle.
> It is time to go into the shop.
> I take my check [a tag] from the gate board and hang it on the department board.
> I change my clothes and get ready to work.
> The starting whistle blows.
> I eat my lunch.
> It is forbidden to eat until then.[5]

Disciplining themselves to whistles, bells, and clocks was not the only time-related difficulty that unskilled workers faced. In a mechanized factory, managers could, almost at a whim, speed up or slow down the work. Workers had a hard time deciding which was worse. A speedup might occur when the business was under economic pressure that required either an increase in production or a cut in labor costs. In some factories, speeding up meant that managers increased the tempo at which the machines were operating or the speed at which conveyor belts were moving. In other factories, it meant that the daily quota of goods was increased or that the piece rate for goods was lowered. Either way, for workers a speedup meant that a bad job became worse; they had to move faster, concentrate harder, eat at their machines, visit the restroom (itself a significant term) less frequently.

Slowdowns were little better, and if a slowdown turned into a shutdown that could, in fact, be worse. A slowdown might occur when business was bad or when the managers of a particular business, unaccustomed to handling both mass production and mass markets, had let inventories pile up. In the meatpacking industry, slowdowns were seasonal; in winter, fewer animals were shipped to market. In his novel *The Jungle*, Upton Sinclair left us a stark description of what a slowdown meant for people such as his protagonist, Jurgis, who were paid either by the hour or by the piece.

> The big packers did not turn their hands off and close down . . . but they began to run for shorter and shorter hours. They had always required the men to be on the killing beds and ready for work at seven o'clock . . . but now, in the slack season, they would perhaps not have a thing for their men to do till late in the afternoon. And so they would have to loaf around in a place where the thermometer might be twenty degrees below zero! . . . Before the day was over they

would become quite chilled through and exhausted, and when the cattle finally came, so near frozen that to move was an agony. . . . There were weeks at a time when Jurgis went home after such a day as this with not more than two hours' work to his credit—which meant about thirty-five cents.[6]

When a factory shut down there were no wages at all, and in those days unemployment insurance did not exist. A factory that still ran on water-power (and many still did in the closing decades of the century) might shut down in the spring if the river had flooded or in the early autumn if it had run dry. Complex machinery often needed repair and, if enough machines had gone on the blink, a factory might shut down for several weeks of maintenance or even retooling. And in a period in which most businesses were small businesses and competition was intense, a factory might be operating at full steam one day and shut down the next because its owner had just plain gone bankrupt. Either way, the workers were out of work—and out of luck; one study done in the relatively prosperous year of 1890 demonstrated that one out of five industrial workers was out of work for at least one month during the year.

Factories were noisy places; the din of the machinery was so intense that workers could not converse with each other. They were also filthy places. Metal shavings, cotton lint, grain chaff, engine oil, sawdust—the waste products of production—were omnipresent; long-term respiratory diseases (tuberculosis, emphysema, black lung) were common. Short-term illnesses were common also, for these factories were dangerous. Before electric motors, every machine was attached by an exposed belt to the engine that powered it, which meant that workers were constantly in danger of being caught in the belting: drawn in and crushed, or thrown across the room. Fingers and hands could be severed by machines that cut and stamped. Knives could miss their mark. Steam engine boilers sometimes exploded. And the rules of antisepsis were largely unknown, so that what might be a minor injury today (the loss of a finger, for instance, or a deep cut on the thigh) might lay a worker up for months or even prove fatal. In 1889, the Interstate Commerce Commission reported that roughly two thousand railwaymen had been killed on the job and ten times as many had been injured in the previous decade—more than one third of the total workforce. Worker's compensation did not exist and neither did medical insurance. An injured worker suffered more than just bodily pain, and so did all of the worker's dependents.

Factory workers eventually learned how to fight back, although, being poor, their weapons were always limited. In the early years, trade unions (such as those that had been established by the carpenters or the machinists) were not interested in unskilled workers; indeed, "not interested" puts the matter too politely. Many trade unions were actually hostile to unskilled workers; from the perspective of the native-born skilled worker, the unskilled worker was a foreigner, a woman, a barbarian, or a scab: someone who was being brought into the workplace to cut wages, break a strike, or steal a job.

At first, lacking organization and also lacking political sophistication, unskilled workers found both individual and communal solutions to their predicament when they could. Workers who lived in large cities, where employment opportunities for the unskilled might be numerous, frequently voted with their feet. Lacking vacation time, they took vacations by claiming to be sick; lacking internal career ladders, they tried to improve their situations (or relieve their boredom) by quitting one factory and going to work for another. As a result, absentee and turnover rates were enormous—the individual's way of going out on strike. The Ford plants, being among the most mechanized, were also among the most severely plagued by this form of worker opposition. On an average day in 1913, about 10 percent of the workforce was absent; and in that year, the turnover rate in Ford's main factory was a staggering 370 percent—managers had had to employ 52,000 workers in order to maintain a workforce of 13,600.

Where alternative workplaces were not numerous (in small towns, for example, or in one-plant towns) soldiering—deliberate restriction of output—and sabotage were common. When they could reasonably do so without lowering their own wages, workers attempted to control the work processes in which they were embedded, sometimes by ostracizing workers who worked too fast or produced too much: pity the man who broke, for example, the bricklayers informal agreement about how many rows could be mortared in a day or the kiln tenders agreement about how many trays of pottery would be fired in a day. Sometimes workers also engaged in various forms of sabotage, for example, by hiding semifinished products (if one person in a sequence was producing more than the others wanted to handle), by threatening a foreman's family, or by deliberately, but not permanently, breaking a machine.

Unskilled workers also learned to band together, both on the job and off, in their own defense. Lacking what we would now call a benefits package, they created fraternal organizations and insurance pools that could help a family over difficult times, provide funeral expenses, or pay doctors' bills. In an effort to make the shop floor a friendlier place, they often persuaded managers to hire their friends and relatives; in the cigar factories, unskilled immigrant workers were able to reestablish the custom that had originated in their home countries of having one worker read aloud to the others as they worked.

Eventually, through the efforts of progressive reformers and the radical leaders of the first so-called industrial unions (such as the IWW, the Industrial Workers of the World, founded in 1905), the worst abuses of the industrial system were somewhat mitigated. Workmen's compensation laws were appearing on the books in various states by World War I. Some halfhearted attempts were being made to regulate the most egregious health conditions in factories by the 1920s. The Depression years brought with them various programs of unemployment insurance, as well as the creation of the Congress of Industrial Organizations (the CIO, founded in 1938), the first successful national coordinating organization for the labor unions that represented unskilled, rather than skilled, workers.

But these improvements did not happen quickly, easily, or universally. In the years between 1870 and 1920, millions of people were victimized by mechanization, some for only short periods of time, others for the whole of their working lives. We who have benefited from their work either directly or indirectly would do well to remember this when we attempt our own accounting of the costs and benefits of industrialization.

Housewives and House Servants

The single largest group of workers in the United States in the years between 1870 and 1920 were housewives and house servants. The census counted 7.5 million American households in 1870 and 24.3 million in 1920. In virtually all of those households, at least one person was doing housework during some portion of every week; in 1870, just under a million of those people were domestic servants (one for every seven or eight households); in 1920, 1.7 million (or, roughly, one for every fourteen or fifteen households). The vast majority of those houseworkers were female. Some worked without pay (the housewives) and some worked with pay (the servants), but in either case, the work that they did was unquestionably work, and like all work in this period, it was affected by industrialization.

Housewives and house servants, like all other groups of workers, were very diverse. Some housewives were rich, some were reasonably comfortable, but in this period most were poor; so were all servants. Some housewives lived in large cities, others in small towns, some on isolated farmsteads; the vast majority of servants were urban. Some houseworkers cared for small families living in large domiciles, others for eight children living in two rooms. In analyzing the transformations that industrialization wrought on housework, however, the most important distinctions are those that have to do with employment status (housewife or house servant), wealth (rich, comfortable, or poor), and ethnicity.

In the colonial period, when most households were rural and rooted in European traditions, the labor of both men and women was required to bring any substantial meal to the table; adults expected either their children or their servants (or both) to assist in the labor. Women cooked at the hearth: peeling potatoes, cutting vegetables, kneading bread, stirring pots, whatever was necessary for what might be called the final processing of foodstuffs. But nothing could be cooked by women until men had supplied the fuel: chopping trees, cutting, hauling, and splitting logs. The raw foods that went into a meal were produced by gender-specific labor processes: men butchered animals and preserved them (by smoking and pickling, for example), and women butchered poultry; men were responsible for most of the work needed to grow such grains as corn and wheat, while women were responsible for vegetables and herbs. Even beverages were gender typed (or, better, gender roled): women brewed beer; men pressed apples to make cider. Children learned through doing: carrying water, collecting kindling, assisting in all aspects of both food processing and fuel gathering. Servants, both male and female, were employed in most households except for the very poorest; servants were expected to engage in the same gender-appropriate work as their masters and mistresses: the hired hand helped with butchering of cattle; the maid servant assisted when the vegetable garden needed hoeing.

Colonial kitchens were organized around open hearths; very large fireplaces (usually made from brick or from stone), which were used for cooking food, heating water, and warming the room. In the early decades of the nineteenth century, an industrialized replacement for the open hearth began to appear on American markets, the cast-iron stove. In their earliest and simplest forms, these stoves were nothing more than footed metal boxes that enclosed a fire; they had vents for letting air in and pipes for letting combustion gases out. However, by 1820, several people had patented modifications to the heating stove that would make it useful for cooking; special holes were cut in the stove top so that pots could be put directly over the firebox, additional compartments were built into the sides of stoves so that breads and meats could be baked; increasingly complex systems of vents were designed so that cooks could exercise some control over the heat of the fire. Several iron founders developed techniques for manufacturing stove plates inexpensively and, as a result, in the 1830s, the prices for cast-iron stoves began to fall

Cast-iron stoves had many advantages over hearths. In the first instance, because they were enclosed, they were fuel efficient—a factor that must have impressed householders in regions where wood was becoming scarce. They were also fuel flexible. With slight modifications, coal could be used as easily as wood in a stove. As the railroads began connecting coal-mining regions with urban areas, coal prices fell. Coal could not be used easily in an open hearth (in part, because of some of the noxious gases that it produced when burned), but it could easily be used in a stove. Unlike hearths, stoves were space flexible: they did not have to be part of a wall. Stoves could project into a room, which meant that air could circulate around them, heating the room better. And stoves could be purchased disassembled, a crucial factor for emigrants, who could buy disassembled stove plates and pipes on the East Coast and then build the cooking and heating apparatus themselves when they arrived at their destination. The cast-iron stove was the first do-it-yourself consumer durable, the first mass-produced appliance intended for use in people's homes.

Although patterns of stove diffusion differed in different parts of the country (the Southeast and Southwest being slow to adopt them), many American houses had converted from hearth to stove for cooking and heating by 1870 or 1880 at the very latest. The implications of the stove for the work processes of cooking were complex. Where wood was still the fuel of choice, stoves lessened the need for male labor at home because they were fuel efficient. Families that had converted from wood to coal experienced an increased need for cash. Wood could be obtained directly by the labor of the men in the household, but coal traveled through many middlemen on its way to the consumer and, as a result, could not ordinarily be purchased except with cash. Thus families that had converted from wood to coal needed to have someone who was working for wages, and oftentimes the person whose household labor could most easily be spared was an adult male. The net result was that a household with a stove was one in which it was both easier for, and more necessary that a man be employed outside the home.

For women (both housewives and servants), stoves meant more work rather than less. Unlike hearths, stoves had to be meticulously cleaned and coated because they were subject to rust. A stove also made the job description of a housewife more complex because a stove made complex meals easier to prepare. With a reasonably well-designed stove, a woman could bake and boil with the same fire or could bake several different items at one time (using ovens at various distances from the firebox)—enterprises that would have been considerably more difficult on a hearth. Partly because of the introduction of stoves the American diet became more varied: an average meal might have two courses instead of one or there might be a cake for dessert (a century earlier, cakes had been served only on very special occasions or in the homes of the very rich, people who had many servants and more than one hearth). Thus the impact of the stove on women was the opposite of what it had been on men. The standard of living of the household went up, but so did the amount of work that women had to do.

Several other changes in household technology had the same net effect. Factories started to make boots and shoes (the manufacture of boots and shoes was one of the ten leading industries in the United States in 1860), which meant that men could keep their families shod without working on leather at home. Factories also started to make manufactured cloth and, as a result, children no longer needed to card, women to spin, or men to weave. Manufactured cloth was relatively cheap in those days, but someone still needed to make it up, by cutting and sewing it appropriately, into clothes, and this had always been work allotted to women and girls; thus one aspect of women's work was eliminated but another was increased. Similarly, the advent of manufactured cloth and larger wardrobes meant that there was more laundry to do (much manufactured cloth was cotton, and therefore washable; the woven fabrics used earlier, linen and wool, were rarely laundered), also women's work. Candlemaking became a lost art in the second half of the nineteenth but making and melting various waxes was replaced by the daily and weekly chore of cleaning the glass globes of oil and gas lamps (electricity did not make much of a dent on household work until after 1920). Wastewater systems (otherwise known as flush toilets) eliminated the chore of collecting slops but added the chore of cleaning toilets; home canning equipment (glass jars especially) made it possible to preserve more fruits and vegetables for consumption during the winter but vastly increased women's workload in the summer and fall.

In short, the first phase of industrialization introduced into American homes various innovations that greatly increased the average American's standard of living but did not substantially decrease the amount of work that women had to do in their homes. Men and children could be spared— and were being spared—to the schools, to the factories, and to the offices of the burgeoning economy. Adult women, on the other hand, could not as easily be spared (although in cases of extreme poverty, even they had to find work that paid wages). Even when they had access to cast-iron stoves and tin washtubs, piped water and gas lamps, carpet sweepers and treadle sewing machines, women found that there was a great deal of backbreaking labor still left to be done at home: meals had to be cooked, sick children had to be tended (little was done to alleviate the scourge of illness before 1920), clothes had to be made, mended, and laundered. A woman's work, it seemed, would never be done, industrialization notwithstanding.

As a result, the so-called servant problem continued to plague wealthy and comfortable American households. Despite the many changes in the work processes of housework occurring between 1870 and 1920, the maintenance of a fairly comfortable domestic standard of living required the labor of at least two adult women. As late as 1925, Americans still believed that a middle-class standard of living could not be achieved without employing at least one full-time household servant.

But good servants were hard to find. Indeed, good servants had always been hard to find in North America, even in colonial times. Householders defined good servants as people who were willing to work hard and maintain a subservient demeanor, at all hours of the day or night, over the course of many years, at very low wages. Except over the very short term and in only a few places (the Southeast both before and after Emancipation, for example, or the Northeast after the Irish potato famine), there has never been a very large supply of such people in North America—a fact foreign visitors have been quick to notice. In the early decades of the nineteenth century, land was too cheap and other employment opportunities too numerous for people, whether male or female, to stay in domestic service for very long (unless they were enslaved). Native-born daughters of poor farmers might go into service, but only until they had managed to lay aside a nest egg and met someone to marry. Immigrants might become servants after they had just arrived, but only as long as it took to learn the language or find a better job or locate a helpful relative. In the land of the free, no one was eager to become a servant, and few people, even if forced into service, were willing to be subservient. "We came to this country to better ourselves," one Irish-born cook is said to have remarked, "and its not bettering to have anyone order you around."[7]

As industrialization proceeded and the nineteenth century turned into the twentieth, the labor market for servants became tighter, except in the Southeast (where domestic work was often the only work available to

More people did housework than any other kind of work in the nineteenth century. Despite the changes wrought by industrialization, the full-time labor of at least two women, a housewife and a domestic servant, was needed to maintain a middle-class household such as this one in Topeka, Kansas. (Photograph of the S. H. Fairfield family, Topeka, Kansas, Courtesy Kansas State Historical Society.)

African American women whose families were not succeeding as farmers). Outside the Southeast, industrialization, as we have already seen, meant an increase in the number of employment opportunities for unskilled women. As dreary and ill paying as factory jobs were, many young women preferred factory work to domestic service. When the factory hand left her job, her life was her own; the domestic servant was expected to be on call twenty-four hours a day with only one day off every two weeks. A factory hand could work with other people her own age, male and female; a domestic servant was isolated. A factory hand could have a social life after work; a domestic servant was not allowed to entertain visitors. And domestic service was hard work, the servant was expected to do anything that the housewife herself considered too difficult or demeaning: floor scrubbing, laundering, heavy cleaning. With hindsight, it is not hard to see why American young women—white, Asian, Latina, or black; native born or recent immigrant—preferred the dark satanic mills to the dark satanic kitchens.

In another kind of social feedback loop, the growing number of alternative employment opportunities for unskilled women made domestic service increasingly unattractive to them. As the labor market for servants became tighter, the number of one-servant households increased, thereby increasing the burden on the servant that remained: a form of speed up. Just as much cleanliness had to be produced, just as many meals, just as many orderly rooms, just as many well-managed dinner parties—but there were fewer workers to do the job. None of the new technologies then on the market (such things as carpet sweepers and canned hams, gas lamps and crank-driven washing machines) were adequate to make up for the loss of other household workers—and they didn't stop for a pleasant chat at dinnertime, either.

For all these reasons the servant problem persisted into the twentieth century; indeed, World War I made matters even worse by ending immigration across the Atlantic while simultaneously creating more factory jobs for women. The number of households in the nation continued to grow and the number of servants continued to decline: there were 188 servants, for example, for every 1,000 households in New York City in 1880; 141 in 1900 and 66 in 1920. This decline in the relative number of servants further aggravated the situation in which comfortable (and now, even rich) housewives found themselves, for there was still work to be done and fewer and fewer people to help with it. As a result (albeit for different reasons), neither the housewife nor the servant perceived that the great nineteenth-century advances in household technology had done much to lighten the burden or ease the trials of housework. Many (and this includes both housewives and house servants) were grateful (if contemporary accounts can be believed) for the improved standard of living that new products had made possible (more and cleaner clothing, larger and cleaner homes, more varied meals, more and less expensive decorative items for the home) but few would have referred to those products as laborsaving devices for women.

Conclusion: Was Industrialization Good or Bad for Workers?

Thus although industrialization wrought profound changes in people's working lives, it is not easy to generalize about precisely what those changes were. For some people, work became less physically burdensome; for others, more so. Some people prospered, others were reduced to penury. Some people worked longer hours for less pay; others, shorter hours for more. Some people observed no change in their working lives at all, but considerable change in their children's employment prospects. Some people observed no change in their working lives at all, but considerable change in their spouse's situation.

Some of these changes were good for people and some bad, but we cannot possibly weigh the good against the bad. "Good" and "bad" meant different, sometimes contradictory, things to different people, and there are no conceivable units of measurement that will make the different goods and bads commensurable. How bad is it that an individual Ford worker spent ten hours a day at a mind-numbing job? And how good is it that his job paid more, allowing him to purchase more than his father would have thought conceivable? And if his father had been struggling to make a living as a tenant farmer or a sharecropper, how much better was the regular wage that Ford could provide? How can we weigh the good for a farmer who was able to turn a profit after buying a harvester against the bad for his neighbor who was driven into tenancy? Indeed, can we even assess whether people found it more or less soul-deadening to have their daily lives governed by the rhythms of the machine than by the rhythms of nature?

In the end, we are left with the generalization with which we began: although industrialization undoubtedly wrought many changes, any statement about the impact of industrialization on the working lives of Americans that does not distinguish them at the very least by occupation, gender, race, and class—and is not very specific about which Americans and which occupations are at issue—is bound to be wrong.

Notes

1. "Resolution passed at a meeting of Gentlemen Engaged in Building," April 21, 1825, as quoted in J. R. Commons and V. B. Phillips, *A Documentary History of American Industrial Society,* Vol. 6 (Cleveland: A. H. Clark, 1910), p. 80.

2. These figures come from reports of the United States Commissioner of Labor (1898) as quoted in Robert A. Christie, *Empire in Wood: A History of the Carpenter's Union* (Ithaca, NY, 1956), p. 26.

3. Both quotes are from the *Report on the Conditions of Employment in the Iron and Steel Industry* (Washington, DC: Government Printing Office, 1911), as quoted in Michael Nuwer, "From Batch to Flow: Production Technology and Work-Force Skills in the Steel Industry, 1880–1920," *Technology and Culture* 29 (October 1988): 834.

4. Adam Coldigger, "Why I'm for Henry," *Auto Worker* 5 (August 1923): 5, as quoted in Stephen Meyer III, *The Five Dollar Day: Labor Management and Social Control in the Ford Motor Company* (Albany: SUNY Press, 1981), p. 40.

5. From *Harvester World* 3 (March 1912), as quoted in Gerd Korman, "Americanization at the Factory Gate," *Industrial and Labor Relations Review* 18 (1965): 402.

6. Upton Sinclair, *The Jungle* [1905] (Cambridge, MA: Bentley, 1974), p. 86.

7. Quoted in Helen Campbell, "Why Is There Objection to Domestic Service," *Good Housekeeping* 11 (27 September 1905): 255

Suggestions for Further Reading

Allan Bogue, *From Prairie to Corn Belt: Farming in Illinois and Iowa Prairies in the Nineteenth Century* (Chicago, 1963).

Harry Braverman, *Labor and Monopoly Capitalism* (New York, 1974).

David Brody, *Steelworkers in America: The Non-Union Era* (New York, 1960).

Robert Christie, *Empire in Wood: A History of the Carpenter's Union* (Ithaca, NY, 1956).

Patricia A. Cooper, *Once a Cigar Maker: Men, Women, and Work Culture in American Cigar Factories, 1900–1919* (Urbana, IL, 1987).

Ruth Schwartz Cowan, *More Work for Mother: The Ironies of Household Technology from the Open Hearth to the Microwave* (New York, 1983).

Clarence Danhof, *Change in Agriculture: The Northern United States, 1820–1870* (Cambridge, MA, 1969).

Pete Daniel, *Breaking the Land: The Transformation of Cotton, Tobacco and Rice Culture Since 1880* (Urbana, IL, 1985).

Gilbert Fite, *The Farmer's Frontier: 1865–1900* (New York, 1966).

Herbert Gutman, *Work, Culture and Society in Industrializing America* (New York, 1976).

Steven Hahn and Jonathon Prude, eds., *The Countryside in the Age of Capitalist Transformation* (Chapel Hill, NC, 1985).

Jacqueline Jones, *Labor of Love, Labor of Sorrow: Black Women, Work and the Family from Slavery to the Present* (New York, 1985).

David Katzman, *Seven Days a Week: Women and Domestic Service in Industrializing America* (New York, 1978).

Alice Kessler-Harris, *Out of Work: A History of Wage Earning Women in America* (New York, 1982).

Bruce, Laurie, *Artisans into Workers: Labor in Nineteenth Century America* (New York, 1989).

Stephen Meyer III, *The Five Dollar Day: Labor, Management and Social Control in the Ford Motor Company, 1908–1921* (Albany, NY, 1981).

David Montgomery, *Fall of the House of Labor: The Workplace, the State, and American Labor Activism* (Cambridge, 1987).

J. Sanford Rikoon, *Threshing in the Midwest, 1820–1940: A Study of Traditional Culture and Technological Change* (Bloomington, IN, 1989).

Leo Rogin, *The Introduction of Farm Machinery in Its Relation to the Productivity of Labor in the Agriculture of the United States during the Nineteenth Century* (Berkeley, 1933).

Fred Shannon, *The Farmer's Last Frontier, 1865–1897* (New York, 1945).

Marc Stern, *The Pottery Industry of Trenton: A Skilled Trade in Transition, 1850–1929* (New Brunswick, NJ, 1994).

Susan Strasser, *Never Done: A History of American Housework* (New York, 1982).

Carroll D. Wright, *The Industrial Evolution of the United States* (Meadville, PA, 1897).

9

American Ideas about Technology

ONE WINTRY DAY IN JANUARY 1903, a group of Boston physicians gathered in a seminar room to witness the demonstration of a new medical device. The demonstration was conducted by two young medical researchers: Harvey Cushing, professor of surgery at the Johns Hopkins University Medical School, and George Washington Crile, professor of medicine at the medical school of Western Reserve University. The device that Cushing and Crile were demonstrating that day had been invented by an Italian physician. It had an awkward name, "sphygmomanometer," and an even more awkward appearance: a piece of rubber tubing (which was supposed to be tied around a patient's arm); a hollow metal disk, similar to a pocket watch on which the face had been replaced by a sensitive rubber diaphragm (this was supposed to be pressed against the patient's wrist); and a barometer connected to the metal disk by more rubber tubing. The tubing and the disk were both filled with mercury.

The story of the sphygmomanometer—what we would call today a blood pressure cuff—can teach us a good deal about the various ideas, and combinations of ideas, that Americans held about technology in the years in which industrial society was being shaped. Cushing and Crile believed that the sphygmomanometer was an invaluable aid to diagnosis, which is why they had taken the time out from their busy schedules to travel to Boston for the demonstration; convincing the doctors at the country's most prestigious medical school that this new medical device was useful was a matter of some importance. The two young researchers believed that the syphgmomanometer indicated how a patient's heart was functioning because it measured the fluid pressure of the blood that left the heart with each contraction. They thought multiple benefits might ensue from such measurements: surgeons would be able to monitor their patients' heart during operations; physicians might be able to keep their patients from going into shock; at the very least, general practitioners might have a quantitative base line against which to assess changes in their patient's health.

Cushing and Crile were aware that diagnostic instruments were somewhat strange to most of their medical colleagues and that some of their colleagues would be opposed to the introduction of such instruments in medical practice. In 1903, most doctors estimated the severity of a fever by feeling a patient's forehead; they estimated the adequacy of lung function by pressing their ears against a patient's chest. At Harvard, however, Cushing and Crile expected to find progressive, forward-thinking physicians who were willing to consider the advantages of change. Both of them were somewhat surprised to discover that some of the physicians in this most prestigious of American medical schools reacted quite negatively to the notion that every physician's office and every hospital's operating room ought to have a sphygmomanometer.

From the perspective of the doubting physicians in that room and elsewhere, the trouble with the sphygmomanometer was that it replaced another, infinitely more natural instrument: the finger. For centuries upon centuries, doctors had been touching people's wrists with their fingers, feeling for pulses: diagnosing diseases, quite literally, by the laying on of hands. Pulse diagnosis was central to the ancient medical arts of China, Egypt, and Greece, and it had continued through many centuries to be a crucial aspect of medical training. Nineteenth-century medical students were taught to notice the hardness of the artery in the wrist, as well as the power, rate, and rhythm of the pulse. Pulse diagnosis was a complex business, and doctors needed many years of experience before they could think of themselves as proficient at it, for every person had a somewhat different normal pulse and every disease had a different way of making the pulse feel abnormal. Proficiency at pulse diagnosis was one of the skills that were absolutely central to the craft of medicine as it was being practiced at the time Crile and Cushing began their campaign for acceptance of the sphygmomanometer.

Thus one doubting physician complained that with the sphygmomanometer "we pauperize our senses and weaken clinical acuity." Another remarked that the machine would "deal a death blow to the painstaking study of the pulse." When one proponent argued that objective measuring instruments might have the same effect on doctors as on factory workers, "creating a wholesome appetite for precision," an opponent countered that doctors were supposed to be professionals, not mechanics: relying on a measuring gadget might be as bad for a doctor's reputation as "to be seen carpentering, painting or displaying other common-place or out-of-place talents."[1] Those physicians who wanted their profession to become more scientific, such as Crile and Cushing, favored the use of the measuring device because it was, at one and the same time, objective, precise, and quantitative—widening the physical and emotional distance between doctor and patient. Those who wanted their profession to remain an art preferred the experienced finger, which was at one and the same time more sensitive, more versatile, more portable (a major consideration when doctors were still making house calls)—keeping the physician actually in touch with the patient. Advocates of the new machine tended to be young men who had learned new skills of precise measurement in physiology laboratories; they wanted to export their skills from the lab to the bedside. Opponents of the new machine were physicians who had spent years perfecting their expertise in pulse diagnosis; they feared, with some justification, that they were going to be deskilled.

The outcome of this battle was clear within a decade; Crile and Cushing and other proponents of what came to be called scientific medicine carried the day. Contrary to what the machine's opponents predicted, a physician today is not considered unprofessional if he uses a sphygmomanometer—or any other machine. Quite the reverse; the use of complex machinery has become a central feature of modern medical practice; the more complex the machine, the more professional the physician is thought to be. Blood pressure cuffs are now so much a part of medical care that people are willing to pay to use them in supermarkets, drugstores, and airports.

People have many different—ofttimes surprising—ideas about technology and many different—ofttimes surprising—reactions to technological change. Which of us living in the latter years of the twentieth century would have imagined that physicians who were living at the beginning of the century would have objected to a device as simple as the now common blood pressure cuff? And who would have imagined that those doctors would have seen a connection between scientific medicine and factory labor? And who would have imagined that in those early decades of the twentieth century, when the medical profession was just beginning its rise to social prominence, some physicians would fear being deskilled by an apparatus that purported to measure a patient's blood pressure? But fear it they did and object to it they did, and in order to understand why, we must understand something of the ideas that people had about technology in the years in which Cushing, Crile, and the other physicians were living.

Technology and Associated Ideas

As odd as it may seem today, the word "technology" does not have a very long history. During the eighteenth century, "technology" had occasionally been used in Europe in its literal sense, "knowledge of the arts." As knowledge of the arts, technology referred to the contents of a certain kind of handbook or encyclopedia, popular in the latter decades of the century, in which the knowledge once transmitted orally by craftsmen to their apprentices had been written down for anyone who was literate to consult. However, no American seems to have used the term in print until the publication of Jacob Bigelow's very popular book, *Elements of Technology*, in 1829.

Bigelow was a professor at Harvard; he was the Rumford Professor of the Physical and Mathematical Sciences as Applied to the Useful Arts. *Elements of Technology* was based on his lecture notes; according to the subtitle, Bigelow understood technology to be "the application of the sciences to the useful arts." The contents of the book, however, suggest a larger meaning for the word, as the historian Howard Segal has explained:

> The "elements" of technology included numerous materials, both natural and man-made . . . ; equally numerous machines and structures . . . ; the processes of discovering or inventing and refining and producing all of those materials and machines and structures; the technical knowledge, skills and equipment needed to carry out all of those processes; and the history of the development of all of those processes and techniques.

In 1865, Bigelow, who had been asked to present a major address at the opening of the Massachusetts Institute of Technology, remarked that the word which was such an important part of the new institution's name had not been "in use, nor was it generally understood" thirty-six years earlier, when he had first adopted it.[2]

Even before the word was part of the general vocabulary, people had ideas about technology, and their ideas were, as all ideas are, ideas about technology in relation to something else. The physicians who were opposed to the use of the sphygmomanometer, for example, were thinking about that particular technology in relation to *nature*; the finger, they said, is more natural than the machine. They were also concerned about this particular technology in relation to *social status* (using a measuring device, they thought, might turn professionals into artisans) and also in relation to *skill* (the person who can diagnose the pulse, they argued, is more skilled than the person who can read the scale of the blood pressure machine). Nature, social status, and skill are important categories in relation to which ideas of technology ought to be assessed, but they do not exhaust the possibilities. Technology has meaning in relation to *gender*, for example; we think of some technologies as being inherently male (steel girders) and others as inherently female (cooking stoves). Technology also has meaning in relation to ideas about *God* (were humans created with opposable thumbs because God intended them to be tool users?) and about *politics* (as in such important expressions as "the machinery of government," "the balance of powers"). All of these ideas—about nature, social status, skill, gender, God, and politics—are connected to each other and all are component parts of the intellectual history of technology in the early years of industrialization (even before the word "technology" was regularly used), during the period when industrial society was developing (which will be the focus of attention in this chapter), and even today.

Precursors to Industrialization

The natural world with which most Americans were acquainted in the early decades of the nineteenth century was one that had actually been subjected to considerable human modification. Over the course of the previous century, much of the eastern coastal forest had been destroyed along with most of the game animals that had once lived there. In place of the forests, the settlers had created farms and pastures and towns; in place of venison and fowl, they now ate wheat and corn and ham and beef. Nature had been domesticated on the Atlantic seaboard.

However, most Americans who expressed themselves on the subject did not consider that modified landscape *unnatural*. Americans were, after all, a biblical people, well versed in the texts of both the Old and the New Testaments. In Genesis, they read that God had intended humankind to "have dominion over the fish of the sea and over the fowl of the air, and over the cattle, and over all the earth, and over every creeping thing that creepeth upon the earth." Furthermore, God had instructed people to "be fruitful and multiply, and replenish the earth, and subdue it." And God had been pretty clear about the way in which that subduing should be done: "God said: 'Behold, I have given you every herb yielding seed, which is upon the face of all the earth, and every tree, in which is the fruit of a tree yielding seed—and to you it shall be for food." Clearly God had intended human beings to be farmers and shepherds, to feed themselves by subduing—domesticating—the earth, the fish, the birds, and the cattle.

The rural, pastoral landscape—the Europeanized domesticated landscape that the colonists had created on the Atlantic seaboard—was thus quite natural to early nineteenth-century Americans, in part because they (and their forebears) had been accustomed to it for centuries and in part because they believed that it was the landscape God had intended them to inhabit. God had made raw nature, many of them thought, and had equipped human beings to subdue it; God had intended them to work hard at the subduing. He had also provided the ability to fashion tools—plows and rakes and knives and scythes—that would at least ease the work a little bit. Raw nature and human beings were thus understood to be parts of God's creation, while subdued nature (the pastoral landscape) and hand tools were understood to be part of God's plan. Thus hallowed both by religion and by tradition, all four—raw nature (which still existed on the frontiers), subdued nature, human beings and what we would now call hand-made tools—were understood to be entirely *natural*.

Early in the nineteenth century, many Americans also believed that there was a very special connection between the American landscape and the institutions they had created to govern themselves as an independent nation. The European monarchies had developed during the Middle Ages when, under the social arrangements that we now call feudalism, military leaders had rewarded the loyalty of their followers with large quantities of the single most important economic commodity at their disposal—land. This meant that from the Middle Ages until the end of the eighteenth century, the people who had governed in Europe, the monarchs and the members of the aristocracy, were also the people who owned all the land: everyone else was either a tenant or an employee. In this European social and political culture, subservience was the keynote of social relations: farm laborers were subservient to tenant farmers, tenant farmers subservient to their aristocratic landlords, the aristocracy subservient to the monarch. Everyone except the king was expected to bow and scrape in front of someone else.

Most of the men who wrote the Declaration of Independence and who composed the Constitution were determined to do away with all that bowing and scraping. The United States was going to be an independent nation composed of independent, self-confident citizens. There would be no monarch and, of course, no aristocracy; indeed, if all went well, there would be no tenants or employees either. The political equality that the Constitution promised to all the citizens of the United States would be grounded (quite literally) in the land. If all went well, all American citizens would experience the economic equality that came from owning and working their own land and the political equality that came from having been granted equal rights as citizens. "Those who labour in the earth are the chosen people of God," Thomas Jefferson wrote,

> whose breasts He has made His peculiar deposit for substantial and genuine virtue. . . . Corruption of morals in the mass of cultivators is a phaenomenon of which no age nor nation has furnished an example. . . . Dependence begets subservience and venality, suffocates the germ of virtue, and prepares fit tools for the designs of ambition.[3]

The virtuous state had to be composed of virtuous citizens, Jefferson thought, and farming was the only kind of work that inclined an individual toward virtue.

Those virtuous citizens, of course, would also be males; Jefferson and others who thought as he did were firm in their conviction that *husband*ry would form the economic basis of the republic, not hus*wife*ry. Men and women are different, Jefferson thought, and suited to different kinds of work. Their working lives should be kept as separate as possible, Jefferson wrote in a letter to a friend in 1787, since too much attention to the ways and the affairs of women would lead a virtuous male citizens into a "spirit of female intrigue destructive of his own and others happiness." [4]

Jefferson's vision of a pastoral republic animated thousands, perhaps millions, of people in his own day and continues to have almost mythic force even into our own, but this set of ideas was difficult to square with the level of industrialization that the United States had reached even by 1870. Indeed, even in Jefferson's day, not everyone had agreed with him; in the early decades of independence (see Chapter 4), some Americans thought that the factory system of manufacture was a very good thing indeed—and that it ought to be imported into the new nation quickly.

Alexander Hamilton, Secretary of the Treasury under Washington, and

Tench Coxe, coauthor with Hamilton of the famous *Report on Manufactures* of 1791, were among the dissenters. "Providence has bestowed upon the United States of America," Coxe said in a speech in 1787, "means of happiness, as great and numerous, as are enjoyed by any country in the world." But God intended Americans to exploit those great resources, and not just through agriculture, but also through "manufactures and commerce." When he spoke of "manufactures," Coxe did not just mean, as Jefferson had, artisans who worked with hand tools. Coxe wanted industrialization; he wanted machines to do the work, and he thought that God had intended machines to do the work that had previously been done by hand. "Unless business of this kind is carried on, certain great *natural powers* of the country will remain inactive and useless. Our numerous mill seats . . . would be given by Providence in vain." And since God does nothing in vain, God must have given us all those waterfalls (the mill seats) so that we could learn how to use them to power machinery. The mill seats, Coxe continued,

> If properly improved they will save us an immense expence for the wages, provisions, clothing and lodging of workmen, without diverting the people from their farms. . . . [And since] Fire, as well as water, affords . . . a fund of assistance, that cannot lie unused without an evident neglect of our best interests. . . . It is probable also that a frequent use of steam engines will add greatly to this class of factories.[5]

Coxe and Hamilton and others who thought the way they did obviously believed that industrialization was entirely natural and entirely ordained by God. They also believed that they knew who God had intended to work in all those factories that were going to have machines powered by water and steam: women and children. Why? Because farmers would still be needed, even in an industrialized economy. Husbandry was man's work; so was building factories and inventing machines. Operating machines, however, was appropriate work for women and children, who could, after all, be put to such labor without diverting the essential (male) workers from agriculture or who could work at night when their assistance was not needed in the fields.

Thus in the early years of industrialization, some people thought that machinery was natural (proponents of industrialization) and some that it was unnatural (proponents of pastoralism), although both agreed that the domesticated rural landscape and the hand tools with which it had been subdued were natural. Both the proponents and the opponents of industrialization believed that their favored economic system was ordained by God. Both sides also agreed that the United States should be a democracy, that all citizens should have equal political rights, and that citizenship should be reserved for those who owned property. Because of the nature of their work neither employees nor tenants were thought to be capable of acquiring the basic political education or the basic political interests that virtuous citizens ought to have. Which was why, in the end, both sides agreed that factory work was not manly work: if there had to be factories, then at least let us only employ women and children in them since women and children do not also have to carry the responsibilities of citizenship. Thus in the years just after the founding of the United States, the concepts of God, of nature, of politics, of gender, and of social status were already associated with notions about technology, even before the word had come into general use.

Technology and Romanticism

Subsequently, in the years between 1870 and 1920, many ideas about technology were influenced by an intellectual movement that had begun somewhat earlier: Romanticism. Romanticism started in Germany in the 1780s and 1790s and then, in the early decades of the nineteenth century it spread first to Britain and, somewhat later, to the United States. The Romantics celebrated passion at the expense of reason; they preferred creativity to discipline, anarchy to order, spontaneity to practicality. Romantic heroes were poets, artists, and musicians, not scientists, lawyers, and politicians. The German Romantics were fascinated by nature, by which they very specifically meant raw nature, not the domesticated rural world: terrifying, grand, sublime, untamed, anarchic, spontaneous.

In Germany in the 1780s and 1790s, industrialization had not yet begun; the young men and women who adopted Romanticism there were rebelling against what they saw as the overly militaristic and hierarchical character of their society and the overly analytic and abstract character of the education that they had received. In Britain and the United States, however—where industrialization was well under way—Romanticism became, not surprisingly, a critique of industrialization. The industrial world enshrined everything that the Romantics despised: the parts (components) rather than the whole; machines rather than organisms; reason rather than feeling; science rather than poetry; profit rather than community; the future rather than the past; progress rather than tradition; means rather than ends; objectivity rather than subjectivity.

The Romantic impulse touched Americans during the nineteenth century in many different ways. In the early 1850s, when Henry David Thoreau retired from the hurly-burly of life in Concord, Massachusetts, to live for a year by the shores of Walden Pond, he was thinking and behaving in a way that was perfectly in tune with Romanticism, and opposed to industrialization. Thoreau hoped to observe nature closely and thereby get to know himself better. The trouble with life in Concord, according to Thoreau, was that the pace was too hectic; there was no time for self-examination, for contemplation, for repose. The pace was too hectic, as Thoreau saw it, because the tempo of life in Concord was set by the machine.

In Concord, Thoreau wrote, "the laboring man has not leisure for a true integrity day by day; he cannot afford to sustain the manliest relations to men. . . . He has no time to be anything but a machine."[6] Thoreau believed—along with many other Romantics—that when men worked with machines, they became like machines, and to be like a machine was to be inhuman and, as he says, *unmanly.* Thoreau also believed that when people used hand tools on nature they altered it, which was a bad thing to do, but at least justified by necessity. When they used machine tools on nature, however, they destroyed it, which was considerably worse, especially as it was justified only by profit, not by necessity. Machinery is destroying nature, Thoreau and other Romantics thought, and also destroying that which is best about human beings: their artistry, their creativity, their sensitivity.

Thus it was the Romantics who first began to worry about what mechanical technology was doing to *skill.* Focusing, as they did, on creative artists, the Romantics defined creativity as the ability to take an inspiration—an idea, something immaterial—and give it some kind of form, to give it material reality. Creative people, according to the Romantics, can be artists, writers, or composers, but they can also be craftspeople—carpenters, masons, stonecarvers, and weavers—anyone who is inspired to give form to that which is unformed. The familiar expression "arts and crafts" has its linguistic origins in the Romantic movement precisely because of this association between what the artist does and what the artisan does; indeed, one aspect of American Romanticism was the arts and crafts movement, which was popular between 1870 and 1920. Proponents of arts and crafts produced and purchased handmade paper, handbound books, handcrafted furniture, handthrown pottery—anything not made by a machine.

Skills, according to the Romantics, were what made people capable of creating; they were the means by which people breathed inspiration into words or notes or wood or paint. To the Romantics, being skilled meant being free. A free person is someone who has unrestrained use of his or her skills, a person who has what we would now call *freedom of expression.* And, according to the Romantics, only the free person is truly and fully human because being creative is the highest, most human, way to be.

The Romantic critique of industrialization, therefore, asserted that machines were robbing people of their skills, skills that had been developed over the course of centuries, that had been handed down from one generation to the next. The machines were also robbing people of opportunities to be creative; ultimately, this meant that machines were also robbing people of their freedom—turning free men into slaves—and of their humanity—turning men into, as Thoreau put it, machines.

Acceptance of Romanticism by Advocates of Industrialization

Students of intellectual history know that there is a common pattern in the history of ideas: when ideas are very powerful, they are sometimes adopted even by opposing sides in the same debate. Thus in the decades before and after the Civil War, some advocates of industrialization adopted some ideas of their Romantic opponents. We can see this pattern developing, for example, in the ways in which Americans began talking about inventors. Several successful inventors—Whitney, Fulton, Morse, Edison, and Bell— had literally became national celebrities, almost on a par with such political celebrities as Washington, Jefferson, and Lincoln. Poems applauding their achievements were published in newspapers; streets, towns, and babies were named after them. A surprising number of people talked and wrote about inventors using the language, and the concepts, of Romanticism. A contributor to *Scientific American* in 1849 wrote that "inventions are the poetry of physical science and inventors are the poets. . . . [T]he same spirit of inspiration dwells in both." Thomas Ewbank, Commissioner of the Patent Office (and precisely the sort of person a good Romantic would have detested), asserted in the 1850s that a steamship is a "mightier epic" than the Iliad. One author believed that inventors looked handsomer—by Romantic definitions of handsomeness—than other men: "It is notorious," he wrote, "that nearly all poets and philosophers, nearly all theologians, too, have something mean and little about them, all inventors something heroic and grand." Later in the century, another author struck the same Romantic note, referring to the process of invention as "the dreamings—the wakeful nightly dreamings . . . [that] create some ballad or produce some epic in machinery."[7] (See the painting on page 137.)

The same trend—acceptance of some Romantic ideas by proponents of industrialization—is apparent in the ways in which the people who liked industrialization, and who wanted it to continue, addressed the issue of deskilling. While accepting the negative things that the Romantic critique suggested about "mechanicalized men," defenders of industrialization began arguing, in the latter half of the nineteenth century, that industrialization would eventually liberate men, not enslave them. The connection between industrialization and liberation would be prosperity.

That connection was a major theme of many of the technological utopias—fictional accounts of societies in which technology has solved all of humanity's problems—that were written in the waning decades of the nineteenth century. Some of those utopias, such as Edward Bellamy's *Looking Backward: 2000–1887* (1888), were extremely popular, reprinted dozens of times; others, such as King Camp Gillette's *The Human Drift* (1894), were less well received (Gillette was the inventor of the safety razor and founder of the company that still bears his name)—but all the utopias were similar in equating industrial technology with prosperity and prosperity both with happiness and with democracy. They were also similar in the ways in which they linked industrial technology with gender and with politics.

In the future that modern technology will create, according to one utopian, "Our sanitary arrangements and lavatories are of the best. . . . Our roads are well paved; smoke, cinders, and ashes are unknown because electricity is used now for all purposes for which formerly fires had to be built." Nature will have been completely tamed by technology; even the climate will be completely controlled. Population centers will have been integrated into giant built environments, megalopolises, which will be linked to the natural world, most especially to farms, by excellent transportation facilities. Workplaces will be beautiful: "lofty, airy halls, walled with beautiful designs in tiles and metal, furnished like palaces." Work there will be, but it will be so pleasant and easy that it will be almost like play: "willing work, in fields fitted to the capacity of the worker." Everyone will want to work, but because of the work of machines, work will not occupy most of people's lives; they will have enormous amounts of leisure to do with as they please and they will spend most of their time educating themselves and pursuing creative activities. People will understand themselves to be "cogs in a great machine," and because they understand this, they will work in a "spirit of cooperation." The military will be abandoned (since everyone is cooperating at work and happy in leisure), and in its place there will be a democratic civilian army in which every member "no matter what station he fills . . . is a public servant." The most crucial criterion in choosing political leaders will be technical expertise since the only decisions that leaders will have to make will be about technical matters. "Permanent success will depend . . . on the civil engineer; not on the shrewd guesses of the so-called business man, but on the accurate knowledge of the manager." Citizenship—and technical expertise, in most of these utopias—would be the province of men only. Women would be entirely domestic and entirely lacking in technical expertise; they would happily submit to the rule of their fathers and husbands. As people will all love each other, there will be a "religion of humanity." "Our highest conception of a personal god," a citizen of one of the utopias says, "is embodied in the perfect man." Science will become a religion, and scientists will become its "new priests."[8]

The late nineteenth-century technological utopians, in short, envisioned a perfect world in which mechanical ingenuity (which the Romantics generally despised) would subdue nature (which the Romantics did not want to do) in order to bring prosperity and leisure, which would help ever larger numbers of men (but not women) to reach the Romantic goals of creativity and free expression. These technological utopias were, of course, fictions, but in the last two decades of the nineteenth century, they had real-life correlates in the economic and social movement called "scientific management."

Scientific management originated in the work of Frederick W. Taylor. Taylor was the scion of a very prosperous Philadelphia family. Refusing to go to college as his parents' wished, he had worked, first as a machinist and subsequently as a foreman, in a metal products factory. Appalled to discover that the production process in the factory was disorganized and inefficient and also that some of the people who worked for him deliberately set a slow pace for themselves (soldiering; see Chapter 8) Taylor had an inspiration that he thought would solve both problems. By analyzing the production process into its component parts, by carefully studying the time that an average worker took to perform each part, and by rearranging all the equipment that the worker needed so that it was easily at hand, Taylor believed a manager could figure out precisely how much an average worker was capable of producing under optimum conditions in an average day. Once this was determined, piece rate wages could be set that would provide a fair rate of pay to the worker, an incentive for the worker not to soldier, and a reasonable profit to the owners of the plant.

Taylor began publishing about scientific management in the 1880s, and by the turn of the century, he had developed quite a following. Much of his thinking (and much of what his movement wanted to do) would have appalled a true Romantic, but not all of it. Taylor and his disciples believed that when industrial work processes were redesigned and workers were paid piece rates, they would become like artisans of times past, mini-entrepreneurs acting in their own interests, managing their own time to suit their own interests; they would be skilled. Greater efficiency, scientific managers also thought, would make industrial work easier and pleasanter for the workers, who would then be able to realize the Romantic goal of expressing their individuality: being creative on the job, cutting loose from the mass of passive workers, increasing their wages and their status. According to proponents of scientific management, under such a system factory work would become manly work, and factory workers would understand what it meant to have an economic stake in their own society. Production would then increase, leading to prosperity for the workers; once people were freed of exhausting toil and constant worry about their finances, they would have enough peace of mind to think about important matters and enough leisure to shoulder their responsibilities as citizens.

Scientific management was an immensely popular movement for several decades, spawning what one historian has called an efficiency craze. Taylor's books became fundamental texts for the new profession of industrial engineering (later also called systems management) that began emerging in the first decades of the twentieth century, but his ideas spread far beyond the factories and construction sites where engineers were employed. Conservationists such as Gifford Pinchot, who became Chief Forester of the United States, and Theodore Roosevelt, who appointed Pinchot, were fond of speaking of the work that they did as "efficient management of natural resources." Progressive reformers carefully studied many aspects of urban life—collecting data, for example, on the number of people living on each square block or the amount of money spent to install every mile of sewer pipe—all in the hope of making the management of cities less corrupt (as Taylor's soldiering workers had been corrupt), less wasteful (as his disorganized shop floors had been wasteful), and thereby more efficient. Home economists, members of a profession that started at the same time as industrial engineering, undertook time studies of housewives in the hope of doing for housework what Taylor had done for factory work. Many home economists were also suffragettes; they believed that if the drudgery of housework could be minimized, housewives would be able to educate themselves and function as good citizens. Underlying all these various efforts was the utopian conviction that if work could be made more efficient, if government could be made more efficient, if households could be made more efficient, then, and only then, would the true promise of democracy be realized. All of these people recognized that, in the industrial society emerging during their lifetimes, most people were doomed to be employees of someone else, but they had a utopian and somewhat Romantic expectation that technology, and the technological frame of mind, would provide both leisure and affluence for Americans.

Technological utopianism and scientific management were responses to the reality of working conditions in industry and to the ongoing Romantic critique of those conditions. Some of that response was cast in terms that the Romantics had first popularized and connected: work, pride, creativity, artistry, self-esteem, skill, freedom, manliness, and independence. Where the Romantics thought that only artists and artisans (who worked with hand tools) could live out that connection, the late nineteenth-century advocates of industry thought that factory workers could, too, if only technology were allowed free rein and factories made efficient. Then, they argued, industrial technology would create prosperity and material comfort for ordinary people, a set of ideas that is sometimes referred to as the American Dream.

Technology and Art

The impact of Romanticism on the debate about technology and industrialization is also reflected in the history of American art. Even though cities were flourishing, factories expanding, and the railroad network growing denser, nineteenth-century American painters and sculptors seemed oblivious to these facets of industrialization; even more, they positively avoided the subject. The most Romantic of all American painters were the members of what is sometimes called the Hudson River School—Asher Durand, Albert Bierstadt, Frederick Edwin Church, and Thomas Cole—who flourished in the decades around the Civil War. Cole, Durand, and Bierstadt painted enormous canvases in which they depicted raw Nature at sublime moments: sunlight breaking through storm clouds was a favorite; so were vast mountain ranges and dense woodlands. Other nineteenth-century painters, usually called genre painters, were fond of depicting everyday life, but only the everyday life of the rural population: William Sidney Mount, for example, is famous for his depictions of rural life on Long Island; George Caleb Bingham for his nostalgic portrayal of trappers and boatmen on the Missouri and Mississippi Rivers; Eastman Johnson for his focus on the activities of farming families. In the latter decades of the nineteenth century, Impressionism rather than representational techniques dominated American painting; such American Impressionists as Mary Cassatt, John Singer Sargent, and Childe Hassam focused their attention almost entirely on scenes from suburban middle-class life.

In the first decades of the twentieth century, however, numerous American artists, devotees of the movement generically called Modernism, dis-

In the decades between the two world wars, many artists were inspired by the industrial landscape and by the promise of technology. *Pittsburgh* by Elsie Driggs, 1927. (Photograph Copyright © 1996: Whitney Museum of American Art, New York.)

covered the visual possibilities of machinery, of factory vistas, and of cities. Some early twentieth-century American painters began experimenting with precise lines and geometric compositions, while sculptors began using machine parts to design three-dimensional collages. Joseph Stella began painting factories and gas tanks around 1915; his most famous painting, *Brooklyn Bridge,* was completed in 1918. Morton Schamberg constructed a sculptural assemblage out of plumbing fixtures in 1918 and sardonically entitled it *God.* Charles Demuth began a series of paintings called *Machinery* in the 1920s and specifically compared one of his factory landscapes to the pyramids of old by entitling it *My Egypt.* Charles Scheeler, who had once concentrated on the geometry of barns, painted *Classic Landscape,* based on the automobile factory at River Rouge, in 1931.

Ironically, these artists who depicted industrial subjects understood themselves as the true American heirs of Romanticism. Stella, for example, had been born in Italy and had, after immigrating, wholeheartedly embraced the set of ideas that was earlier referred to as the American Dream: the notion that industrialization would improve people's lives. Stella thought that the industrial landscape was vastly more inspiring, and more humane, than the frontier landscape that had so enthralled the members of the Hudson River School. He wrote of New York, for example, as "massive dark towers dominating the surrounding tumult of surging skyscrapers with their gothic majesty sealed in the purity of their arches," and of the Brooklyn Bridge as "the shrine containing all the efforts of the new civilization, *America*—the eloquent meeting point of all the forces arising in a superb assertion of their powers, an Apotheosis"—perfectly Romantic ways of talking.[9] Thus the American modern artists of the early twentieth century thought of themselves as Romantics, all the while painting canvases and modeling sculptures that would have horrified the members of the Hudson River School, who also thought of themselves as Romantics.

Beginning in the 1880s, some Americans also began thinking of engineers as Romantic heroes. Walt Whitman may have started the process in the 1870s and 1880s when, in poem after poem, he used not just the medium most beloved of the Romantics but also their idioms and their ideas to laud the work of engineers.

> Singing my days
> Singing the great achievements of the present,
> Singing the strong light works of engineers,
> Our modern wonders, (the antique ponderous Seven outvied,)
> In the Old World the east the Suez Canal,
> The New by its mighty railroad spann'd,
> The seas inlaid with eloquent wires[10]

Two generations later popular writers began to join in Whitman's chorus. In the twenty years from 1900 to 1920, engineers were the main characters in dozens of books written for the edification of young men (for example, *Tom Swift and His Airship* and *Young Engineers in Mexico*) and equally many novels for adults. Engineers were portrayed, in these books, as both romantic and Romantic heroes. They are unusually tall, handsome, brave men. They work in dangerous places on monumental projects, and they do it under the sway of some visionary inspiration about the future—in much the same way that Romantic poets were inspired by Nature. "The idea took possession of his mind," one author wrote of his main character, who has just discovered a potential railroad route through Alaska, "He had begun to dream."[11]

Unlike Romantic poets, these twentieth-century engineer-heroes want to conquer Nature, not commune with it. Their battle is, in appropriately Romantic terms, Herculean: they never take on small tasks and they always work on some raw frontier. They always win, big, managing to turn raw Nature into pastoral nature every time. "I've won," the Alaskan railroad builder exults,

> I hold the keys to a kingdom. . . . I saw it in a dream, only it was more than a dream. . . . I saw a deserted fishing village become a thriving city. I saw the glac-

In Greek mythology, Prometheus was the god who defied the other gods to give fire to human beings. Romantics admired this myth and tried to emulate it. This advertisement, from *Colliers*, October 1922, depicts the chemical engineer as a Romantic hero. (*Colliers*, 1922; Courtesy DuPont Corporation.)

iers part to let pass a great traffic in men and merchandise. I saw the unpeopled north grow into a land of homes, of farms, of mining camps, where people lived and bred children. I heard the mountain passes echo to steam whistles and the whir of flying wheels.[12]

In these novels, there is always romance as well as Romanticism. The Romantics had idolized a particular kind of woman, a *natural* woman; unschooled, unsophisticated, domestic, demure, like "the coolness of the shadows, the scent of the damp earth, and the faint fragrance of the wild flowers," according to one novelist's description.[13] As the engineer-heroes always succeed in conquering nature, so, too, they always succeed in winning the natural girl. As opposites, they are attracted to each other: he is strong, she is weak; he is analytic, she is intuitive; he builds, she inhabits. Ultimately, he symbolizes technology and she, nature—although usually nature pastoralized rather than nature in the raw. As the romantic frame of mind developed over the course of the nineteenth century, it thus suggested a gender division with regard to mechanization very different from the one with which Thoreau had started out: a good society, according to the early twentieth-century version of Romanticism, was one that kept women in their homes (the *natural* place for a woman) rather than in the factories.

This Romantic vision of engineers also inspired some of the work of a prominent sociologist and social critic of the early twentieth century, Thorstein Veblen. Veblen, who had been born in the Midwest, came to New York in his youth and made his living as a writer, an editor, and eventually, a professor. Veblen believed, along with the Romantics, that people had what he called "an instinct for workmanship." That instinct cannot, however, be given free rein, according to Veblen, if people must work like animals; drudgery kills whatever instinctive pleasure people might find in their work. Engineers use their instincts for workmanship to create and build new technologies, which can liberate other people from drudgery. If the work of engineers has not yet succeeded in doing that for most Americans, Veblen argued in his last important book, *The Engineers and the Price System* (1923), it is only because businessmen and politicians, who do no productive labor, have perverted their creations and their intentions. Veblen's ideas took root in a political movement called technocracy. Technocracy attracted several hundred adherents in the 1920s; it was part of the political momentum that swept Herbert Hoover into office, the first engineer to serve as president of the United States.

Conclusion: The Cultural Meanings of Technology

Thus in the years when industrial society was developing, in the decades between 1870 and 1920, ideas about nature, gender, social status, politics, and skill were still prominently associated with ideas about technology. In the early decades of the nineteenth century, when factories were just beginning to dot the landscape, both the proponents and the opponents of industrialization had shared and linked certain ideas. Partly as a result of Romanticism and partly also as a result of greater experience with industrialization, some ideas had become more important than others and some terms had been linked to each other in different ways. Neither the Romantic opponents of industrialization nor the post-Romantic defenders of it focused, as their predecessors had, on God or on His Providence, but they all worried (as the opponents of the sphygmomanometer worried) about whether machine technology was natural. By 1900, many Americans had come to associate manliness with industry and engineering, something that might have been incomprehensible a hundred years earlier. Like their predecessors, both the opponents and the proponents of industrialization still believed that there were natural differences between men and women and that these differences were somehow connected to the place that each gender should occupy in society, but even proponents had ceased believing that factories were natural places for women and wheat fields the natural places for men. The Romantic opponents of industrialization disagreed with the post-Romantic proponents on the question of whether artisans or engineers and inventors should be celebrated as archetypes of creativity, freedom, and manliness, but they all agreed that there was some natural and necessary connection between working with your hands, being skilled, being independent, and being a good man. Some Americans had started to prefer raw nature to domesticated nature and some had even started to prefer cities and factories to the great outdoors in any of its manifestations, but all continued to justify their preferences in terms of what they considered natural to the human condition.

The history of ideas about technology is, like the history of all ideas, both very complex and very subtle. Objects, technologies, even technological systems—railroads and bridges, electric lights and fractional horsepower motors—carry, as a result of this complexity and subtlety, many levels of cultural meaning. As the story with which this chapter began, the story of the sphygmomanometer, was meant to suggest, those meanings can sometimes be more potent to people than the social and economic functions those objects and technologies and technological systems were designed to perform.

Notes

1. All these quotes, and the story of the sphygmomanometer, can be found in Hughes Evans, "Losing Touch: The Controversy over the Introduction of Blood Pressure Instruments in Medicine," *Technology and Culture* 34, no. 4 (1993): 784–807.

2. Both quotes in this paragraph are from Howard Segal, *Technological Utopianism in American Thought* (Chicago: University of Chicago Press, 1985), pp. 80–81.

3. This passage comes from "Query XIX: The present state of manufactures, commerce, interior and exterior trade," from Jefferson's *Notes on the State of Virginia* [1781], ed. William Peden (Chapel Hill: University of North Carolina Press, 1955), p. 165.

4. Jefferson to John Bannister, Jr., October 15, 1785, as quoted in Leo Marx, *The Machine in the Garden: Technology and the Pastoral Ideal in America* (New York: Oxford University Press, 1964), p. 131.

5. Tench Coxe, *A View of the United States of America* (Dublin: Wogan, 1795), pp. 12–13.

6. Henry David Thoreau, *Walden* [1854] (New York: New American Library, 1942), p. 9.

7. The quotations in this paragraph can found in John Kasson, *Civilizing the Machine: Technology and Republican Values in America, 1776–1900* (New York: Penguin, 1977), pp. 146–53; for the sources see Kasson's endnotes, p. 253.

8. All of the quotations in the paragraph above are from one or another of the twenty utopian novels discussed in Segal, *Technological Utopianism,* pp. 23–31.

9. Joseph Stella, "Brooklyn Bridge, A Page of My Life" (1929), as quoted in Irma Jaffe, *Joseph Stella* (Cambridge, MA: Harvard University Press, 1970), p. 57.

10. Walt Whitman, "Passage to India" [1868], in *The Complete Poetry and Prose of Walt Whitman,* ed. Malcolm Cowley (Garden City, NY: Garden City Books, 1948), p. 361.

11. Rex Beach, *The Iron Trail: An Alaskan Romance* [1913], as quoted in Cecilia Tichi, *Shifting Gears: Technology, Literature and Culture in Modernist America* (Chapel Hill: University of North Carolina Press, 1987), p. 119.

12. Rex Beach, *Iron Trail,* as quoted in Tichi, *Shifting Gears,* p. 121.

13. John Fox, Jr., *Trail of the Lonesome Pine* [1908], as quoted in Tichi, *Shifting Gears,* p. 124.

Suggestions for Further Reading

William Akin, *Technocracy and the American Dream: The Technocrat Movement, 1900–1941* (Berkeley, 1977).

Joseph J. Corn, *The Winged Gospel: America's Romance with Aviation, 1900–1950* (New York: 1983).

Samuel Haber, *Efficiency and Uplift: Scientific Management and the Progressive Era* (Chicago, 1964).

Samuel P. Hays, *Conservation and the Gospel of Efficiency: The Progressive Conservation Movement, 1890–1920* (New York, 1969).

John F. Kasson, *Civilizing the Machine: Technology and Republican Values in America, 1776–1900* (New York, 1976).

Carolyn Marvin, *When Old Technologies Were New: Thinking about Electric Communication in the Late Nineteenth Century* (New York, 1988).

Leo Marx, *The Machine in the Garden: Technology and the Pastoral Idea in America* (New York, 1964).

Daniel Nelson, *Frederick W. Taylor and the Rise of Scientific Management* (Madison, WI, 1970).

David E. Nye, *Electrifying America: Social Meanings of a New Technology* (Cambridge, MA, 1990).

Daniel T. Rodgers, *The Work Ethic in Industrial America, 1850–1920* (Chicago, 1978).

Howard P. Segal, *Technological Utopianism in American Culture* (Chicago, 1985).

Cecilia Tichi, *Shifting Gears: Technology, Literature and Culture in Modernist America* (Chapel Hill, NC, 1987).

Alan Trachtenberg, *Brooklyn Bridge: Fact and Symbol* (New York, 1965).

Rosalind Williams, *Notes on the Underground: An Essay on Technology, Society and the Imagination* (Cambridge, MA, 1990).

Julie Wosk, *Breaking Frame: Technology and the Visual Arts in the Nineteenth Century* (New Brunswick, NJ, 1992).

III

TWENTIETH-CENTURY TECHNOLOGIES

THIS LAST SECTION EXAMINES FOUR TECHNOLOGICAL SYSTEMS that have dominated twentieth-century history: automobiles, and their attendant roads and fuel sources; aircraft, spacecraft and also rockets; electronic communication devices, from wireless telegraphy to personal computers; and finally, biotechnologies, new foodstuffs, medications, and contraceptives. All of these systems had their origins in previous centuries and none of them are solely American. All of them have shaped, and been shaped by, the history of the twentieth century. Indeed, when we call ours the century of technology, when we wonder over the rapidity of technological change, when we rail against the ways in which technology has changed our lives, we are usually thinking about one or all of these technological systems.

The last chapter of this section introduces a new concept: *technoscience*. In the twentieth century, it has proved very hard to distinguish between technology and science. For most of this century, technological development has been conducted using scientific methods and scientific research has been conducted, and funded, for technological reasons. Chapter 13 explains that technoscience has been prevalent in agricultural research, which produces new things for us to eat, and also in medical research, which produces new diagnostic, therapeutic, and reproductive technologies. Chapter 11 (about aerospace) explores the funding relationship that developed between the government, businesses, and universities in order to use technoscience for military ends. Implicitly, Chapters 10 (about automobiles) and 12 (electronic communications) reflect the results of technoscience as well, for every electronic device we use is based on discoveries made by such nineteenth-century scientists as Maxwell and Hertz, and the internal combustion engine was created and improved by people who either knew a lot or were hoping to learn more about thermodynamics. The transformative power of twentieth century technology is derived from the investigative and creative power of technoscience.

Blessing or Curse?

But to what ends have people used this power? For good or ill? Enhancement or destruction? Pleasure or torment? These are the ultimate questions that a history of technology must ask.

People are complex, so are the technological systems that they have created—and so is history. Because history is complex, there can be no simple answers to these questions. Chapter 10 uses the story of the sorcerer's apprentice to make this point about automobiles: that in the process of solving all kinds of transportation problems that Americans wanted solved, the automobile created additional problems—pollution, congestion, environmental degradation—that those very same people did not want to create. There doesn't seem to be any way out of this impasse; it is not unique to the twentieth century. Indeed, the impasse may be part of the human condition; the story of the sorcerer's apprentice is, after all, very ancient.

Chapter 11 is about aircraft and rockets; it illustrates another, very similar, paradox. In all times and all places, military defense—the efforts that societies make to protect themselves from assault—have been a major stimulant of technological change. As a result, and whether we like it or not, some of the technological systems from which we derive the most pleasure are the very same systems that have been designed, and often used, to torment the lives of other people.

The chapter on electronic communications is focused on a related question, "Who gets to control the powerful systems that technoscience helps create?" The answer is ironic. In the United States in the twentieth century, brilliant inventors, powerful business executives, and confident politicians have all tried, and failed, to control electronic communications. Something about the nature of electromagnetic waves, something about the hams and nerds and hackers who have been fascinated by them, something about American politics, and something about the mechanisms of the free market have all conspired against giving any individual or organization ultimate control of communications.

Finally, Chapter 13 explores some surprising examples of the *unintended* and *unexpected consequences* of technological changes. The history of twentieth-century foods and medications reveals that even as mundane a technology as hybrid corn—summer's delectable treat—has social and ethical implications that none of its creators had anticipated, implications that touch intimate aspects of everyone's life. One crucial lesson of this chapter—and this book—is that because of technoscience we cannot trust the experts to make decisions about these issues because none of them is disinterested. As members of the species *homo faber,* we are all, for good or ill, enmeshed in technological systems from which we cannot escape—and about which we need to be informed.

10

Automobiles and Automobility

THE STORY OF THE SORCERER'S APPRENTICE is a story about the paradoxes of technology; although the story is about a mop and although it is set in the Middle Ages, it could just as well be about the automobile in the United States in the twentieth century.

Like all apprentices, the sorcerer's apprentice has some menial jobs, like mopping the floor, to perform. One day while the sorcerer is away, the young man tries to find (against his employers' expressed instructions) a special incantation that will command the mop to clean the floor itself—and he succeeds. Within minutes, the mop is off to the well to retrieve the water that it will need; a few minutes more, and the mop has finished cleaning.

The apprentice is ecstatic, relishing a vacation for himself. But his pleasure soon turns to apprehension. The mop has returned to the well; instead of resting after its labors, it has retrieved another bucketful of water and has started mopping once again. The apprentice begins to worry: will he need another special incantation to *stop* the mop? Apparently, he will; he searches frantically through all the sorcerer's dusty tomes but to no avail. Meanwhile, the mop has been diligent; bucketful after bucketful has been applied to the floor; the water level is slowly rising. Unless the apprentice can stop the mop, the shop will soon be flooded. He tries breaking the mop in two, but then both halves of the mop continue the work; the water is now rising twice as fast! What to do? Will the wonderful device destroy the sorcerer's shop and the apprentice's future?

Like the sorcerer's mop, the automobile is a desirable technology that has gotten out of hand. As its numbers have doubled, and then doubled again and again, a flood of problems have been created: traffic jams, fatal accidents, smog. Dozens of social policies were created in an effort to find the incantation that would stop the nation's favorite technology from destroying the nation's environment—and its future. To foreigners, the automobile has come to represent everything quintessentially American; to many Americans, the automobile has come to represent both the blessings and the curses of technology.

Who Invented the Automobile?

People often ask historians of technology, "Who invented the automobile?" There is no easy answer to that question (except in the negative form: "*Not* Henry Ford") because the automobile is not so much a technology as a technological system, and not a fixed technological system but one that has changed over time. Many nineteenth-century inventors took up the challenge of creating a self-propelled road-running vehicle, an engine-powered substitute for horse drawn wagons and carriages. Like Oliver Evans (whose "Amphibious Digger" is discussed in Chapter 4), several early nineteenth-century inventors patented designs for "auto-mobiles" ("self-movers") that were powered by steam engines, but these turned out to be little improvement over the horse. The fuel was heavy and bulky, and the engines needed lengthy warm-up times. Later in the century, there was a flurry of interest in electric vehicles, and several companies actually went into commercial production of electric cars, but the batteries then available were very heavy (some weighed over a ton) and had to be frequently recharged or replaced.

By midcentury, various inventors had come to the conclusion that a successful self-propelled vehicle would have to be powered by some kind of internal combustion engine that used a liquid or gaseous fuel. A steam engine is an *external* combustion engine: the fuel is burned in a container (the boiler) which is separate from the engine and a gas (the steam) communicates energy between them. The science of thermodynamics, which had developed in the first decades of the nineteenth century, suggested that an *internal* combustion engine, in which the explosion of a flammable fuel powers the drive train directly, would be much more efficient because there would be no need for a mediating fluid.

An internal combustion engine presented, however, numerous problems: how to control explosions within the confined space of a cylinder so that the shaft driven by the piston moved smoothly; how to consistently achieve just the right mixture of air and fuel in the cylinder; how to ignite the mixture and, having ignited it once, repeatedly ignite it at just the right split second of the piston stroke. Such practical questions proved intractably difficult for many years; the first commercially successful internal combustion engine was not developed until 1860, several decades after scientists had first predicted that it would be a more efficient type of engine.

The first workable internal combustion engine was developed by a Belgian mechanic, Etienne Lenoir, but Lenoir's engine (which could develop about two horsepower and weighed several hundred pounds) could not possibly have pulled a vehicle. The first step in the direction of a more powerful engine was taken in 1876 by a German inventor, Nicholas Otto. Otto's engine, which is the direct ancestor of all current automobile engines, achieved greater power by compressing the fuel–air mixture before it was ignited, a four-stroke cycle for the piston. The first (up) stroke drew in the mixture; on the second (down) stroke, the mixture was compressed; the third stroke, which was the power stroke, resulted from ignition of the compressed mixture; the fourth stroke exhausted the spent gases from the cylinder.

Many people realized that a fast running, lightweight Otto-type engine with a reliable ignition timing mechanism would be able to power a roadworthy vehicle. Many inventors butted their skills against this task, and by the 1880s, three men had succeeded, all in Germany. Gottlieb Daimler and Wilhelm Maybach had together developed an engine that could run at 600 to 900 revolutions per minute by 1883; at about the same time, Karl Benz, working independently of Daimler and Maybach, developed an electric ignition system. As a result of this pioneering development work, motor vehicles were already in commercial production in France, Germany, and Britain in the early 1890s; aside from a few steamers and electrics, most were based on some combination of the devices that Daimler, Maybach, and Benz had patented.

Americans were latecomers to automobile production, and most of the work of the early American vehicle builders was entirely derivative of what had been accomplished in Europe. The very first American horseless carriage seems to have been the brainchild of Charles E. and J. Frank Duryea, who were bicycle mechanics. The Duryea brothers copied a published description of Benz's automobile and succeeded in running it on the streets of their hometown, Springfield, Massachusetts, in September 1893. A trio of engineers and mechanics—Elwood Haynes, Edgar Apperson, and Elmer Apperson—put together an automobile that they tried out in Kokomo, Indiana, in 1894. Hiram Percy Maxim, son of the inventor of the machine gun and a graduate of MIT, watched an Otto engine at work and then built himself a copy, which he tried attaching to a tricycle in 1895. That year, on Thanksgiving Day, the *Chicago Times Herald* sponsored the first American automobile race (there had already been several in Europe), which was won by J. Frank Duryea, who managed to cover the 55-mile, snow-covered course at an average speed of 8 miles per hour.

The crucial turning point for the American automobile industry seems to have come in the two years between 1895 and 1897. A trade journal, *Horseless Age*, began publication in November 1895. By the end of the year, the editors had counted roughly three hundred companies or individuals who were building and testing experimental vehicles in the United States, and three department stores had begun selling imported European cars. In 1896, the Duryea brothers actually sold their first vehicle and produced twelve more of the same design. "All over the country mechanics and inventors are wrestling with the problems of trackless traction," the editors of *Horseless Age* remarked, "all signs point to the motor vehicle as the necessary sequence of methods of locomotion already established and approved. The growing needs of our civilization demand it; the public believe in it, and await with lively interest its practical application to the daily business of the world."[1]

These early American companies were essentially assemblers rather than manufacturers. They purchased parts from suppliers and assembled vehicles, one by one, in barns and sheds and warehouses: a body from a local carriage maker, an engine from an importer in New York, wheels from a bicycle manufacturer, tires from a rubber plant in Akron. Very little capital was required and no fixed plant was needed, parts were generally bought on credit and cars were sold to dealers or customers for cash. A few early manufacturers invented new devices (for example, James H. Packard, who was an engineer, went into business in Warren, Ohio in 1900, after having invented the H-slot gearshift), but most usually purchased patented devices or manufactured devices after paying license fees to the patent holders.

By 1899, the first year for which the *United States Census of Manufactures* compiled separate statistics on the automobile industry, roughly thirty companies had entered into commercial production of motorized vehicles and among them had produced 2,500 vehicles. A year later, ALAM (the Association of Licensed Automobile Manufacturers) had been formed to administer the sharing of patent licenses. Colonel Albert Pope, who had made a fortune manufacturing bicycles, turned his company to the manufacture of automobiles, declaring that his bicycle agents throughout the country were "fairly howling" for automobiles to meet an "enormous demand."[2] By 1910, some 458,500 automobiles were registered in the United States, and the industry had already jumped from 150th to 21st in the value of its product—an unprecedented rate both for growth and for diffusion, especially for an item that was, even at the bottom of its price range, expensive. In the United States, the automotive age and the twentieth century arrived more or less in tandem.

By 1910, it was clear that there were two ways of manufacturing cars and two rather different markets for them once they were manufactured. One way was to hire skilled mechanics and carpenters (much of the car body in those years was made out of wood, just as carriages had been), who would, as a team, assemble one vehicle at a time. The cars that resulted from this form of manufacture were high-priced and fairly heavy, touring cars, which appealed to economically comfortable people, the kinds of people who might once have kept a horse and carriage, who might be able to afford a chauffeur (whose job it was not only to drive the car but to clean and fix it). Touring cars could be designed precisely to a customer's specifications. The other way was to develop and adapt the techniques of mass production (which had already been used—see Chapter 4—in the manufacture of such wood and metal objects as clocks, guns, sewing machines, and bicycles) to produce a car that was at once lighter in weight and lower in price, a standardized car for the mass market, a car that would appeal to all kinds of people—who once might have owned a horse and buggy or a horse and wagon or no horse at all.

Henry Ford and the Mass-Produced Automobile

Henry Ford is generally credited with developing the first high-quality mass-produced car intended for the mass market, which is why so many people think of him as having been the inventor of the automobile. What Ford pioneered was not the car itself, but a new way to make cars—and many other things. This new method of assembly became the base from which the American automobile industry grew, until, for a time, it was the central industry of the American economy; it is also the base on which Ford's fame properly rests. What Ford invented was not the automobile, but the characteristically American automobile.

Ford began tinkering with automobiles in 1896 and developed something of a reputation as a designer and driver of racing cars. After two unsuccessful attempts to go into business as an automobile manufacturer, he finally managed to form the Ford Motor Company in 1903: he had one financial backer, a dozen or so workmen, an assembly plant that measured 250 by 50 feet, and paid-in capital of $28,000. Ford was committed to the production of light, low-priced, high-quality cars. The Model N, which he brought out in 1906, cost $600. The Model T came out two years later; it cost a bit more ($825–$850), but could generate twenty horsepower, which was unprecedented for a car in its price range. Ford advertisements claimed that "No car under $2000 offers more, and no car over $2000 offers more except the trimmings"—and they were right. Demand for the Model T was enormous, and as a result, Ford's staff designed and built a new plant for producing it—sixty acres in Highland Park, Michigan.

Before 1910, Ford had been able to build high-quality low-cost cars by using interchangeable parts and by putting a lot of money into his tool department, where specialized machine tools were created by very highly skilled mechanics. Overhead conveyors were used to transport particularly heavy items and Ford production engineers were in the habit of using time and motion studies (see the discussion of Taylorism in Chapter 9) so as to relocate tools and workers, making production more orderly and therefore more efficient. The Highland Park plant, when it opened, incorporated all of these innovations and more; it was well lighted and well ventilated.

Between 1912 and 1914, however, the Highland Park plant was modified to incorporate continuous conveyor belts at arm level that could move all the parts to be assembled. The moving assembly line had been conceived by a team of Ford engineers back in 1908, but had Ford waited for all the experiments that were needed to make it work effectively, he would have had to delay introduction of the Model T, which he did not want to do. So he decided to introduce the assembly line piecemeal. (Ford later claimed to have invented the basic idea of the assembly line himself. This is unquestionably not true; both the idea and the tedious work needed to make it a reality came from a team of his engineers. Ford encouraged them, approved their innovations, and made the decision to invest company resources both in development and in construction.)

By the summer of 1913, magnetos, transmissions, and motors were being constructed on assembly lines. Within a few months, production of these subcomponents threatened to swamp final assembly of the chassis, and so that was converted to assembly-line production also; between October and December 1913, the time required to assemble the chassis fell from 12 hours and 30 minutes to 2 hours and 40 minutes. On February 27, 1914, a new line, built on rails, was constructed for final assembly of the dashboard, the front axle, and the body: the Ford assembly line was complete. "Every piece of work in the shop moves," Henry Ford remarked a few years later. "It may move on hooks or overhead chains . . . it may travel on a moving platform, or it may go by gravity, but the point is that there is no lifting or trucking of anything other than materials."[3]

"The point" was considerably more than that, as Ford knew perfectly well. Assembly-line production resulted in substantial cost savings, which were partially passed on to customers in the form of lower prices; having come on the market at about $850 in 1908, the price of a Model T had dropped to $360 by 1916 (and this during the inflation caused by the outbreak of war in Europe) and to $290 by 1927 (also an inflationary period). Ford's cars had been popular from the very beginning because they were relatively cheap and remarkably easy to repair. With assembly-line production, they became even cheaper, and the Ford Company began to dominate the automotive market. Just under 6,000 Model T's were sold in 1908, just over 577,000 in 1916.

Other automobile companies were quick to realize that they were going to have to cut prices in order to compete; cutting prices meant that they, too, would have to build assembly lines. Within a decade, assembly-line techniques became standard in the auto industry, forcing many small manufacturers to close since they were unable to afford the high cost of retooling. The end result, by 1930, was that Ford's manufacturing technique had changed not only his own business, but the entire automotive industry: fewer and fewer companies were making more and more automobiles, with greater standardization, at lower and lower cost.

But even that isn't the whole of the story. The assembly line made it possible for Ford to hire a different kind of worker, replacing skilled craftsmen with unskilled laborers (see Chapter 8). Assembly-line work was simple and repetitive; virtually anyone could learn to do it successfully in a matter of hours. In order to retain workers and compensate for the awful monotony of the labor, Ford began offering selected workers $5 per day and an eight-hour day in 1914, a far higher wage rate in those days than that available to unskilled workers. Ford wanted to price his cars at a level that every worker could afford and to pay his workers at a level at which they could become his best customers. The Ford Company hired large numbers of immigrants from southern and eastern Europe (where they were accustomed to earning six cents an hour for equally mind-dulling work), large numbers of African Americans (by 1923, some 5,000 black men worked for Ford, more than worked for any other American corporation), a substantial contingent of ex-convicts, and in 1919, over 9,000 disabled men (amputees, the blind, the deaf, the epileptic, the mentally retarded, and the tubercular). On the other hand, Ford would not hire workers who had any connections to, or interest in, radical political associations or labor unions—and he created a special unit, the Sociological Department, to weed out such potential troublemakers. Some people hated Ford for his labor policies (he also created a company police force to spy on and intimidate his workers), but other people were grateful for the employment that his company offered.

After 1920, "Fordism"—the production of inexpensive goods by assembly-line methods—spread throughout American industry. Like the automobile itself, Fordism is a blessing when looked at from one perspective and a curse when looked at from another. It is not, however, the only contribution that the automotive industry made to American industry in general: there is also "Sloanism."

Alfred P. Sloan and the Mass-Marketed American Automobile

After he graduated from MIT Alfred P. Sloan, Jr. had gone to work for the Hyatt Roller Bearing Company, which manufactured a tapered roller bearing used by many automobile makers. The inventor of the roller bearing and the founder of the company was, as it happened, an indifferent businessman, and Sloan eventually became the president and co-owner of the business. One of Sloan's most important customers was Henry Leland, founder and president of the Cadillac Motor Car Company. In 1910, Leland had sold his company to the General Motors Company, which was then headed by a wealthy financial speculator, William C. Durant. Durant had formed General Motors in 1908, hoping to make it a multicompany combination of both car manufacturers and parts suppliers. Its first several years had been very rocky, but in 1916 Durant purchased two manufacturers, Buick and Chevrolet, and two crucial parts suppliers. One of these was Delco (the Dayton Electric Company), which made the electric ignition systems patented by its president, Charles F. Kettering. The other was the Hyatt Roller Bearing Company, owned and managed by Alfred P. Sloan.

In 1918, General Motors grew even larger when Durant purchased the Fisher Body Company, the leading manufacturer of automobile bodies; a year later, he created the General Motors Acceptance Corporation to finance automobile sales. He also purchased a small manufacturer of electric refrigerators, the Frigidaire Company. One year later, during a short downturn in the nation's economy, Durant was relieved of the presidency of General Motors, having hopelessly overextended his own, and his company's, capabilities. A few years later, in 1923, Sloan became the leader of the massive combination that Durant had assembled.

During his first decade at the helm of General Motors Sloan pioneered organizational and marketing innovations that were to have as much impact on the American economy—and on American culture—as Ford's assembly line. First, he reorganized General Motors—which under Durant's stewardship had been essentially ungovernable—along methodical, almost military lines. The production divisions, both for parts and for finished automobiles, were given virtually complete autonomy, except that their heads met together, for overall policy-making decisions, in an executive committee. Certain nonproduction functions were centralized: separate agencies were created for research (Kettering became the chief of the General Motors Research Division), for advertising, and for product planning. These central agencies had an advisory relationship with the production divisions; if conflict arose (which it frequently did), crucial decisions could be made only by the president, with the advice of the executive committee.

Sloan's decentralized organizational scheme was the exact opposite of the dictatorial way in which Henry Ford controlled the company that bore his name—and it worked exceedingly well. Within three years, General Motors had overcome Ford's considerable lead in the automotive market; in 1927, for the first time, a division of General Motors, Chevrolet, sold more cars than the whole of Ford. Within a few years, other very large companies had also adopted Sloan's management style, and it has since become characteristic of many of the most successful multidivisional and multinational American companies. By the time Sloan retired from the presidency of General Motors to become chairman of the board in 1937, GM was the largest privately owned manufacturing company in the world, and its various divisions had come to dominate the American automobile industry.

If Sloan's management policies were the opposite of Ford's, so too, were his marketing strategies. Stubbornly adhering to his original vision, Henry Ford made one car (until 1927, the Model T; between 1928 and 1938, the Model A), almost all of them in one color (black) for one basic market. Under Sloan, General Motors made several cars for several different markets, with Chevrolet at the bottom of the price range and Cadillac at the top. Sloan also pioneered the notion of an annual model change, with new models carefully designed so that one year's model could be distinguished from another; customers were also given a choice of colors. Toward the end of the 1920s, Sloan also had come to believe that the seller's market in automobiles was a thing of the past and that it would be necessary to pay very close attention to the retailing of his company's products. As a result, General Motors worked very hard at establishing good relations with its dealers, offering them, for example, discounts as high as 24 percent, when Ford dealers were getting only 17 percent.

By the end of the 1920s, the American automobile industry had settled into the form that it would retain for most of the rest of the century. A few huge corporations had emerged out of the several hundred that had started in the business thirty years earlier. Ford had achieved prominence first on the basis of its decision to build a low-priced car for the mass market and then on its development of assembly-line manufacture. General Motors had been created as a speculative venture by the purchase of several assemblers and suppliers, and it had succeeded by innovations both in management and in marketing. By 1950, the "Big Three" (Walter P. Chrysler created the multidivisional company that still bears his name in 1925), commanded more than 90 percent of the American market—and a goodly portion of the overseas market as well.

The characteristically American automobile had also taken shape by the end of the 1920s. A product of the assembly line, this automobile was, generally speaking, cheaper than its European cousins. It was also easier to repair because it was more likely to have been made with interchangeable parts. Its engine was larger (in many European countries, car purchasers paid a tax based on the horsepower rating of the engine) and, as a consequence, its maximum speed was higher. The shift from four to six cylinders began in the 1920s; from six to eight in the 1930s. In addition, American automobiles were more likely than European cars to be made entirely out of metal and their bodies tended to be roomier than those made in Europe, perhaps because the price of steel was lower in the United States than it was in Europe. Hypoid gearing (introduced by the Packard Motor Company in the late 1920s) made it possible to begin building low-slung cars (which were less likely to flip over on curves). All cars had four-wheel brakes by 1930, and the more expensive models had hydraulic brakes. Automatic transmissions finally appeared in the late 1930s, after more than fifteen years of developmental work based on earlier patents for the epicyclic gearbox, the torque converter, and hydraulic coupling.

This characteristically American automobile was (and is) the result of creative engineering work carried out by many people, working in many different institutional settings over the course of several decades. Early in the twentieth century, assemblers acquired new technologies either by purchasing them from a manufacturer (who was usually the original inventor) or by purchasing the manufacturer. Once Ford started designing new machine tools and new assembly lines, and once General Motors had created its Research Division, all the larger companies employed highly trained engineers and skilled craftsmen to search through the technical literature and conduct laboratory and field experiments designed to solve specific automotive problems. Thus after 1920, most innovations in the mechanics or design of American automobiles were the result, essentially, of managed development, rather than invention.

Automobility and the Road Sysetm before 1945

The American people welcomed the automobile enthusiastically, putting the sorcerer's mop to work for them without much more than a moment's hesitation. By 1920, when reliable national statistics first became available, there was already one registered automobile for every thirteen Americans—and by 1930, the figure had risen precipitously to one for every five.

The very first automobile owners were relatively wealthy people who liked driving as a sport, but within a very short time people lower down the economic scale were putting the motor car to work. Doctors purchased automobiles to visit their homebound patients; housewives learned to drive to run errands; farmers wanted cars and trucks to haul produce to market, then they wanted tractors to pull plows and harvesters across their fields; retailers used cars to make deliveries to customers; rural school districts began buying buses to transport children to school.

Motorcar drivers very quickly became aware—as bicycle riders had been a decade or two earlier—that the nation's roads were not equipped to handle new kinds of traffic. Since the advent of the railroad seventy-five years earlier, very little effort had been expended to maintain, let alone extend, either the local or the national road systems. By 1900, the National Road between Washington and St. Louis, most of which had been built before the Civil War, had fallen into disrepair; rural farm to market roads were unpaved (manageable, although not in all seasons, with a horse and wagon, but murderously destructive to the automobile); city streets, paved with cobblestones, were congested with slow moving pedestrians and horse-drawn vehicles.

Local automobile clubs, voluntary associations of owners, began forming early in the twentieth century (they created a national association, the Automobile Association of America, the AAA, in 1905), and one of their first items of business was to pressure city, state, and national governments to improve the roads: to widen city streets, to smooth cobblestones with asphalt, to build limited-access highways with concrete road surfaces, to pave existing rural roads.

Progress was slow, however; road building and road paving projects were expensive, and in the years before 1932 governments were unaccustomed to such large-scale expenditures for public works. The first federal Road Aid Act was passed in 1916 and another followed in 1921; both allocated monies to state legislatures so that local authorities could build new roads and pave old ones. The highways that were funded under these two acts (among them were the two north–south roads labeled U.S. 1 and the east–west roads labeled U.S. 40 and U.S. 66) were a hodgepodge created by the location of older roads and the interests of local politicians. These roads meandered from town to town; the flow of traffic was repeatedly interrupted by sharp curves, narrow roadbeds, crossroads, and traffic signals.

During the 1920s several states began to tax gasoline sales in order to raise additional revenue for the construction and maintenance of roads, but it was not until the Depression years, when the federal government began massive payments to the states for public works projects, that new roads were designed and built specifically to accommodate automotive traffic. These were limited-access roads (built so that there were very few crossroads at grade level) with fewer sharp curves than the older roads had had. In New York City, the elevated West Side Highway; in Westchester County, New York, the Bronx River and Taconic Parkways; in Los Angeles, the Arroyo-Seco Parkway; in Pennsylvania, the Pennsylvania Turnpike—these are all examples of Depression-era roads, many built with funds provided by the Works Progress Administration (WPA).

As admirable as these new roads were, in 1940 they amounted to only a tiny fraction of the total road mileage in the United States, slightly more than half of which was still unpaved at the outbreak of World War II. In those years, Americans used their cars principally for local driving—to get to work, to run errands, to visit relatives, to make deliveries, to go out on dates. High-speed driving over long distances traversed by express highways—the kind of driving that is considered routine for most Americans today—did not arrive until the decades immediately after the end of the war.

During the 1920s, some people had already begun to notice that automobiles were creating some unanticipated problems, but they were also convinced that with some coherent planning and some modest expenditure of funds most of the problems could be solved. At the turn of the century, most of the older American cities had been extremely crowded places; with trolley cars on tracks running down the middle of the streets, horse-drawn wagons and carriages on either side of the street, pushcarts at the curb, and pedestrians occupying many of the spaces in between. Congestion was often a very serious problem. Traffic jams caused unnecessary expenses and considerable ill temper. Between the squealing of the streetcars and the rumbling of the wagon wheels, noise levels were sometimes unbearable. So were pollution levels—as hard as it is to believe this today. Horses urinated and defecated on the streets, and when they died, their bodies were often left to rot at curbside. In New York City, one reformer estimated, 2.5 million pounds of manure and 60,000 gallons of urine were deposited on the streets of the city *every single day*; in the course of twelve months, 15,000 dead horses had to be carted away at city expense. On top of all this, accidents were frequent; adults and children were regularly hit by streetcars or trampled by horses hooves or run over by wagon wheels.

Some politicians and city planners thought that the automobile would solve the problem of urban congestion. Unlike streetcars, the new motorcars did not operate on fixed rails, which meant that they could use all the streets of a city instead of just a few, and they could pass one another at will. Urban planners also thought that motorcars would enable more people to move their residences into the suburbs while still holding jobs in town. The streetcar companies had helped begin the exodus to the suburbs by extending their lines from the central cities into previously undeveloped districts, but in streetcar suburbs, the only useful residential property was on or within walking distance of the rail line. The automobile would make it possible to develop suburban property that had previously been inaccessible—to fill in, as it were, the spaces between the spokes of the streetcar lines. As more and more people moved out of the center of the cities, planners thought that congestion would ease and the accident rate would fall. In addition, pollution derived from horses would, of course, disappear; although the fumes from automobile exhausts smelled noxious, they had the virtue, at least, of dissipating quickly—no one in those very germ-conscious days thought that the fumes posed anything like the threat to human health that the feces and rotting bodies of horses did.

By 1920, some of these predictions had already begun to seem fantastical. Traffic congestion in many of the nation's central cities was getting worse, not better. When not in use, cars had to be parked somewhere, and if that somewhere turned out to be at curbside, then streets became impassable. Automobile suburbs were indeed developing, but suburbanites were still populating the central cities during the daytime working hours and were now bringing their cars with them. One traffic expert calculated that automobiles required twenty times as much street space per passenger as the streetcars: "It is a good deal as if our ladies and our men, also, wore a hooped skirt arrangement ten or twelve feet in diameter and went through the sidewalks."[4] And traffic fatalities continued to increase. Automobile manufacturers seemed to worry more about styling than about safety and cars were consistently built with the capability of operating at much higher speeds than permitted by law. In 1924, automobile accidents accounted for 23,600 deaths (of which 10,000 were deaths of children), over 700,000 injuries, and close to a billion dollars in property damage.

A traffic jam in midtown Manhattan in 1917. Notice the electric streetcar in the middle of the street and the horse-drawn cart in the foreground. (Reproduced from *Motor Magazine* © 1917 by permission of the Hearst Corporation.)

Despite these problems, Americans remained enamored of automobiles. In the 1920s, numerous social prophets were looking forward to the day when virtually every American would be able to travel easily and independently wherever she or he wished to go, at high speeds, in a privately owned vehicle, at times and along routes of his or her own choosing. In the prosperous 1920s, it was fairly easy for people to assume that, when that day came, the country's roads would be redesigned and its factories, offices, and residences relocated so that individual ownership of automobiles would become not a luxury or a convenience but a necessity.

In the 1920s, no one was prophetic enough to have predicted that the Depression and another world war would intervene to delay the automotive future. When the economic crisis began, the rate of increase in automobile ownership slowed precipitously—and it collapsed altogether during World War II. Many of the smaller automotive manufacturers went out of business; only the giant companies had pockets deep enough to survive. Then between 1940 and 1945, virtually all the manufacturers that had survived the Depression converted from civilian to military production. Household incomes increased significantly during the war, but since the supply of consumer goods, including automobiles, was severely restricted, most people saved their money. At least temporarily, unexpected events had conspired against the enthusiastic predictions of the 1920s; the automotive age was peeking over the horizon in the 1920s, but it did not fully arrive until the 1960s.

Automobility and the Road System after 1945

Even before peace was declared in 1945, politicians and industrialists had been laying plans for the postwar reconstruction of the economy; one cornerstone of those plans had been the assumption that there would be a seller's market for automobiles for quite some time. The assumption turned out to be correct. Production figures climbed steadily after the war, and sales kept pace with production, despite the fact that prices were also climbing. In 1940, there had been roughly 27 million registered passenger cars in the country; by 1950, that figure had almost doubled; a decade later, it had trebled. By 1970, there were more cars than there were households in the United States, and in Los Angeles in that year, there were more cars than there were people. The promise of almost universal ownership had finally been fulfilled.

Postwar American cars were enormous; they were bigger than the cars built before the war, bigger than the cars being built in Europe (and later Japan) in the same decades. Engine capacities increased with every passing model year: so did compression ratios, horsepower ratings, wheelbase lengths, body weights, interior volumes, and potential speeds. The intermediate-sized cars of 1958 were larger than full-sized cars had been ten years earlier; the least expensive 1968 model could deliver more horsepower than the most expensive Cadillac of 1950. In the early 1960s, the average American automobile had an eight-cylinder engine which was capable of generating roughly up to 300 horsepower; its body measured about 130 inches across the wheelbase and 210 inches down its length, with enough interior space to seat six tall people in sometimes lavish comfort. In those years, automobile advertising campaigns encouraged the notion that increased power, size, speed, and price signified increased social status—and postwar Americans, many of whom were just beginning to enjoy the fruits of their hard-won affluence, seem to have been convinced by the ads. Europeans spoke with derision about Ford's tail fins, about General Motors' boats, about Chrysler's gas-guzzlers—but Americans bought them, and enjoyed them, in record numbers.

This increase in size and capacity was the result of deliberate industry policy. Early in the 1920s, manufacturers had learned that, by using the cost-effective assembly-line techniques, larger automobiles generated more profit than smaller ones: affluent consumers would pay more for a larger car, but production costs were not significantly higher. Not surprisingly, once the economic constraints of the Depression and World War II were lifted, industry executives, ever motivated by the bottom line, were eager not only to increase the size of most of their automobiles, but also to convince Americans that larger, more powerful cars were worth the increase in their price tags. As gasoline prices remained low through most of this period, many Americans, even those who weren't affluent, seemed to have been happy to go along for the ride. Those who weren't happy purchased one of the imported cars that began coming on the American market in the 1950s—the small Hillmans, Renaults, Morrises, and, most especially, Volkswagens—or one of the compact cars that American manufacturers began to produce in the early 1960s in order to fend off the imports. Ironically, even the American compacts were larger and less gas efficient than the foreign cars they were meant to supplant, and throughout the 1960s, the big car remained the national standard.

Whether big or small, foreign or domestic, automobile ownership had become an economic necessity for most Americans by the 1960s. In the postwar years, ownership figures rose not because automobiles had become status symbols but because an increasing number of Americans could neither get to work nor do their work without them. The tailfin era was also the era in which the suburbs expanded. Along with new cars, new housing had been scarce during the fifteen years of depression and war. In the immediate postwar period, this pent-up demand for housing, rising family size (the baby boom), rising affluence, and the federal policy of providing low-cost mortgages for veterans all combined to create an unparalleled increase in the nation's stock of single-family houses. All over the country, developers began buying farmland within easy reach of cities, planting dwellings where corn had once grown and cows had once grazed.

These new suburbs were predicated on the automobile rather than the streetcar—not just automobile accessibility but automobile dependency. Both developers and purchasers assumed that the families living in those new houses would have cars; either driveways or garages were part of the house plan or streets were made wide enough to accommodate parked cars. Schools, hospitals, and shopping districts were dispersed across the landscape—too far to walk—on the assumption that residents would prefer car ownership to congestion. Zoning laws were passed to prevent commercial and industrial facilities from impinging on residential districts, as no one wanted the new suburbs to go the way of the older cities. Since mass-transit facilities were deficient in the suburbs, this meant that everyone who lived in a suburb would need a car to get to work, even if work was in the suburb itself. As a result, a family that moved to the suburbs without a car soon purchased one—and within a few years, discovered that it was going to need two. By 1970, more Americans were living in suburbs than in any other demographic region, and study after study of automotive usage had demonstrated that, even in families with more than one car, the vast majority of automotive miles were devoted not to leisure travel but to necessities. Universal ownership arrived by the end of the 1960s, and the built environment was redesigned to accommodate that fact; the motorcar had become indispensable.

High-speed travel had also become the norm. In 1944, in an effort to anticipate postwar needs, Congress had passed the Defense Highway Act, which allocated, as soon as hostilities would cease, $1.5 billion for the construction and maintenance of roads, 40,000 miles of which were to be in interstate routes. The wartime experience, during which the railroad network had proved inadequate to the job of moving troops and munitions effectively, had convinced Congress, in the name of national defense, to overlook what had earlier been thought to be the constitutional barriers to funding roads. Using federal monies and smaller amounts of their own matching funds, several states built toll roads in the postwar years, having discovered, through the experience of Pennsylvania, that motorists and trucking companies would be willing to pay a fee for the use of a multilane, limited-access, high-speed highway. The New York, New Jersey, Connecticut, Ohio, and Indiana turnpikes were just a few of these postwar toll roads; they were the first roads to make extensive use of the now-famous cloverleaf design for entrance and exit ramps. The Indiana Toll Road Commissioners calculated that by using these toll roads on a round-trip between New York and Chicago, the average motorist could save 10 hours, 834 gear shifts, and 741 brake applications, all the while traveling at an average of 14.66 miles per hour faster.

Yet even these roads were not sufficient to meet the nation's desire—commercial, military, and civilian—for high-speed, far- ranging automotive travel. The postwar inflation had decreased the value of Congress's original allotment, which meant that fewer roads had been built than originally anticipated. In any event, some of the older roads were too small for the amount of traffic they were carrying; the tollbooths were becoming the cause of considerable congestion; and some of the less populated states, having a much smaller tax base of their own, had no express highways at all.

Consequently, in 1956, Congress passed another massive road-building measure, the Interstate and National Defense Highway Act, providing for the construction of 41,000 miles of toll-free expressways, with 90 percent of the cost to be borne by the federal government. This act also imposed excise taxes (on fuel, tires, new buses, trucks, and trailers) as well as a use tax on trucks, all of which was to be put in a highway trust fund to underwrite the building and maintaining of interstate superhighways into the far distant future. The net result of the nation's continuing prosperity, fears generated by the cold war, Congress's generosity, state initiative, and the special taxes paid, in some form, by virtually every American was that sometime in the 1960s, the automobile and the express highway became the transportation mode of choice for most of the nation's commercial and personal needs.

The Unexpected Consequences of Automobility

Ironically—or perhaps not so ironically—just at that historical moment, the enthusiasm that had so motivated the social prophets of the 1920s began to dissipate. The sorcerer's apprentice discovered, just at the moment when he had finished congratulating himself, that the helpmate he had created was taking on a rather unpleasant life of its own. Similarly, just when most Americans had become completely dependent on their automobiles, the nation began to discover that auto dependency had unintended, unexpected, and unpleasant consequences. Fortunately for the apprentice he had only one problem with his mop: how to get it to stop carrying and dumping water. Between 1965 and 1975, Americans began to understand, somewhat to their surprise and often to their horror, that they had three serious problems with their cars: safety, environmental degradation, and the climbing price of fuel.

Traffic fatalities and injuries had always been a problem, even in the age of carriages and wagons, but as the level of automobile ownership began to rise, so, too, did the level of death and damage created by automobiles— and the level of public concern about safety. According to the statistics compiled by the automobile industry itself, in the two decades between 1945 and 1965, the average annual number of deaths caused by an automobile rose from 30,000 to 50,000. In the mid-1950s, in response to complaints from the public, Congress authorized the creation of an industry-wide research effort, the Vehicle Equipment Safety Compact; without specific congressional authorization, this effort would have been a violation of the antitrust laws. But five years later, the participants in the compact had agreed on only one safety standard, for tires, and the research effort still did not even employ a single full-time staff member. Convinced that safety doesn't sell, the automobile industry was clearly dragging its heels, and various members of Congress and the public decided to take matters into their own hands. Investigatory hearings were held; safety standards for automobiles purchased by the government were set; a muckraking book, *Unsafe at Any Speed*, by the consumer activist Ralph Nader, achieved wide readership; and finally, in 1966, the National Traffic and Motor Vehicle Safety Act was signed into law.

This act created the National Highway Traffic Safety Administration, which was authorized to set safety standards for new cars—whether purchased by the government or not—beginning in the 1968 model year. Passage of the act meant that for the very first time federal regulators were going to play a determining role in the design of American automobiles. The first standards adopted by NHTSA included safety belts in the front seat, padded visors and dashboards, recessed control and instrument knobs, safety door latches and hinges, impact-absorbing steering columns, dual braking systems, and standard bumper heights—essentially minor changes, requiring no great technological sophistication.

During the congressional hearings apologists for the automobile industry had argued that Americans would never voluntarily pay for or use these new safety options. They had also argued that safety equipment was unnecessary since, as a result of improvements in the design of highways, the ratio of deaths to vehicles on the road or vehicle miles driven was actually declining in the postwar years. In the public mind, however, all of these statistical arguments paled to insignificance when compared with the mounting toll of dead children, paralyzed teenagers, totaled cars, and failed brakes. In addition, evidence uncovered during the hearings had clearly indicated that the managers of automobile companies could have easily introduced safety equipment at no great expense several years earlier—and could also have taken steps to improve quality control on automotive assembly lines.

The public image of the automobile industry was further damaged, two years after the first safety standards went into effect, when it was discovered that the rate of decline of those ratios (deaths to vehicles on the road and deaths to vehicular miles driven) had fallen precipitously, just as the reformers had argued that it would. Public faith in the credibility and the reliability of the American automobile manufacturers was badly shaken by this course of events, despite the fact that, at first, many drivers and passengers chose not to use—just as the manufacturers had predicted—the seat belts with which their cars were now coming equipped.

In the same years that Americans were starting to worry about whether they could safely drive their cars, they were also starting to worry about whether they could safely breath their air. By the early decades of the twentieth century, residents of industrial cities had become aware that their clothes and windowsills were frequently soiled by unsightly and destructive ashes, but it was not until the late 1940s that the detrimental effects of air pollution on health became apparent. In the late 1940s and early 1950s, there were major air pollution crises in a number of cities when atmospheric inversions stilled the winds that might ordinarily have dispersed industrial pollutants: 20 people died during an inversion in Donora, Pennsylvania (a steelmaking center), in 1948; 4,000 people in died in London during a two-week period in 1952; a year later, 200 people died in seven days as a result of intense pollution in New York City. Congress enacted the National Air Pollution Control Act in 1955 in an effort to fund research that might uncover some way to eliminate these detrimental effects of industrialization.

At the time that the act was passed, no one realized the extent to which automobiles were contributing to pollution; the chemicals that were causing the trouble were thought to come principally from burning coal and refining petroleum. However, in the late 1940s, the citizens of Los Angeles had begun to notice that when a strange haze appeared, their eyes would tear. Thinking at first that their smog (the word "smog,"—combining "fog" with "smoke,"—was coined in the 1950s) had the same origins and content as the smogs that had appeared in London and New York, the Los Angeles City Council banned the use of coal and fuel oils for industrial purposes, but the problem continued to get worse. Within a few years, however, researchers at the California Institute of Technology in nearby Pasadena had identified some unusual components in Los Angeles smog. The smogs of New York and London were gray, not yellow, and they were composed largely of sulfur dioxide; Los Angeles smog was yellow, and its major component turned out to be nitrous oxide. The culprit was the city's cars, not its smokestacks.

Further tests revealed that, in addition to nitrous oxide, the pollution caused by automobile emissions contained carbon monoxide, hydrocarbons, oxidants, peroxyacyl nitrate, and lead. In just small doses, this smog was known to cause eye irritation and to reduce visibility—so it was even increasing the frequency of automobile accidents. Automotive smog also damages vegetation (the Los Angeles area is no longer, as it once was, a major producer of ornamental flowers for market); it eats away at rubber, textiles, and dyes. In large doses, smog is immediately toxic to people whose respiratory systems are already impaired; over long periods of time, even small doses can cause cancer.

As early as 1953, Los Angeles officials began asking automobile companies to develop emissions control systems, but a decade later, having received no concrete response, the California legislature required the installation beginning with 1963 models, of crankcase blowby devices (which returned unburned gases to the engine) on all cars sold in the state and then, beginning in 1966, of exhaust emission devices. One of the most promising of these devices, the catalytic converter, could change nitrous oxide into a harmless by-product, but it could not function well in cars that burned leaded (antiknock) gasoline. As a result, the California legislature found itself forced to require the sale of unleaded gasoline at all the state's filling stations.

Other states and the federal government quickly followed California's lead (no pun intended), as it became apparent that the smog produced by automobiles was both a multicity and an interstate problem. In the decade between 1965 and 1975, city after city, state after state, congressional com-

Los Angeles was not the only city plagued by smog. If there had been no smog on the day this photograph was taken in 1973 residents of Denver, Colorado, would have been able to see the Rocky Mountains, just beyond the city skyline. (Adapted from a photograph in *Fortune*, 1973, by Daniel C. Kramer.)

mittee after congressional committee held hearings and suggested legislation in an effort to control the air pollution that cars were producing. Through it all, the manufacturers continued to drag their feet. "First, industry spokesmen denied that the problem existed" says the historian John Rae, "they then conceded that it did exist but asserted that it had no solution; finally, they conceded that it could be solved but that the solutions would be very expensive, difficult to apply, and would require a long time to develop."[5] The discovery that automobiles were poisoning the air (and that automobile manufacturers were reluctant to do anything about it) dealt another blow to Americans' enthusiasm for their cars.

But air pollution was not the only environmental problem. Vast stretches of the countryside had been scarred by the interstate freeways. Meadows had been covered with asphalt to make parking lots. Steam shovels had dumped tons of dirt to convert streambeds into roadbeds. Once flourishing urban neighborhoods had been destroyed to ease traffic congestion at bridges and tunnels. Beaches were fouled when oil tankers ran aground or offshore oil wells exploded. Rivers and streams were fouled by the chemicals used to convert petroleum to gasoline. Scenic views had become favored locales for billboards; and scenic areas had turned ugly when motels were constructed to house the tourists who came to visit them. In the 1930s, progressive federal and state governments had busied themselves creating roads into wilderness parklands; forty years later, those roads had to be blocked and automobile traffic banned to preserve the wilderness. The technology that had opened more and more of the country's natural beauty to more and more of the country's citizens was also, in the process, the technology responsible for destroying that beauty.

And to make matters worse, the technology appeared to be increasingly expensive because gasoline prices were creeping upward. Most Americans did not notice or become upset by this increase until 1973, but the conditions that led to the rise in the price of gasoline in that year had been building for some time. The oil industry in the United States had begun in the decades after the Civil War (see Chapter 7). Extensive underground supplies had been discovered, by geologists and by mining engineers, first in Pennsylvania in the 1860s then, a few decades later, in Ohio and Indiana. In 1900, virtually the entire oil industry in the United States belonged to the Standard Oil Trust, which had been created by John D. Rockefeller. In 1906, however, the federal government had begun antitrust action against Standard Oil, and in 1911, the Supreme Court had ordered the trust dissolved. Rockefeller and his associates broke up Standard Oil into several competing units: Standard Oil of New Jersey (which became Exxon), Standard Oil of New York (which became Mobil), Standard Oil of California (which became Chevron). When new oil fields were discovered in California, west Texas, Oklahoma, and Kansas early in the new century, several other companies, not controlled by Rockefeller, were formed: Sun Oil (later Sunoco), Texas Oil (or Texaco), and Gulf Oil (which hasn't changed its name). Between them, these companies controlled most of the American oil industry for most of the twentieth century; when Sun dropped out and Shell (which is a Dutch–British conglomerate) and British Petroleum were added, they form the so-called seven sisters, who have dominated the international oil industry for most of the century.

In the 1920s, as more and more cars were produced and purchased, some experts began to worry that the demand for oil would be greater than the domestic supply. Lead additives were developed in that decade, partly to prevent engines from misfiring (knocking) but also partly to make cars more fuel efficient, stretching the supply. In addition, the American oil companies began developing foreign sources of supply—Mexico, Venezuela, Turkey, Saudi Arabia, Mesopotomia (later called Iraq)—in order to meet the anticipated growth in domestic demand; much of the oil that was being pumped into American gas tanks in the interwar years was, unbeknownst to the American public, refined from a mixture of domestic and imported oils.

During the 1920s (and partly as a result of their experiences in World War I), military experts became aware that the country's defenses were becoming increasingly dependent on internal combustion engines (in tanks, military trucks, aircraft, and ships), which meant that they were becoming increasingly dependent on oil supplies. The federal government began to worry about whether domestic supply would be sufficient to meet military demand and whether foreign supplies could be relied on. As a result, during the 1920s, the needs of the oil companies started to become important factors in both American domestic and foreign policy.

During the Depression, worries about an oil famine were transformed into worries about an oil glut. Huge oil reserves were discovered in east Texas at the very same time that the general economy was collapsing, and the demand for gasoline was, as a consequence, falling. Officials in Texas began to worry about what would happen to the oil companies (and their state's economy) if oil prices fell as the result of oversupply. Similarly, officials in Washington began to worry about what would happen to the national economy and the national defense if the stability of the oil industry was threatened. And so both governments, stimulated by the interventionist attitudes of the New Deal, began, in different ways, to regulate the level of domestic oil production and the rate of foreign oil imports. The price of gasoline at the pump was no longer only responsive to the market, to the laws of supply and demand alone.

In addition, after World War II, imported oil began playing a larger role in the American oil market. Between 1947 and 1990, there were several years in which the total amount of oil imported into the United States—from Central America, Latin America, Africa, and the Middle East—was greater than the total amount produced domestically. Some of that imported oil came from wells that were dug on land owned by American oil companies; some was sold to American refiners by foreign oil companies; some came from under land owned by foreign countries yet leased to American producers. But whatever the legal and financial arrangements may have been (and they were quite varied), the supply and the price of the oil being refined for use in American cars was becoming an international rather than purely a domestic matter—and a matter over which neither the government of the United States nor the American consumer could exert very much control. In one year, a revolution in one country might lead to the nationalization of wells that had previously belonged to an American company; in another year a war between two oil-producing foreign countries might lead to the total cessation of their export production; in another year, the discovery of oil reserves in a new locale might depress the price worldwide.

Thus in the postwar years, as the American economy became increasingly dependent on the internal combustion engine, the price and the supply of the fuel for that engine became increasingly dependent on the vagaries of international and domestic politics. Beginning in the Truman years, successive American governments became successively more enmeshed in the conflicting politics of oil. Sometimes the interests of one company conflicted with the interests of another. Sometimes the legislators from oil-producing states advocated policies that were detrimental to the interests of oil-refining states or oil-consuming states. Other times the State Department would want the president to do something to further its foreign policy objectives that was directly contrary to what the Commerce Department wanted done in order to further its domestic policy agenda.

Through it all, by a combination of special subsidies and special policies and special tariffs and special tax abatements, state and federal governments and the oil companies were able to collaborate to keep the supply of gasoline relatively high, the pump price relatively low, and both price and supply relatively stable. Gasoline prices began rising somewhat in the mid-1960s, but as all prices were going up in a general worldwide inflationary pattern, and as petroleum supplies continued to seem almost infinite, few policy makers and even fewer consumers became particularly alarmed about what was happening—until the autumn of 1973.

The inflationary spiral of the previous decades had devastated the economies of undeveloped nations because the prices that they had to pay for manufactured goods (like gasoline) were rising faster than the prices that they were getting for raw materials (like petroleum). In an effort to wrest control of the market for petroleum, the undeveloped nations that were oil producers had banded together in a loose economic union, which they called the Organization of Petroleum Exporting Countries—OPEC.

Several of OPEC's member states were Arab nations in the Middle East, and in those years the politics of the Middle East was dominated by the conflict between the Arabs and the Israelis. The Arab–Israeli conflict, which had begun in 1948 when the United Nations had agreed to the creation of an independent Jewish nation in the Middle East, had become part of the larger conflict then dominating the whole of international politics, the cold war between the Soviet Union and the United States. The Arab nations of the Middle East were producing much of the oil that was being imported in increasingly large quantities into the United States, and they were all implacable enemies of the state of Israel. Israel was, however, an ally of the United States, and the United States was supplying both economic aid and military hardware to the Israelis as a bulwark against Soviet incursion in the area. Thus if the domestic politics of oil was complicated, the international politics of oil was even more complicated: sometimes, for some reasons, the American government decided to act in alliance with the Israelis; other times, for other reasons, it decided to act alliance with their enemies. Continuance of a low pump price for gasoline in the United States hung on an exceedingly fragile thread that was being pulled tauter and tauter with each passing year.

In October 1973, when Egypt and Syria invaded Israel in a coordinated surprise attack on the eve of Yom Kippur (the most solemn religious holiday in the Jewish year), the thread finally broke. During the summer, the government of Saudi Arabia had been pushing OPEC to consider doubling the price per barrel of petroleum, and it had also warned the United States that serious consequences would ensue if it continued its policy of aiding Israel militarily. The OPEC member states were meeting in Vienna when the Yom Kippur War began; they quickly agreed on a substantial price increase. Military reverses early in the war led Israel to request armaments from the United States, and on October 19, President Nixon authorized a $2.2 billion weapons airlift. A few days later, OPEC announced a boycott on shipments of oil to the United States and all other countries who were supporting the Israelis; the radical curtailment of production that occurred in the next few weeks led to further increases in the price refiners were willing to pay for oil.

In the United States, the results of the oil embargo were felt immediately. Refiners and retailers began rationing supplies of gasoline. Long lines of irritable motorists began forming at filling stations—and prices began to skyrocket. Between October and December, the price of a barrel of crude oil increased almost 130 percent—the average pump price more than doubled; some gasoline stations changed their prices daily. Accusatory fingers pointed in six directions at once: consumers said that the gas stations were hoarding gasoline; refiners said that the government should have acted more decisively; many economists argued that the oil companies had engaged in profiteering; some politicians blamed the situation on the diplomats who had favored Israel; others railed at the diplomats who had been courting the Arabs; the oil companies complained that the auto manufacturers should have stopped building gas-guzzlers a decade earlier; the manufacturers complained that consumers had never been enthusiastic about the compacts. At some level, all of these accusations were correct. Dependence on the automobile was culturewide and complex; every aspect of American life—and every American—was, somehow, caught in its web and, consequently, also caught in the web of oil prices.

The Yom Kippur War was over within a few weeks, and the oil embargo ended a few months later; in real dollars (that is, dollars that have been adjusted for inflation), the price of gasoline eventually fell back to its earlier levels and even (in the 1980s) dropped below them. American consumers, however, reeled from the shock of discovering how dependent on oil (and foreign oil at that) they really were. The last banner year for sales of big American cars was 1973, the last year in which the domestic manufacturers of those cars—GM, Ford, Chrysler, and American Motors—were able to control the character of the American automobile market. After 1973, Americans began deciding that they wanted smaller cars, cars that would have better gasoline mileage. When they went out shopping, they discovered that there were some American-made cars—compacts and subcompacts—that answered that description, but for the same money they could get a Japanese car (both Toyota and Nissan had begun exporting cars to the United States in the 1960s) that had more safety features, better emissions controls, a better maintenance record, and even better gas mileage. Just over 60 percent of the cars that Americans purchased in 1970 were domestic full- and mid-sized models. By 1980, the figures were reversed, just over 60 percent of the automobile sales in that year were for small cars, and 27 percent of those had been manufactured outside the United States.

Thus by 1980, partly as a result of the 1973 oil embargo and partly as a result of distress about both environmental and safety problems, many Americans had become disillusioned not only with what had once been the nation's favorite technology but also with what had once been the nation's leading industry. The characteristically American automobile was transformed by this disillusionment and the characteristically American romance with the automobile began to sour.

Conclusion: The Paradox of Automobility

The tale of the sorcerer's apprentice ends happily for society, if not for the apprentice. The sorcerer returns from his trip, evaluates the situation, utters his special incantation, and stops the mop. The apprentice loses his job, but at least the shop and the town are saved.

As the twentieth century draws to a close, no one can predict whether the tale of the automobile will end the same way. Before the sorcerer returned to save the day, the apprentice tried frantically to find an incantation that would stop—but not destroy—his automatic mop; he wanted to have his cake and eat it, too. Since 1980, American policy makers, politicians, manufacturers, and consumers have all been behaving in roughly the same way—with roughly the same lack of success—trying to develop solutions to the problems without giving up the automobile. Politicians and policy makers have tried to encourage carpooling and to resurrect commuter railroads; they have raised emissions control standards, lowered maximum speed limits, and created heavier penalties for drunken driving. Some scientists and engineers have been trying to develop synthetic fuels that would end American dependence on petroleum. Others have been trying to develop electric and even solar engines that would end dependence on the combustion of hydrocarbons. Yet others have been trying to design increasingly safer, less obtrusive highways. Consumers, meanwhile, have learned to buckle their seat belts, to stop disconnecting the catalytic converters on their cars, and to consider good gas mileage as an important criterion in the choice of a new one. Even American manufacturers have mended (some of) their ways, shrinking the average size and weight of their cars; returning to six-cyclinder and even four-cylinder engines; developing new, less damaging gasoline additives.

Unfortunately, none of these attempted solutions—like the apprentice's frantic search through the sorcerer's books of magic—has succeeded in solving any of the complex problems that automobile dependence has created. Technological systems, once they are in place, have enormous staying power. The technological system that some people call "automobility" was originally built out of inventors' hopes and entrepreneurs' dreams, but it has now been set in the concrete of several hundred thousand miles of highway and several million suburban subdivisions and in habits that Americans have been developing for several generations. No simple, single set of incantations will make it go away.

Notes

1. "Salutatory," *Horseless Age* 1 (1895): 1, as quoted in James J. Flink, *The Automobile Age* (Cambridge: MIT Press, 1988), p. 22.

2. "Fairly Howling," *Motor World* 1 (1900), as cited in Flink, *Automobile Age,* p. 27.

3. Henry Ford and Samuel Crowther, *My Life and Work* (Garden City, NY: Doubleday, 1922), p. 59.

4. Charles R. Harte, "Discussion," *Proceedings of the American Society of Civil Engineers* 55, no. 1 (1929): 574, as quoted in Mark S. Foster, *From Streetcar to Superhighway: American City Planners and Urban Transportation, 1900–1940* (Philadelphia: Temple University Press, 1981), pp. 62–63.

5. John B. Rae, *The American Automobile Industry* (Boston: Twayne, 1984), p. 134.

Suggestions for Further Reading

Warren J. Belasco, *Americans on the Road: From Autocamp to Motel, 1910–1945* (Cambridge, MA, 1979).

Michael L. Berger, *The Devil Wagon in God's Country: The Automobile and Social Change in Rural America, 1893–1929* (Hamden, CT, 1979).

George H. Daniels and Mark H. Rose, eds., *Energy and Transport: Historical Perspectives on Policy Issues* (Beverly Hills, 1982).

James J. Flink, *America Adopts the Automobile, 1895–1910* (Cambridge, MA, 1970).

James J. Flink, *The Car Culture* (Cambridge, MA, 1975).

Mark S. Foster, *From Streetcar to Superhighway: American City Planners and Urban Transportation, 1900–1940* (Philadelphia, 1981).

David Gartman, *Auto Slavery: Labor Process in American Automobiles* (New Brunswick, NJ, 1986).

Kenneth T. Jackson, *Crabgrass Frontier: The Suburbanization of the United States* (New York, 1985).

Stuart W. Leslie, *Boss Kettering* (New York, 1983).

W. T. Lhamon, Jr., *Deliberate Speed: The Origins of a Cultural Style in the American 1950s* (Washington, DC, 1990).

Martin V. Melosi, *Coping with Abundance: Energy and Environment in Industrial America* (New York, 1985).

Robert C. Post, *High Performance: The Culture and Technology of Drag Racing* (Baltimore, 1994).

John B. Rae, *The American Automobile* (Chicago, 1965).

John B. Rae, *The American Automobile Industry* (Boston, 1984).

John B. Rae, *The Road and Car in American Life* (Cambridge, MA, 1970).

Mark H. Rose, *Interstate: Express Highway Politics, 1941–1956* (Lawrence, KS, 1979).

Virginia Scharff, *Taking the Wheel: Women and the Coming of the Motor Age* (New York, 1991).

Bruce E. Seely, *Building the American Highway System: Engineers and Policy Makers* (Philadelphia, 1987).

Joel Tarr and Gabriel Dupuy, eds., *Technology and the Rise of the Networked City in Europe and America* (Philadelphia, 1988).

Sam Bass Warner, *Street Cars Suburbs: The Process of Growth in Boston, 1870–1900* (Cambridge, MA, 1962).

Daniel Yergin, *The Prize: The Epic Quest for Oil, Money and Power* (New York, 1991).

11

Taxpayers, Generals, and Aviation

THE UNITED STATES GOVERNMENT, particularly the various branches of the military, has always been an important sponsor of technological change. Technology, government, and warfare have been multiply linked at least from the time in 1798 that Oliver Wolcott, Secretary of the Treasury, first wrote a contract with Eli Whitney to develop a new means of manufacturing muskets (see Chapter 4). The government, especially when it was involved in or preparing to go to war, has sometimes spent money to forge previously untrod technological paths, paths that civilian industries had deemed too expensive, or too risky, to explore. As a result, military technologies have often been the most advanced technologies of their day.

Many people believe that the linkages between technology, warfare, and government profoundly changed character in the years during and after World War II, that the links became firmer and the chain larger—much larger—growing to the point where it threatened to dominate (some people say succeeded in dominating) all facets of social, economic, and political life in the United States. In 1959, President Dwight David Eisenhower cautioned Congress to beware of the growing power of the phenomenon that he called the military-industrial complex; in the following years, many people outside of government also began to notice the connections between the military functions of government and the presumptive progress of technology. During the cold war, as the links between military procurement and high-technology industry grew firmer (just as Eisenhower predicted that they would), the problems posed by the military-industrial complex (which by then had actually become the military-industrial-academic complex) became part of an often acrimonious public debate. The relationship that had seemed natural and normal in the world of muskets and cannon had become suspect and fearful in the world of hydrogen bombs and intercontinental ballistic missiles.

There are at least seven modes through which governmental agencies affect technological change: *patenting, tariffing, regulating, educating, researching, building*, and *consuming*; the military has been particularly involved in the latter four. One good way to understand the importance and complexity of the military role in technological change, as well as to evaluate the extent to which that role changed after 1940, would be to examine the history of the aerospace industry. Aerospace has been crucial both to the nation's defenses and to its economy for much of the twentieth century; aerospace is also one the industries in which technological change has been both conspicuous and celebrated.

The Early Days of Aircraft and the Aircraft Industry

Most people think of manned airflight as having begun in December 1903 when Wilbur and Orville Wright, after several years of patient study and experiment, sailed into the air (59 seconds, 852 feet) over the sand dunes of Kitty Hawk, North Carolina. In fact, the Wrights (as they well knew) had both predecessors and contemporary competitors. This is not said with any intention of demeaning the Wright's accomplishment (they were the first men to build and then fly a self-propelled vehicle that could be controlled and could land safely), but only to indicate that no technological device as complex as an airplane and no industry as vast as aerospace can have one, and only one, point of origin.

For more than a century before 1903, Europeans and Americans had been experimenting both with balloons and with gliders. The perfection of the internal combustion engine (in France and Germany) during the 1880s gave these aviation pioneers reason to believe that powered flight might some day be possible. One of these men, Samuel Pierpont Langley, has the distinction of having been awarded the first United States government contract for aircraft development.

Langley (1834–1906) was a distinguished astronomer and the head of the famous Smithsonian Institution in Washington, D.C. He had been experimenting with internal combustion engines attached to gliders for many years, but the expense of building his prototype vehicles was becoming too great to be borne by his salary alone. Langley tried to convince the Navy Department to underwrite his costs on the grounds that a powered aircraft would be even more useful for surveillance than the balloons that the Union Army had used occasionally during the Civil War. Langley's pleas fell on deaf ears for a number of years, until an assistant secretary in the department, Theodore Roosevelt, finally became interested and wrangled $50,000 (a very large sum in those days) to support the experiments. On two separate occasions in the autumn of 1903, the plane that Langley built with that money (its five-cylinder engine weighed 125 pounds and could generate 53 horsepower) was catapulted off a houseboat on the Potomac River. Both attempts were failures; both times the plane had to be fished out of the river. The last attempt was on December 8, 1903, and as Roger Bilstein, an eminent aerospace historian, puts it, "[n]ine days later . . . the Wright brothers flew into history."[1]

Initially the Wright brothers financed their experiments themselves. And initially their successes were ignored, misinterpreted, or greeted with skepticism. The Wrights, in fact, hesitated to publicize their achievements until their patents were secure, but they did have a rough idea of how they were eventually going to turn a profit. "It is our intention," Orville announced, "to furnish machines for military use first, before entering the commercial field."[2] Thus it came to pass that their fledgling company (among the initial investors were Cornelius Vanderbilt, Howard Gould, and August Belmont), after entering into negotiations with the governments of Great Britain, France, and Germany, delivered its first airplane, in 1909, to the War Department of the United States for $25,000 (a sum that covered not just the plane itself but also a year's worth of developmental testing).

The Wrights were not alone in the sky for very long, however; interest in powered airflight was so high that they soon had many competitors, both in the United States and abroad. One of these competitors was Glenn Curtiss, an experienced engine builder, who had received a contract in 1904 from the U.S. Army to build a power plant for its first dirigible. On June 20, 1908, Curtiss made a flight of 1,266 feet in an airplane belonging to the Aerial Experiment Association, a private company that had been formed by Curtiss and several investors, including Alexander Graham Bell.

By 1911, almost a dozen American firms had begun manufacturing aircraft, but none were very profitable. Most of the planes they produced were purchased either by wealthy men and women or by the military; the Army and Navy intended to use them only for surveillance. Built out of wood and canvas, these early planes—called biplanes because of their double wing structure—had open cockpits and only one engine; none were large enough to carry freight or more than one passenger. In order to stay in business, the early manufacturers created traveling air shows; pilots and planes traveled from town to town—barnstorming—charging admission for the privilege of seeing feats of aerial daring. Research and development were virtually impossible under such economic conditions. In the years before World War I, American aircraft design changed at a snail's pace. To make matters worse, the two companies with the largest production capacity, the Curtiss company and the Wright company—were locked in a legal battle over who had the right to a patent on aileron control (ailerons are the hinged flaps on airplane wings that make banking and turning possible).

World War I changed everything. In 1914, the British, the French, and the Germans entered the war with small numbers of aircraft that they intended to use for reconnaissance. By the time the United States became a combatant two and half years later, air battles had become commonplace, aircraft production had been accelerated, and machine guns had been adapted to aerial warfare. In 1916, all of the American manufacturers together built and sold 411 airplanes. Five months into 1917, the French government requested the United States, its new ally, to furnish 4,500 planes; the War Department estimated that its own forces were going to need almost 20,000. Congress appropriated $640 million to do the job; in one fell swoop, the size of the American aircraft industry was transformed. In addition, the government flexed its muscles as the industry's most important customer and boldly demanded that the Curtiss and Wright interests settle their patent battle and agree to a cross-licensing arrangement.

During the next year, new engines were designed; 20,000 pilots and mechanics were trained; new companies were formed; subcontractors began specializing in the production of aircraft parts; both the Army and the Navy created air services; bombs were altered so that they could be dropped from aircraft. World War I was the first air war and also the first total war: the war in which governments learned how to mobilize civilian enterprises, such as aircraft manufacturers, on a massive scale; the war in which, for the first time, civilians far from the front lines were endangered by attack from the air. During World War I, the government used its power as a very large consumer to aid in the development both of the aircraft industry and of aircraft technology. The birth of the aircraft industry had made total war possible; the advent of total war made the growth and development of that industry both a defensive and an offensive necessity.

By the time peace came in 1918, a strong link had been forged between the American military and the American aircraft industry. Most knowledgeable people in the military services had come to believe that the aircraft industry had to be kept healthy in order to provide for the nation's defense. In fact, more than just health seemed to be required; American aircraft and aircraft engines were going to have to be improved if they were going to compete successfully with European designs.

How the government would participate in doing this during peacetime, and with peacetime budgets, was resolved during the 1920s in two different ways. First, the federal government built an airmail service. Regular airmail (between New York, Philadelphia, and Washington) began on May 15, 1918, while hostilities were still raging in Europe; the War Department supplied both the planes and the pilots. After the war, airmail service was transferred to the Post Office Department, which at first hired its own pilots (many of whom had gotten their training in the Army or Navy) and purchased its own aircraft directly from the manufacturers.

By 1930, however, the Post Office Department's airmail business was used not only to support the aircraft manufacturers but also to support the newly created commercial airlines. In the United States, commercial air transport began in the early 1920s; entrepreneurs tried to create businesses that would fly passengers and freight from one city to another. In those early years, air travel was difficult, expensive, and dangerous. Bad weather wreaked havoc with schedules; nighttime flying was virtually impossible; long distances required frequent stops to refuel the small engines. The fledgling airlines struggled to survive. In 1925, in order to alleviate their plight, Congress passed legislation requiring the Post Office to contract its airmail services with the commercial airlines. This legislation, the Kelly Act, salvaged many of the struggling passenger airlines.

The Kelly Act also encouraged aircraft manufacturers to modify their World War I designs so as to be able to carry greater weight and provide both freight and passenger compartments. With assured government contracts for airmail, there was no doubt that airlines could become long-term, large-scale consumers of aircraft, thus keeping the aircraft manufacturers in business in peacetime. In the 1920s, air cooling replaced water cooling for engines; single-engine craft gave way to double- and triple-engine designs; fuselages were streamlined; the biplane was replaced by the monoplane with a cantilevered wing; wood and canvas construction gave way to all-metal construction. Much of the developmental work resulting in these new designs was trial and error: build a prototype motor a new way and see how efficiently it runs; try a different type of steel sheeting for the fuselage and test it to discover how much stress it can endure. This developmental work produced incremental increases in the speed, size, and durability of aircraft; much of this work was both financed and undertaken by the aircraft manufacturers themselves, with varying degrees of both technological and economic success.

Much but not all of the development work was financed by the manufacturers, however, which brings us to the second mode of government support: research and development. Early in the history of American aircraft design, several pioneers had expressed the hope that the federal government would finance an aeronautical research laboratory. The experimental facilities that would be required would be very expensive, beyond the means of either individuals or start-up companies. In any event, the argument went, the federal government had long been in the habit of conducting research that was deemed important to the national defense: the Naval Observatory and the Army Signal Corps, for example, had been doing meteorological research for decades. In 1915, a rider was attached to a naval appropriations bill that provided for the creation of the National Advisory Committee for Aeronautics (NACA) "to supervise and direct the scientific study of the problems of flight with a view to their practical solution."[3]

For the first five years of its existence, the NACA appropriation was only $5,000 a year. The committee members met twice a year; they isolated problems worth studying and allocated funds to a government agency or a university laboratory to undertake the work. During World War I, NACA answered queries for the armed services and ran experiments at the naval laboratories; it also laid plans for the construction of its own laboratory to assist in airframe and engine design, which was to be housed on the

The wind tunnel constructed at Langley Research Center in the late 1920s was so large it could be used, as in this picture, for testing full scale models of aircraft. (Courtesy National Aeronautics and Space Administration.)

grounds of a newly established Army airfield in Virginia, named after air flight pioneer Samuel Langley. When the base was dedicated in 1920, the annual NACA appropriation increased significantly; the NACA facility at Langley consisted of an atmospheric wind tunnel, a dynamometer laboratory, an administration building, a warehouse, and a staff of eleven people.

During the 1920s and 1930s, the NACA staff at Langley grew; many of the people who were employed there had degrees in mechanical or civil engineering—or, later, aeronautical engineering. Scale models of new wing designs were tested in the wind tunnels; new instruments were invented in order to detect pressure distributions on the exterior of aircraft during flight; contract work for the Army and Navy led to the development of standards against which to measure the controllability, stability, and maneuverability of new aircraft. (Controllability, stability, and maneuverability are of particular importance in the development of aircraft that are specially modified for bombing missions and airborne fighting, fighters and bombers.) A large air tunnel built in 1927 made it possible to test not just scale models but whole airplanes; in this tunnel a new cowling for air-cooled engines was tested, which allowed a new record for aircraft speed to be set (177 mph); the Langley cowling subsequently became a standard feature of engines manufactured both for commercial and military aircraft.

Like all operators of federal laboratories, NACA was a public agency, and the results of its research had to be made available to all interested parties, even if the item being tested belonged to a manufacturer or an individual. Langley engineers published the results of their tests in newsletters, in professional journals, and in trade magazines. Once a year, representatives of the manufacturing companies and of the military air services were invited to a conference in which the results of the year's work were displayed.

NACA-Langley was not the only government-funded aeronautic research facility; each of the armed services had also established an air research operation after World War I. And the government was not the only player in the field of aeronautical research. The aircraft companies themselves did some of it, as we have seen, and so did a growing number of university laboratories. The university laboratories tended to do fundamental, basic research in aerodynamics rather than applied, practical, or trial-and-error research into the features of aircraft components. Many of these laboratories were funded by private foundations, the most important of which was the Daniel Guggenheim Fund for the Promotion of Aeronautics. The university laboratories (at such institutions as Cal Tech and MIT) also served as the basis for undergraduate and graduate degree programs in aeronautical sciences and engineering.

In the 1930s, incremental change in aircraft and engine design continued, much of it based on research conducted in some combination of all three types of laboratories. The variable-pitch propeller was introduced, as was the automatic piloting system; stressed skin construction became more widespread; streamlining was explored; landing gear became retractable; new engines required less frequent overhaul; pressurization of cabins made higher flying feasible; new fuel additives led to greater fuel efficiency.

As a result of all this developmental activity, new air travel records were set. In 1927, Charles Lindbergh made the first nonstop solo trip across the Atlantic. The prize-winning speed in 1929 was 194 mph; by 1937, Howard Hughes had managed to reach 352 mph in a single-engine monoplane. In 1931, Wiley Post flew around the world in nine days; seven years later, Hughes did it in half the time. The first versions of the Pratt and Whitney Wasp engine needed to be overhauled every 150 hours of flight time; by 1929, this had been increased to 300 hours; by 1936, to 500 hours. One model of the Wright Cyclone engine could generate 550 horsepower in 1930 and 1,100 horsepower by 1939. During the 1930s, two of the most famous American commercial airliners were introduced: the Boeing 247 and the Douglas DC-3. Because of their several technical and commercial advantages, these aircraft were sold widely both in the United States and abroad, helping American, rather than European, companies dominate the market for both civil and military aircraft. Even with the benefit of hindsight, it is probably impossible to say which of these three components of the aeronautical research community—government, industry or university—was the most significant factor in the development of new knowledge and new designs. For our purposes, it is sufficient to note that the government laboratories were regarded as significant by their contemporaries, so much so that for many decades they attracted some of the best graduates of the degree programs in aeronautical engineering.

Nonmilitary aspects of government were also important to the history of aviation in the years between the two world wars. American airlines prospered in those years, even during the Depression: 475,000 passengers traveled in the air in 1932; 4 million in 1941. Local and federal governments—acting in their roles as both builders and regulators—were, at least in part, responsible for this prosperity. Realizing the economic potential of air travel, local governments and private investors began building and maintaining airports. Fifty years earlier, local governments had tried to attract railroad lines; now the most forward-looking of those governments were trying to attract airlines by building and maintaining airports, flat places (or terminals on the water for hydroplanes) where the aircraft of more than one airline could arrive and depart, essential parts of the growing air travel network.

The federal government also increasingly became involved in the effort to ensure the safety of airline passengers and air freight. The Air Commerce Act of 1926 established the authority of the federal government to regulate routes and rates and also to provide for passenger safety. The activities of the Bureau of Air Commerce were frequently controversial, especially in light of the differences that existed between the American business community and the New Deal administration, but by 1938, the federal government had taken over the responsibility of operating air traffic control centers. (The first air traffic control center had been created in Newark, New Jersey, in 1935; it was funded by a consortium of the airlines and utilized positional information transmitted on radio broadcasts between pilots and controllers.) Late in the 1930s, the federal government also began to issue weather forecasts for pilots and to investigate and assign blame in air accidents. All of these activities were costly; the fact that the government was willing to undertake them, financing them from tax revenues, meant that the airlines did not have to, adding to their prosperity.

Thus during the late 1930s, when the United States began to rearm for what was to become World War II, the multiple links between aviation technology, government, and warfare had already been well forged. Some of those links connected the military with the manufacturers of airplane bodies and the operators of commercial airlines; others, just as strong, connected civilian government agencies (such as the Post Office and NACA) with both arms of the aviation industry.

World War II: A Turning Point

World War II was the first war in which military strategists assumed that supremacy in the air was going to be a key—some said *the* key—component of victory. Yet in 1939, when the war began in Europe, the United States was not prepared to fight, and win, in the air. In 1939, the American Air Corps had 26,000 personnel; the British Royal Air Force had

100,000; the German Luftwaffe, 500,000. "Our existing [air] forces are so utterly inadequate that they must be immediately strengthened," President Roosevelt declared, six months before the war actually began, two years before the United States entered it.[4]

The president and the Congress began to increase appropriations for the purchase of military aircraft (bombers, fighters, transports), and those appropriations kept increasing for the next six years. Shortly after the United States declared war in 1941, new plants were built, and later, new production techniques were introduced. By the end of 1943, over two million workers (many of them women) were employed round the clock, seven days a week, producing aircraft and the parts for aircraft. Production capacity soared. In 1939, the entire United States Air Corps possessed 800 modern planes (the British had twice that number; the Germans eight times as many). During 1944–45, the Ford plant at Willow Run, just one of dozens of factories producing aircraft in that year, produced 5,476 B-24 bombers. At the war's end, the entire American aircraft industry was capable of turning out 110,000 aircraft a year.

But the nation's air defenses required more than improved quantity; quality as defined in specifically military terms, needed to be improved as well. Most of the aviation successes of the 1920s and 1930s had been in civilian, rather than military, aviation; a concerted research and development effort would be needed to match the military capabilities of the Luftwaffe. In the late 1930s, in order to meet the mounting threat of war, the NACA installation at Langley Field turned from basic to applied research. In addition, two new NACA research facilities were opened, one of which was devoted specifically to work on aircraft engines. Both the Army and the Navy increased the research activities in their own laboratories and both services began letting contracts to universities for additional research.

Cognizant of the important role that research and development were going to play in the war effort, President Roosevelt authorized the creation of a new organization in 1941. The Office of Scientific Research and Development (OSRD) was to be a civilian organization, staffed by scientists and engineers, which would coordinate all the research activity (in many fields) that the government—most particularly, the military—was sponsoring. World War II was thus the first war in which the nation's scientists and engineers were almost completely mobilized for the war effort, in which a large-scale, nationwide research effort was mounted to solve a series of very particular and very practical problems. Industry and the academic world cooperated in all of these endeavors, but the government—which was managing the research, and paying for it—was in charge.

In aeronautics, much of OSRD's coordinating effort proved successful; between 1940 and 1945, changes in aircraft design and flight technology occurred at a dizzying pace. Some of the changes were incremental. Bombers and transports got bigger; their payload capacity increased; so did their geographic range. Fighters became more maneuverable. Other changes were more revolutionary. Radar, the use of microwave reflection to locate objects, was first developed and then miniaturized for use on aircraft. The helicopter was introduced—the first successful rotary winged aircraft. Jet engines, based on gas flow through a turbine, were tested and manufactured. So were systems for gyrostabilization and radio control of bombs. The long-range rocket, both with liquid and with solid propellants, was a wartime innovation. Not all of these revolutionary developments were the result of American, or solely American, research and development—but many were. And not all had their greatest impact during the war years; some (for example, jet engines and rocket propulsion) were introduced rather late in the war and had their greatest impact once the hostilities were over.

These wartime research projects were huge: dozens, in some cases, hundreds or even thousands, of people were put to work, often at several different locations, usually under conditions of extreme secrecy and time pressure. Scientists and engineers who had previously been accustomed to working alone, at a leisurely pace, often on projects of their own creation, discovered that they had to change their working habits overnight. Some became managers who supervised both other people's work and complex physical facilities, tasks that they had not been trained to undertake, but that were thrust on them by the exigencies of wartime.

Many people believe that the techniques these scientists and engineers developed during the war, the techniques for managing large research and development projects, were the most revolutionary wartime technological changes of all because they had such a profound impact on the postwar years, the years when the "hot" war turned into the "cold" war. Managed research is a social technology in which the end goal is a specific artifact, but many different social systems have to be successfully manipulated in order to both invent and produce the artifact. There may be one or two initial inventor-entrepreneurs, but a managed project is one that is too big to be carried out successfully by just one person: a development team has to be assembled. Some members of this team must be able to operate in the political realm (since public funds are almost always required), others must be able handle economic matters (since budgets must be prepared, adhered to, audited), yet others have to know how to locate and supervise the scientists and engineers who will cooperate in doing the work (which may mean that contracts and subcontracts will have to be given to universities, to independent research companies, or to the research and development facilities of corporations); yet others have to know how to locate and supervise the manufacturers (prime contractors and subcontractors) who will produce the object that is required at the end.

The most justifiably famous of the wartime managed research projects, and the one that taught the country and the world about the awesome potential of managed research, was the Manhattan Project, the enterprise that created the atomic bomb. Much has been written, and will continue to be written, about the Manhattan Project, but for our purposes it is sufficient to note just three of its key features. First, the initial discoveries about the potential release of huge amounts of energy in uranium fission were made in the 1930s by scientists whose aim was to understand the fundamental atomic structure of matter: they were doing what they, and others, called pure science. Several of these scientists understood the military potential of their discoveries and decided, after the outbreak of the war, to bring this potential to the attention of their governments. The Manhattan Project was, therefore, a case in which a scientific discovery was applied for a practical purpose. Not all subsequent managed research projects had this characte—some were examples of applied technology (old technologies are applied to reach new goals) and others were examples of managed science (scientific research was stimulated by a desire to reach some new practical goal)—but all involved, somewhere along the way, discoveries that had been made by scientists.

Second, the practical work—the research and development that actually produced the bombs—was a huge, complex undertaking, and its complexity was both technical and social. Both research and development were required: in the Manhattan Project for example, research to learn how to separate the various isotopes of uranium that are found naturally in uranium ore, and also to ascertain whether a nuclear chain reaction could be controlled, and development to build the production plants in which large quantities of fissionable material could be obtained and then to build and test a portable device (the bomb) that could deliver that material to a target. This work was coordinated by two government agencies, at first OSRD and subsequently the Army. Some of the work was contracted out to universities (the first controlled chain reaction occurred in a reactor that was built under the football stadium at the University of Chicago) and some was done in facilities built especially for the government (for example, the reactors at Oak Ridge, Tennessee, and Hanford, Washington, and the bomb development facility at Los Alamos, New Mexico). Some of the materials that were needed were produced by the project itself and some were purchased from private industry. Some of the people employed on the project were scientists and engineers; others were secretaries, clerks, technicians, and janitors. Overall supervision of their work fell partly to an Army officer, Colonel Leslie Groves, and partly to a scientist, J. Robert Oppenheimer.

Finally—and crucially—the ultimate goal of the Manhattan Project was military, not technological. The atomic bomb was only a means to an end and that end was military: the defeat of the Axis powers. Because the end was military, control of the project was, ultimately, military: Oppenheimer, the scientist, reported to Groves, the military officer. When the time came to make a decision about whether to use the bomb, the scientists and engineers, much to their dismay, were not consulted. The experts were forced to recognize that they were employees, not policy makers; their job was to get the bombs ready to go; politicians and military leaders decided whether, and where, they would be used. Those who paid the piper called the tune—a fact of life that, in the postwar years, many scientists and engineers would come to find extremely disconcerting.

The Military-Industrial-Academic Complex

The cold war began even before the hot war had completely ended. Throughout World War II, the alliance that was fighting the Axis had been a rather tentative affair, its weakest link the relationship between the Soviet Union and the United States. By the spring of 1945, the Soviet Union had already made it clear that it was going to extend its hegemony over much of the territory in eastern Europe that it had wrested from the Nazis: it was not going to allow Poland, Yugoslavia, Czechoslovakia, or any other incipient nation-state in eastern Europe to become independent or democratic. Thus at the very same time that President Truman and his advisors were worrying about how to end the war against Japan in Asia, they were also starting to worry about how to counter the Soviet threat in Europe, about how the expansionist intentions of the Soviet Union were going to be contained. Four years later, in 1949, these worries became even more pressing when the Soviet Union tested its first atomic bomb, making it clear that there were now two superpowers in the world, each potentially in a technological position to destroy the other.

Even before the war had ended, many knowledgeable Americans, particularly those associated with OSRD, had come to believe that science and technology were essential to maintaining the nation's military security. In 1944, while the war was still raging, President Roosevelt had asked Vannevar Bush, who was then head of OSRD, to prepare a report on how the federal government might promote scientific and technological progress after the war "for the improvement of the national health, the creation of new enterprises bringing new jobs, and the betterment of the national standard of living."[5] In the finished report, *Science, The Endless Frontier*, Bush pointed out that there was a fourth domain, the nation's offensive and defensive capability, to which scientific and technical research could, and should, contribute. In the ensuing decades, as the cold war became both more intense and overtly technological, a very large number of people came to agree with Bush.

Although estimates are difficult to make, some scholars have concluded that in the four years that the United States was involved in World War II (1941–45) the federal government spent roughly $1 billion *just* on research and development. By 1950, that figure had risen to $1 billion *every year*. Some of that research and development money was spent by agencies of the Department of Defense (the new name given in 1947 to the War and Navy Departments). Other funds were spent by civilian agencies of the government; the most important of these was the Atomic Energy Commission, which had been created to oversee the production of nuclear armaments and nonmilitary applications of nuclear technology.

With the outbreak of the Korean War in 1950, Congress began allocating even more money for research and development—$3.1 billion in 1953 alone. Some of this was allocated for new research facilities owned and operated by the government, such as the Brookhaven National Laboratory on Long Island. Yet another portion of it ($350 million in 1949–50) spilled over into research laboratories owned and operated by private industry. According to estimates made by defense industry analysts, the federal government paid for about a quarter of all industrial research and development in the early years of the cold war, but by the mid-1960s that figure had risen to one third of all R&D costs, and in the defense-related industries—aerospace and electronics—it was closer to three quarters. In addition, many firms, recognizing a growth sector when they saw one, began creating research facilities precisely so as to be able to attract government funds. One of the first of these was the RAND Corporation (its name derives from the phrase "research and development"), which was originally a part of the Douglas Aircraft Corporation; in the 1960s, it employed roughly 800 people and did $13 million worth of business with the Air Force every year. Some men and women who had been consultants to the Defense Department formed their own companies for the purpose of conducting research for the government.

Another substantial segment of federal research funds went to universities. In the postwar years, some universities became the operators of research facilities that belonged to the government. The University of California at Berkeley, for example, operated the Lawrence Radiation Laboratories and the Los Alamos Laboratory for the Atomic Energy Commission, leading Oppenheimer to refer to it as "a great, liberal university that is the only place in the world . . . that manufactures, under contract with the United States government, atomic bombs."[6] The California Institute of Technology operated the Jet Propulsion Laboratory for the Army. Other universities created their own laboratories for the express purpose of attracting contract research for defense-related agencies: MIT Research Laboratory in Electronics was one of these, and so was the Applied Physics Laboratory at Johns Hopkins and the Stanford Research Institute. Other universities that were not willing to create wholly new departments began allowing individual faculty members to accept federal contracts for research. As the years wore on and the benefits of this largesse became apparent, faculty members were actively encouraged not just to accept but also to solicit such contracts—and to forgo their teaching obligations in order to undertake such research. Thus, under various guises, American universities were becoming the government's partner in the arms race.

In the fall of 1957, when the Soviets launched Sputnik (the first manmade orbiting satellite), the nation's concern about maintaining its technological advantage increased. Federal expenditures for research and development skyrocketed again, literally and figuratively. NACA was transformed into NASA (the National Aeronautics and *Space* Administration), and the country's space program went into high gear. In the next ten years, federal R&D expenditures nearly quadrupled, reaching roughly $15 billion per year. Also as a direct result of the cold war and the launch of Sputnik, Congress passed the National Defense Education Act, and the government began to pay for the graduate education of increasingly large numbers of scientists and engineers, particularly in the physical sciences. The effects of that legislation were unmistakable; in the seven years between 1960 and 1967, the number of doctorates granted by American universities in science and engineering more than doubled, from 6,000 to 13,000. The historian Daniel J. Kevles has estimated that in 1969, one third of the 132,000 science and engineering graduate students in the United States were being funded by the federal government. Competition in certain scientific and technical domains—particle physics, aeronautical engineering, rocketry, microelectronics—became part of the cold war, part of the nation's international strategy.

By 1959, when President Eisenhower warned the Congress and the country about the dominance of what he called the military-industrial complex, the federal government, in the name of military security, was calling a very large number of tunes—so many, in fact, that some industries and universities were becoming almost entirely dependent on federal largesse. By 1964, three fifths of all the scientists and engineers in the United States—whether employed in industry, academia, or government service—were financed in some way by the federal government. A close relationship had always existed between the federal government and the manufacturers who supplied it with armaments; in the years after World War II, that close relationship was extended to many more industries as well as to universities.

Civilian Spin-offs and the Race into Space

Thus as a consequence of the cold war, the United States government—as consumer, as funder of research, and as supporter of education—became the major determinant of aerospace technological change. What the politicians and the military wanted in order to fight both cold and hot wars, and what the scientists and engineers accomplished through managed research, was what the rest of the country got. In order to convince people that there were advantages to this very costly enterprise, federal agencies developed the concept of spin-off to designate the civilian benefits that government-funded research was providing to the general population.

In the years between 1945 and 1990, there were many different kinds of spin-offs. One was cheaper and more convenient air travel; in the postwar years, those Americans who began taking to the skies in greater and greater numbers with greater and greater frequency, both for business and for vacation travel, were the beneficiaries, probably unwittingly, of military contracts.

The companies—Boeing, Douglas, Lockheed, and Convair were among the most important—that had built most of the American aircraft used during World War II continued to be major suppliers of aircraft for the military services after the war, but they also learned how to convert their designs and their production facilities to meet the needs of the commercial airlines. Aerospace historian Roger E. Bilstein describes one of the most widely used of these postwar commercial planes, the Boeing 377 Stratocruiser, as "a collection of parts; wing, tail and landing gear from the B-29 [bomber]; fuselage from the post war Air Force C-97 cargo/tanker aircraft; and engines from an improved version of the B-29 known as the B-50."[7]

Postwar propeller-driven commercial aircraft became both faster (they carried four engines, rather than two or three) and larger (roughly three times the carrying capacity of prewar planes). Because of the size and character of their engines, they had to use 100-octane fuel, but this, too, had become cheaper as a result of processes developed by the oil refineries during the war. Thus virtually all of the important technological changes in postwar propeller-driven aircraft were the results of lessons that had been learned by manufacturers while completing contractual work for the Defense Department. The net result was a reduction in the fares charged to passengers, a reduction in the time spent traveling from one airport to another, and an increase in the number of nonstop flights.

And what was true of propeller-driven craft was also true of airplanes with jet engines: they, too, were spin-offs from military research and development. The gas turbine engine provides the thrust that an aircraft needs to take off and remain in the air by forcing very large quantities of air out of its rear vents at very high speeds. Workable gas turbine engines had been developed in Britain and in Germany during the 1930s. In 1940 when the U.S. Army Air Corp decided to explore the potential of this new engine, it imported a British engine and asked the General Electric Company to make copies of it. During the war, several aircraft companies designed and tested fighter aircraft outfitted with these new engines—one of those planes, the Lockheed P-80 Shooting Star, was flown in a few combat missions in the closing months of the conflict.

Jet aircraft had the potential for greater speed, faster climbs; they could also operate at higher altitudes. In the years immediately after the war, both the Navy and the Air Force (which was separated from the Army in 1947) were eager to convert to jet fleets. Contracts were given to aircraft manufacturers and university laboratories to conduct aerodynamic research that would lead to aircraft designs that could realize the potential of the jet engines. In pursuit of this objective, the Air Force, disappointed that NACA had not stayed abreast of British and German developments, also created its own research and testing facilities—among them the Aeronautical Engineering Development Center in Tullahoma, Tennessee; Wright-Patterson Field, in Dayton, Ohio; and Edwards Air Force Base in California. Both the Navy and the Air Force organized a special corps of pilots to test the new planes and special training programs (which included flight simulators) to teach new skills to their pilots, who were accustomed to propeller-driven aircraft. As a result of intense activity in research, development, testing, production, and training in the decade between 1945 and 1955, the nation's military services entered the jet age with such aircraft as the F-86 fighter, the B-47 bomber (with its swept-wing design), and the B-52, the first intercontinental bomber, capable of carrying uranium or hydrogen bombs nonstop from the United States to the Soviet Union.

The commercial jet airliner was an afterthought, a spin-off. Airline executives had been watching developments in military aeronautics throughout the first postwar decade but had not been convinced that jet aircraft would prove commercially profitable, in part because fuel expenses were very high and in part because new, longer runways were going to have to be constructed at airports. (British turboprop engines—a gas turbine engine with a gearbox linkage to a propeller—were a compromise solution for a few years.) In the early 1950s, however, the Pratt and Whitney Company developed a new jet engine (an axial flow engine) for the B-52s that reduced fuel consumption. At that point several aircraft manufacturers decided to try creating a commercial airliner powered by these new engines. Senior executives realized that the development work would not be difficult; as defense contractors they already had teams of engineers with experience in jet aircraft design, and their machine tool departments, having helped develop the jet fighters and bombers, were already equipped with the tools needed to manufacture the basic component parts. The Boeing 707 jet airliner entered service in 1958; its competitor, Douglas Aircraft's DC-8, was in the air a year later. Both planes and their successors established the superiority of jet travel—for speed, for passenger comfort, and for airline profitability—and also established American predominance in the manufacture of jet airliners. In the United States, civilian jet travel was, one might say, spin-off with a capital "S," for it had important implications not just for the lives of individuals but also for the American economy as a whole.

As important as jet travel was, and still is, to Americans, the advent of the jet did not galvanize their attention or capture their imaginations in quite the same way as the leap into space. The various projects that together constitute the American contribution to space travel were all examples of federally funded managed research. In a technological sense, these projects were all spin-offs from military research; each had spin-offs of its own, some of which were important to scientists and some to consumers. In a political sense, each of these projects was also an important element of the cold war with the Soviet Union. The political impetus behind the race into space was the American government's desire to demonstrate that the United States was scientifically and technologically superior to the Soviet Union. The space race was as much a part of the cold war as the arms race and just as much an example of managed research as the uranium and hydrogen bombs. The international prestige that the government hoped to generate by paying for space exploration was, quite straightforwardly, another kind of spin-off.

Rockets, missiles that receive their motive power from an explosion, are very old technologies (the Chinese were using them as far back as the year 1200), but the idea of building rockets powerful enough to travel beyond the atmosphere appears to be a twentieth-century invention, or perhaps it would be better to say a twentieth-century compulsion. The pioneers of rocketry were a diverse lot, but the one thing they had in common was a passionate interest in space travel. Konstantin Tsiolkovsky (1857–1935) was a Russian theorist who, building on the scientific work done by several European chemists, first suggested that liquid hydrogen and liquid oxygen would be the ideal propellants for a vehicle traveling in space. Robert H. Goddard (1882–1945), an American physicist, succeeded, during more than twenty years of experimentation, in building a rocket that attained an altitude of 7,500 feet. Goddard's work was supported partly by the government-funded Smithsonian Institution and partly by the Guggenheim aeronautical philanthropies; he also pioneered technologies for the guidance and control of rockets, as well as for the special welding, insulation, and pumps that would be needed for liquid fuels, which must be kept at very low temperatures. Hermann Oberth (b. 1894), a German, was an early propagandist for space travel. In the 1920s, he created and presided over the German Society for Space Travel and succeeded in attracting the interest of the German military. The Versailles Treaty, which ended World War I, had restricted German rearmament, but rockets were not covered by the treaty. Oberth's protégé Wernher von Braun, was hired by the German army to direct the rocket research facility in Peenemuende, where the famous V-2 rockets were developed and then launched (against British cities) in the closing days of World War II.

Von Braun and several members of his team were captured by American forces in the spring of 1945, and in what many people later came to believe was disgracefully short order, they became employees of the United States Army, assigned to developing increasingly larger and more powerful rocket engines for increasingly longer-range missiles that would have increasingly more accurate guidance systems. These missiles were intended for the strategic purpose of delivering nuclear warheads to the Soviet Union, but von Braun never lost, and indeed never disguised, his ultimate interest in space travel. Neither did many of the scores of aeronautical scientists and engineers who were set to work—at universities, at consulting companies, at in-house research laboratories, at manufacturing companies—on numerous rocket engine and ballistic missile projects for the other armed services.

The opportunity to work on a space project rather than a warhead project finally arose in 1955, when a committee of scientists managed to convince Congress and President Eisenhower that a good deal of international prestige could be garnered if the United States could succeed in launching an artificial satellite into orbit around the earth during the International Geophysical Year, 1957–58. A blue-ribbon panel was given the task of choosing the launching rocket and designing the capsule of the satellite, but a great deal of time was lost as each of the armed services politicked to have its own missile development team selected. In the end, the Soviet Union beat the United States to the punch—and to the prestige: Sputnik I, the first artificial satellite, was sent into orbit on October 4, 1957. Explorer I, the American equivalent, was not launched for another four months; its success was credited to an Army team under von Braun, which was asked by the Secretary of Defense to step into the breach created by the failure of the Navy team originally chosen by the panel.

Sputnik I sent the nation into a tailspin, inspired in equal parts by fear and embarrassment. Fully aware of the delays that had been caused by military politicking, President Eisenhower proposed to Congress, on von Braun's urging, that a new agency be created, a civilian agency, devoted entirely to space exploration, for peaceful purposes as well as for defense. On July 29, 1958, NASA (the National Aeronautics and Space Agency) was created, replacing NACA. Within a few weeks, NASA announced plans to begin launching communications satellites (about which we will have more to say in Chapter 12) and to put an American astronaut into orbit— Project Mercury. Wernher von Braun and his team (which now numbered dozens of scientists and thousands of engineers and technicians) were reassigned from the Army to NASA.

Project Mercury used an Army rocket (the Redstone) as the boosting rocket for suborbital test flights and an Air Force rocket (the Atlas ICBM) for its orbital efforts. A capsule, which would enable one or more astronauts to live and work in space, had to be designed, as well as launch vehicles that would adapt the military rockets to the capsule. These tasks were carried out at new research and development facilities that NASA created: the Marshall Space Flight Center in Huntsville, Alabama; the Manned Spacecraft Center (later the Johnson Center) in Houston, Texas; the Goddard Space Flight Center near Beltsville, Maryland; and the Cape Canaveral (later Kennedy) Space Center in Florida. Many university and commercial laboratories also acted as contractors to NASA.

However, in the race for manned flight, the Soviet Union once again got there first: Yuri Gagarin completed one full orbit of the earth on April 12, 1961—a full year before John Glenn accomplished the same feat for the United States. The nation once again was both embarrassed and fearful. In the 1960 presidential campaign, the Democratic candidate, John F. Kennedy, had assaulted the Republican party and its candidate, Richard M. Nixon (who was Eisenhower's vice president) for its failure to compete successfully with the Soviets in space. A few weeks after Gagarin's successful voyage, President Kennedy announced that the United States, under von Braun's leadership, was going to land a man on the moon. There was some, but not very much, scientific value in such an enterprise. Kennedy's motives were fundamentally political; he was determined to beat the Russians at what had become by then their own space game.

Eight years later, on July 20, 1969, millions of people all over the world watched their televisions in wonder as Neil Armstrong and Edwin Aldrin planted the American flag on the moon and announced that they had taken "one small step for a man, one giant leap for mankind." The program to set an American on the moon, dubbed the Apollo program, was the largest managed research project of all time. Even the scale of the vehicle itself was outsized; the dimensions of the launch vehicle that stood on a pad in Florida in mid-July of 1969 are difficult to visualize. The Saturn V rocket had a diameter of 33 feet (three moving vans could have been driven, side by side, into the fuel tanks for the first stage) and a height of 363 feet (about the size of a thirty-six-story building). At lift-off, the vehicle weighed 6.1 *million pounds,* and when the five engines of the first stage were fired (the pumps that sent the fuel and the oxygen into the firing chambers generated the power of thirty diesel locomotives), they generated 7.5 million pounds of thrust, almost ten times more than conventional ballistic missile engines of the day. Complex control systems had to be developed for these engines since at liftoff they had to be able to receive three *tons* of fuel per *second* and to fire (by combining the fuel with enough liquid oxygen to fill several dozen railroad cars) simultaneously. Developing the second- and third-stage engines (for powering the command module in which three astronauts traveled and the lunar module in which two of them actually landed on the moon's surface) was an additional monumental task; special tanks and pumps had to be used because the liquid gases had to be kept at −423 degrees Fahrenheit. Special space suits were required, as well as special communications and tracking systems; the instruments that the astronauts would use to collect specimens and perform experiments on the moon's surface also had to be specially designed.

But the managerial enterprise that coordinated the work of all the contractors and subcontractors who developed and manufactured the components for that vehicle was equally gargantuan. New manufacturing techniques had to be developed to achieve new levels of tolerance so that the component parts would fit together perfectly and stay together under extraordinary stress. Wholly new management systems for keeping track of the progress of very abstruse work going on at hundreds of different sites had to be developed as well. Manufacturing and testing of component parts took place in Alabama, Mississippi, Louisiana, and California. Transporting segments of the vehicle required special barges that floated down the Mississippi River and across the Gulf of Mexico, as well as the creation of a special company (aptly named Aero Spacelines) that flew enormous cargo planes (one of which was called the "Pregnant Guppy"). The Apollo program took the techniques of managed research to new heights, literally and figuratively.

During the 1970s and 1980s, Apollo I was followed by several other manned and unmanned missions to the moon. In addition, NASA managed the Skylab program (which sent astronauts into earth orbit in order to perform experiments in space), and the space shuttle program (which developed orbiter vehicles—reusable space transports). Several explorato-

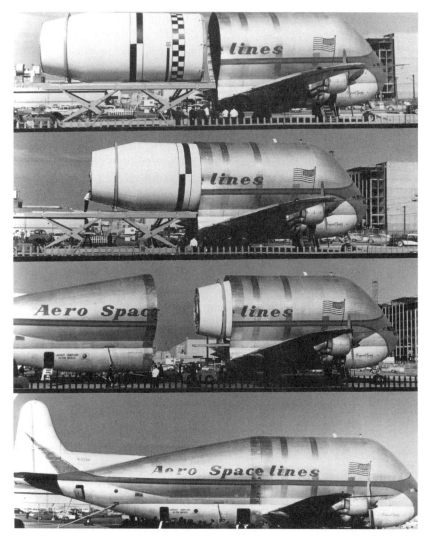

Coordinating the manufacture and transportation of space vehicle components was a complex enterprise. Here, the fourth stage of the Saturn I vehicle is being loaded onto the "Pregnant Guppy" for the trip from the West Coast to Florida prior to an Apollo launch. (Courtesy National Aeronautics and Space Administration.)

ry programs that flew by all the planets followed, as did the Hubble space telescope (which, once repaired, proved capable of "seeing" into space without atmospheric distortion). But none of these programs rivaled the Apollo program either in scope or in size. The various antiwar movements of the late 1960s and 1970s, the debilitating inflations of the 1970s and 1980s, the end of the cold war have all, in one way or another, taken the wind out of NASA's sails. NASA's budgets have been repeatedly cut back, and a good part of the nation's research effort has been diverted away from aeronautics and into medicine. While the military-industrial-academic complex has not by any means disappeared, in the mid 1990s it has lost much of its commanding role both in the American economy and in the scientific and technological community.

Conclusion: Costs and Benefits of Military Sponsorship

In the postwar years, managed research projects brought us the hydrogen (fusion) bomb, the intercontinental ballistic missile, ever larger and more powerful atom-smashing reactors, the nuclear-powered and nuclear-armed submarine, the early warning system (to detect Soviet ICBMs), the rocket-launched orbiting satellites, the manned orbital missions, the moon walk, the space probes, the space telescope, and the space shuttles. All were enormous projects, requiring the coordination of complex components, in which the efforts of scientists and engineers who worked for government, private industry, or universities were directed toward the accomplishment of military or political goals at taxpayer expense.

All these projects had numerous material spin-offs for Americans. A host of consumer goods based on components that were either invented for, designed for, or specially manufactured for the needs of space vehicles—solar energy cells, transistors, silicon chips—have since found their way into calculators, radios, computers, and television sets, as well as pacemakers, artificial hearts, replacement hips, and dental fillings. Yet other spin-offs have to do with manufacturing techniques that were invented to satisfy NASA's extraordinary requirements, but have since been used to produce the multitudinous plastics and fabrics and widgets and gadgets on which our daily lives have come to depend. Thus the space program, paid for with tax dollars, is akin to the numerous military aviation programs of the postwar years, and both are akin to the numerous programs by which—even before World War II—the federal government sponsored technological change through its military services.

Some people are discomfited by the extent to which the military has determined the path of technological change, not only in aeronautics but in other fields as well. These people wonder what technological and social progress we might have made if as much money and brainpower had been expended, for example, in finding a cure for cancer or developing nonpolluting automobiles. They are also more than a little disturbed by the fact that between 1945 and 1990 the economies of several large regions of the country came to depend on the prosperity of defense industries, which has made it difficult (because of the unemployment that would result) for the nation to disarm now that the cold war has apparently ended. Universities also changed profoundly in the decades after the end of World War II. In far too many instances, as far as these critics are concerned, univeristy budgets became dependent on military contracts and government-sponsored research grants. Faculty ranks were swelled by large numbers of scientists and engineers who were paid to do research, not to teach. In addition, because the government feared Soviet spies, scientific work done under many government contracts was frequently designated "top secret," which meant that the professional ethic which requires scientists to make their results public was violated. Particularly in the late 1960s, student and faculty protests were often directed against this situation, which was deemed to distort the mission of educational institutions and compromise their traditional independence from governmental authority.

Other people—over time, probably the majority and certainly the majority of those in positions of power—applaud both the methods and the results that this historically unique combination of military, university, and industrial talent produced. The arms race, they say, was indeed an unfortunate distortion of the American economy, but it kept World War III at bay, and World War III—had it broken out—would have been vastly more distorting. While it would have been wonderful to win the "war" against cancer or the "war" against pollution, neither victory (had there been a victory) would have been worth a hill of beans, these proponents argue, if the war against communism had not been won first. The space race, people believe, wasn't unfortunate at all since it opened a new frontier that the human race could profit not only from exploring but also from exploiting. Military needs have always been important determinants of technological change, the argument continues, but in the twentieth century, as a result of the intensification of this relationship, spin-offs that have improved everyday life have been produced more rapidly, and in greater numbers, than ever before.

Trying to weigh the costs against the benefits when the military acts not just as sponsor but also as determiner of technological change is a difficult exercise. Even if it were possible to get the various contending parties to agree on how to estimate costs and benefits, it would probably never be possible to get them to agree on how to balance those estimates against each other. The history of technology can, however, help us to see that there are some valid points to be made on both sides of the argument.

Notes

1. Roger Bilstein, *Flight in America: From the Wrights to the Astronauts* (Baltimore: Johns Hopkins University Press, 1984), p. 10.

2. Letter to the War Department (June 15, 1907), as quoted in Fred Kelly, *The Wright Brothers* (New York: Harcourt, 1943), p. 216.

3. As quoted in Roger Bilstein, *Orders of Magnitude: A History of the NACA and NASA, 1915–1990* (Washington, DC: NASA, 1989), p. 4.

4. Speech to Congress, February 12, 1939, as quoted in Alfred Goldberg, ed., *A History of the United States Air Force, 1907–1957* (Princeton, NJ: Van Nostrand, 1957), p. 44.

5. FDR to Vannevar Bush, November 17, 1944, as quoted in Vannevar Bush, *Science, The Endless Frontier* [1945] (Washington, DC: National Science Foundation, 1990), p. 3.

6. J. Robert Oppenheimer, *Uncommon Sense*, ed. N. Metropolis, G. C. Rota, and D. Sharp (Cambridge, MA: Birkhauser, 1984), p. 30.

7. Bilstein, *Flight in America*, p. 172.

Suggestions for Further Reading

Roger E. Bilstein, *Flight in America: From the Wrights to the Astronauts* (Baltimore, 1984).

William David Compton, *Where No Man Has Gone Before: A History of Apollo Lunar Exploration Missions* (Washington, DC, 1989).

Joseph Corn, *The Winged Gospel: America's Romance with Aviation* (New York, 1983).

Tom D. Crouch, *A Dream of Wings: Americans and the Airplane, 1875–1905* (New York, 1981).

Virginia Dawson, *Engines and Innovation: Lewis Laboratory and American Propulsion Technology* (Washington, DC, 1991).

A. Hunter Dupree, *Science in the Federal Government: A History of Politics and Activities* (Baltimore, 1957).

Sylvia D. Fries, *NASA Engineers and the Age of Apollo* (Washington, DC, 1992).

Constance M. Green and Milton Lomask, *Vanguard: A History* (Washington, DC, 1971).

James R. Hansen, *Engineer in Charge: A History of the Langley Aeronautical Laboratory, 1917–1958* (Washington, DC, 1987).

Peter L. Jakab, *Visions of a Flying Machine: The Wright Brothers and the Process of Invention* (Washington, DC, 1990).

Daniel J. Kevles, *The Physicists: The History of a Scientific Community in Modern America* (Cambridge, MA, 1977).

Nick A. Komons, *Bonfires to Beacons: Federal Civil Aviation Policy under the Air Commerce Act, 1926 to 1938* (Washington, DC, 1978).

Clayton R. Koppes, *JPL and the American Space Program: A History of the Jet Propulsion Laboratory* (New Haven, CT, 1982).

William M. Leary, *Aerial Pioneers: The U.S. Air Mail Service, 1918–1927* (Washington, DC, 1986).

Stuart W. Leslie, *The Cold War and American Science: The Military-Industrial-Academic Complex at MIT and Stanford* (New York, 1993).

W. David Lewis and W. Phillips Newton, *Delta: The History of an Airline* (Athens, GA, 1979).

John M. Logsdon, *The Decision to Go to the Moon: Project Apollo and the National Interest* (Cambridge, MA, 1970).

Pamela Mack, *Viewing the Earth: The Social Construction of the Landsat Satellite System* (Cambridge, MA, 1990).

Donald MacKenzie, *Inventing Accuracy: A Historical Sociology of Nuclear Missile Guidance* (Cambridge, MA, 1990).

Howard E. McCurdy, *The Space Station Decision: Incremental Politics and Technological Choice* (Baltimore, 1990).

Walter McDougall, *The Heavens and the Earth: A Political History of the Space Age* (New York: 1985).

Seymour Melman, *Pentagon Capitalism: The Political Economy of War* (New York, 1971).

Ronald Miller and David Sawers, *The Technical Development of Modern Aviation* (New York, 1970).

David Noble, *Forces of Production: A Social History of Industrial Automation* (New York, 1984).

Carroll Pursell, *The Military-Industrial Complex* (New York, 1972).

Alex Roland, *Model Research: The National Advisory Committee for Aeronautics* (Washington, DC, 1985).

C. R. Roseberry, *Glenn Curtiss, Pioneer of Flight* (Syracuse, NY, 1991).

Merritt Roe Smith, ed., *Military Enterprise and Technological Change: Perspectives on the American Experience* (Cambridge, MA, 1985).

Robert W. Smith, *The Space Telescope: A Study of NASA, Science, Technology and Politics* (New York, 1989).

John Tirman, ed., *The Militarization of High Technology* (Cambridge, MA, 1984).

Walter G. Vincenti, *What Engineers Know and How They Know It: Analytical Studies from Aeronautical History* (Baltimore, 1990).

Donald R. Whitnah, *Safer Skyways: Federal Control of Aviation, 1926–1966* (Ames, IA, 1966).

12

Communications Technologies
and Social Control

COMMUNICATION IS A FUNDAMENTAL ACTIVITY in which all human beings engage. Electronic devices—radio, television, computers—have had a profound and quite conspicuous impact on American society in the twentieth century because they have altered patterns of communication, the ways in which people convey information to one another. Like all profound social changes, the communications revolution has raised numerous troubling questions; some of those troubling questions concern the issue of social control. Who should be in charge of all the various technologies of communication? Industry? The government? Individuals? And who gets to control the broadcasting of information when one of the media that allows the passage of electronic waves—the earth's atmosphere—belongs, in some sense, to everyone?

Wireless Telegraphy

The history of modern American communications technology begins not in the United States, but in the United Kingdom, Germany, and Italy. A nineteenth-century Scottish physicist, James Clerk Maxwell, demonstrated that light was an electromagnetic wave, and he predicted, on theoretical grounds, that similar waves of different frequencies (either higher or lower than light) could be generated by electric discharges (sparks). In 1887, a German physicist, Heinrich Hertz, created an apparatus to generate and measure both high-frequency (a few centimeters between each wave crest) and low-frequency (a few meters) waves from a device called a spark-gap transmitter.

A few years later, Guglielmo Marconi, a young Italian who was studying electricity, read a description of Hertz's apparatus and noticed a feature of the apparatus that had apparently escaped Hertz's attention: Hertz had sent an electric signal from one place (the transmitter) to another (the measuring device) *without wires*. If it would work across the laboratory, would it work across a field? across a river? between mountaintops? Marconi began experimenting. Over the next two years, he succeeded in sending messages in Morse code (long sparks, short sparks) as far as two miles; he also developed a simple apparatus that would receive the waves (an antenna) and convert them into direct current so that, like telegraph signals, they could be heard by someone listening to the pattern of the current through earphones.

Marconi was not the only person experimenting with "wireless telegraphy" at the time, but he was the person who figured out a way to make money with it. First, he identified a likely market for his invention: ship-to-shore or ship-to-ship communications, he reasoned, were markets that could not possibly be served by the existing systems that depended on wires, the telegraph and the telephone. Marconi's father was a wealthy businessman and his mother was English; the family knew precisely what had to be done. In 1896, Marconi and his mother went to England to obtain a patent on the invention. Through his parents' social connections, Marconi was able to demonstrate his apparatus to officials of the British Navy and to raise the money that was needed to manufacture transmitters, antennae, and receivers. British Marconi was formed in 1897; a subsidiary, American Marconi, followed in 1899.

Wireless telegraphy was an instant success. The British Navy bought wireless equipment to communicate with its imperial fleet, which plied every ocean of the world; the British Army wanted Marconi's invention so that command posts could communicate with troops in the field. The United States had recently acquired the Philippines, Cuba, Hawaii, and Puerto Rico; the State Department wanted wireless systems to communicate with its new overseas possessions, and the American Navy wanted to be able to stay in touch with its newly far-flung fleet. Dozens of companies whose business depended on shipping were also interested. So were the owners of newspapers, who realized (after some of Marconi's more spectacular publicity stunts) that reporters could use wireless transmitters to communicate the news from distant places, places not served by telegraph offices.

Within five years, Marconi had become a very rich man and wireless telegraphy had become a turn-of-the century popular fad. Young people with technical interests were fascinated. They were enthralled by their ability to manipulate mysterious waves that could be neither seen nor felt, yet the equipment needed was not very expensive and the techniques for using (and repairing it) were fairly easy to learn. Amateur wireless operators were soon forming into clubs, astounding their friends with school yard demonstrations, chatting with each other over long distances, volunteering to summon doctors for distressed neighbors—all in Morse code.

Marconi, young and dashing, suddenly both very famous and very wealthy, was a charismatic figure; many young men tried to emulate him. Wireless telegraphy was clearly a growth industry, and the potential for making a great deal of money had already been demonstrated. The apparatus that Marconi was manufacturing had its limitations (enormous spark-gap transmitters, for example, were dangerous, and the larger they got, the fuzzier the signal they sent out). By 1900, young men in all the industrialized countries, many with degrees in physics or electrical engineering, were trying to beat Marconi at his own game—to invent a better receiver, concoct a better transmitter, improve the characteristics of antennae. Investors were willing to back some of those young men in the hope that they might be able to eat away at some of Marconi's share of the market and the profits. As a result, the first control issues that reared their heads in the new century, in the newborn world of electronics, had to do with the control of patents and, through the control of patents, the control of markets.

Wireless Telephony

Wireless telegraphy was invented and commercialized by Europeans, but wireless *telephony* was pioneered by Americans. Among the thousands of young men and women who became amateur wireless operators, there were a few who realized that the next technically interesting and commercially exploitable frontier would be the transmission not of Morse code, but of *real sounds*: voices and music, not just dots and dashes.

One of these young men was Reginald Fessenden. Fessenden had been born in Canada in 1866; after working for several years as a schoolteacher, he had come to the United States in 1886 and had succeeded in landing jobs, first in one of Thomas Edison's manufacturing companies and, then, in the laboratory of the great inventor. Fessenden stayed with Edison for only three years, but that was enough; he had apprenticed himself to the world's foremost master of invention and development.

By 1900, the mechanism of the telephone, which converts sound waves into electric currents, was already well-known. Fessenden reasoned by analogy with wireless telegraphy that wireless telephony could also be possible; sound waves could be converted into electromagnetic waves and then projected across the atmosphere just as they could be sent across wires. The practical problems involved were immense, but in theory, they were all solvable. New kinds of transmitters and receivers would be required. Marconi's success was an inspiration; Fessenden, who was then teaching at the University of Pittsburgh, began experimenting.

In 1901, Fessenden designed a new kind of receiver, a heterodyne receiver, which could convert high-frequency waves produced by spark-gap transmitters into low-frequency waves, the kind that make diaphragms resonate in telephones. Fessenden took out a patent and went into business for himself (and several investors) as the National Electric Signalling Company. By 1902, he had also designed a high-speed spark transmitter, called an alternator, that would generate sparks so fast that the waves it created were almost continuous (voice transmission requires continuous waves; Morse code transmission involves intermittent waves) and had contracted with the General Electric Company to build it.

Fessenden's alternators had a greater range than Marconi's spark-gap transmitters, making them suitable for *very* long distance point-to-point wireless telegraphic communications. The company built such a signaling station on the Atlantic shore, in Brant Rock, Massachusetts, hoping to contract with merchant shipping companies and the Navy. On Christmas Eve 1906, wireless operators on ships hundreds of miles out at sea were startled to hear—instead of the dots and dashes that they were accustomed to—the voice of a woman singing, then a violin being played, then passages being read from the Book of Luke. Using equipment that would otherwise be sending Morse code, Fessenden had succeeded in sending the first "radio" messages.

Fessenden was unable, however, to interest very many people in his new system. His wireless telephonic signals were very weak, amplification was difficult to achieve, and accurate tuning to the transmission frequencies was close to impossible. Point-to-point voice communication on land was handled by the telephone system, and those who wanted ship-to-shore communications had already invested whatever capital they were going to invest in communication in wireless telegraphy, which was adequate for most commercial purposes. Although Fessenden continued to experiment with voice transmission, much of his energy in the next several years was devoted to perfecting his alternator and selling it to organizations that wanted telegraphy services; he was also embroiled in a patent dispute with Lee DeForest.

DeForest was, like Fessenden, a minister's son. He had been born in Iowa, raised in Georgia, and educated at Yale, where he had earned a doctorate in physics in 1899—the same year in which Marconi had broached the American market. DeForest had written his dissertation on features of high-frequency alternating currents. Chronically short of cash, he had taken a series of unsatisfactory jobs after completing his degree, but he had also been bitten by the wireless bug and had started experimenting with improvements on receivers. Sometime in 1901, he made patent applications for a detecting device he called a responder; with this invention and a small amount of capital contributed by friends, he attempted to go into business—just as Fessenden had done—in competition with Marconi. The responder didn't work very well, but in his effort to garner publicity for his new company, DeForest caught the attention of a stock promoter by the name of Abraham White. Before long, White had created a new company, DeForest Wireless Telegraphy, and had brought it public with a $1 million stock offering; by 1904, the capitalization had leaped to $15 million—reflecting both White's public relations skills and the popularity of wireless telegraphy.

DeForest suddenly became very rich; he could afford to build and staff the best laboratories for developing and testing new pieces of equipment. White and his assistants ran the company; DeForest was responsible for research and development. In short order, however, Fessenden sued DeForest for patent infringement; when that suit was settled in Fessenden's favor in 1906, White used it as a pretext to do something he had probably wanted to do for a long time: drive DeForest out of the company that bore his name. DeForest, then desperate to invent something on which he could subsequently begin building another wireless empire, invented and patented the audion.

The audion was a device based on a phenomenon discovered several years earlier by Thomas Edison: if a light bulb contains a filament *and* a metal plate, when an electric current is run through the filament, the current will pass to the metal plate and, if the plate is attached to its own wire, pass out of the bulb. Sometime in 1906, while searching the literature for new ideas for detectors, DeForest had come across a paper published two years earlier by John Ambrose Fleming, a physicist who was then working for British Marconi. Fleming had discovered that when alternating current was applied to the filament of such a bulb, direct current would emerge from the plate. Fleming's device was called a diode: it contained two electrodes, the filament and the plate, with the very special property of being able to convert alternating to direct current.

DeForest immediately began experimenting with Fleming's device and discovered that when a third element was added to the bulb—a small wire grid that could carry its own current—the bulb could operate as a very sensitive wave detector, picking up fairly faint electromagnetic signals at any frequency and amplifying them. By adding a third element to the bulb, DeForest had created a triode, but he called it an audion to signify that it could detect a continuous electromagnetic wave, the kinds of wave that made wireless telephony possible. The audion was, in short, the ancestor of all the vacuum tube technology on which, a decade later, radio broadcasting would be based.

DeForest believed that there might someday be a mass market for wireless telephonic receivers—and therefore a mass market for his audions—but he also understood that a good deal of developmental work was going to have to be done before that day would arrive. He frequently experimented with voice transmissions—playing phonograph records into a microphone in his laboratory or trying to broadcast from the stage of the Metropolitan Opera House—but without much success; listeners just a few blocks away could barely make out what was being sung or played. Had he been able to devote himself full time to the necessary developmental work, it is possible that radio broadcasting would have emerged sooner than it did, but DeForest's business and personal affairs were in a perilous state for several years and commanded most of his attention. In 1912, in fact, at the height of his financial problems, he sold the patent rights for the audion to AT&T, which wanted to use this first vacuum tube as an amplifier to improve its long distance telephone service.

While DeForest was neglecting his audion, someone else was experimenting with it. As a child, Edwin H. Armstrong had been a devoted amateur wireless operator; as a student in the engineering college of Columbia University, he decided to experiment with DeForest's remarkable amplifier for his senior thesis. Neither Fleming nor DeForest had been able to explain why the diodes and triodes worked the way they did, but Armstrong thought that the then new theory of electric currents—that they involved the flow of negatively charged subatomic particles called electrons—might be applicable. During 1912, in the course of his experiments, Armstrong discovered something remarkable: if the current coming off the plate of an audion was fed back into the grid of the tube—creating what was later called a feedback, or regenerative, circuit—sounds could be even further amplified, making it possible to dispense with earphones.

A few months later, still experimenting, Armstrong went one, very crucial step further; he discovered that, under certain circumstances, regenerative circuits could transform vacuum tubes into transmitters of electromagnetic waves. This was even more startling. During the first decade of wireless telegraphy, spark-gap transmitters and alternators had been getting larger and larger in order to meet the commercial demand that electronic messages be able to travel farther and farther. The state of the art was the Alexanderson alternator, named for the engineer who had designed and built it for the General Electric Company: a huge disc that turned 20,000 times every minute (its outer rim reaching speeds of 700 miles per hour) when supplied with 200 kilowatts of energy. Working away in his laboratory at Columbia in the winter of 1913, Armstrong had discovered that a handful of specially adapted light bulbs could do the same job as the room-sized alternator.

Armstrong graduated from Columbia in June 1913. In the fall, he applied for a patent on his regenerative amplifying circuit; in the winter, for one on his transmitting circuit. Because he had failed to interest either AT&T or the Marconi companies in his patents, Armstrong sold the rights to his regenerative circuit, for the then-munificent sum of $100 a month, to the Telefunken Company of Germany. In the fall of 1914, World War I had started. The British had cut the telegraph cables that linked Germany with the United States; wireless was going to have to be used instead. Telefunken needed the best amplifiers it could find.

World War I raged for two and a half years (between August 1914 and April 1917) before the United States entered it on the side of the Allies. During those two and a half years, it became clear to virtually everyone involved in the communications industry that after the war wireless telephony based on vacuum tube technology was going to become both feasible and profitable. A mass market for radio receivers, simple and durable instruments that people could use in their homes, was likely to develop. But who would reap the profits from this potential market? Would it be British and American Marconi, who owned the patents on Fleming's diode, but whose principal business, wireless telegraphy, would be diminished if radio became popular? Or AT&T, who owned the patents on DeForest's audion, but whose principal business, telephone service, might also be diminished? DeForest, who had sold the patent rights to his audion, but who was legally entitled to manufacture it? Telefunken, who owned the rights to the regenerative circuit on which radio amplification would be based? Armstrong, who owned the patent on the transmitting circuit, but who would be able to exploit it only if Telefunken chose not to interfere?

For the duration of the war, none of these questions could be answered. Nonetheless, in anticipation of the war's end and in an effort to control the radio marketplace—which had not at that point been either clearly defined or measured—all the potential recipients of the postwar radio profits had filed patent suits against each other. In April 1917, when the United States declared war on Germany and Austria, neither the outcome of the war nor the outcome of those suits could have been easily predicted.

Government Regulation of Wireless Communication

In April 1912, a luxury ocean liner, the *Titanic*, making its maiden voyage, hit an iceberg in the north Atlantic and sank. The wireless operator on the *Titanic* sent out wireless distress signals, which were picked up by another ship, twenty-five miles away, and this ship changed course in an effort to rescue passengers. (In part because several other nearby ships had closed down their wireless receivers for the night, about half the passengers drowned.) The news of the *Titanic* disaster reached American newspapers by wireless telegraphy (having been transmitted to New York by a Marconi station in Newfoundland), and anxious relatives of the passengers began trying to radio the rescue vessel to get a list of survivors, using any transmitter, amateur or commercial, that they could locate. Ship-to-shore communications became chaotic; every passing hour brought more false good news, false bad news, and totally incomprehensible news. Within days, journalists began demanding government regulation of wireless communication; within a week, Congress began taking testimony.

The Radio Licensing Act of 1912 was intended to cope with what then seemed to be a pressing problem: the frequency spectrum was becoming cluttered with electromagnetic signals, some of which were interfering with each other, some of which were so close together that receivers could not be properly tuned. Thousands of amateur operators, hundreds of commercial systems, and dozens of governmental, educational, and military stations were all sending messages. The American Navy in particular was becoming increasingly dependent on wireless telegraphy and increasingly concerned about interference with and interception of its messages; in one case, perhaps apocryphal, amateurs were said to have sent fake orders to naval vessels at sea.

The Radio Act of 1912 required that anyone with transmitting equipment had to have a license issued by the federal government and that anyone wishing to be a transmission operator had to have passed an examination set by the government. In issuing licenses, Congress also gave the government (through the Secretary of Commerce and Labor) the right to assign a particular frequency and particular time limits for transmission to the licensee; in addition, the licensee was given an identifying call number that had to be used on all transmissions. The frequency spectrum was carved up by function. Commercial transmitters got one range of frequencies; the government got another; amateurs were left, quite literally, with the short end of the stick: short waves of 200 meters or less, then considered the useless part of the spectrum. Amateurs could listen in on any frequency, but transmit on only a few—and heavy fines were exacted for irresponsible transmission or transmissions that trespassed on frequencies allotted to others.

With this act, the United States government asserted its right to control the airwaves—in much the same way that it had earlier asserted its right to control the railroad rights-of-way and the shipping lanes—in the name of public safety; the radio spectrum was understood to be the common property of all the citizens of the United States and therefore under the jurisdiction of the federal government. Licensees would have the same kinds of property rights that renters do: exclusive use of the frequencies they had been assigned without the ability to buy and sell that allocation on the open market. The government would retain control over allocations and over the criteria by which allocations would be made, but not over the content of transmissions (except for the very restricted right to punish people who sent false emergency signals or who trespassed on someone else's frequency).

Amateur operators were not happy with the Radio Act of 1912; they argued that its provisions compromised their right to free speech. Their arguments, however, fell on deaf congressional ears; even newspaper owners and editors, usually avid proponents of free speech, had become dependent on the wireless news services in the previous decade, and they were perfectly delighted to have the amateurs—who occasionally interfered with commercial transmissions—sequestered in a remote corner of the spectrum. The government was not proposing to regulate the *content* of transmissions, most members of the public did not depend on the airwaves as a source of information, nor did they use Morse code to communicate. Public opinion, insofar as there was any on this rather arcane issue, sided with the government; the amateur operators had no recourse.

Another provision of the act allowed the government to commandeer privately owned transmission and receiving facilities in the event of a national emergency, and shortly after the declaration of war in 1914, that was precisely what the government began to do. The United States was initially a neutral party to the conflict, but many people in the government assumed that it would only be a matter of time before the United States would be drawn into the fray, on one side or the other, which meant that it was only a matter of time before wireless equipment was going to be in very short supply. In 1915, the government ordered the end to the dozens of patent suits that were holding up the manufacturing of various electronic components. A patent pool was created for the duration of the conflict, allowing any company with adequate facilities to manufacture anything that the government needed. The Navy began to exercise its right to control transmitting stations and frequency allocations. All amateurs were ordered off the air; their transmitters were to be sealed shut for the duration. In addition, all privately owned transmitting stations on the coasts (many of which belonged to American Marconi) were commandeered for governmental use.

This last was a very sensitive matter. American Marconi was a subsidiary of a British firm. The consequences of having American wireless capacity owned and therefore controlled by a foreign nation were very worrisome during wartime. How could America pretend to be neutral if its ability to communicate with all combatants was compromised by the British owners of its transmitting stations? What would happen if (and during the first year or so of the war, this seemed a possibility) the United States joined combat on the side of the Germans? American Marconi had been the first company to broach the American market, and in 1914 it still controlled a very large share of the market and was still the single largest government contractor for wireless goods and services. Various American companies, most particularly AT&T, had long been pressuring federal officials to support the American competitors of Marconi—and now the advent of war was demonstrating, they thought, the wisdom of that pressure. Between 1914 and 1916, the government began to shift some of its contracts to American firms; it also began to take control, in effect to nationalize, all of American Marconi's transmitting and receiving stations.

During the last months of 1916, as it became clear that the United States was going to enter the war as an ally of the British and the French, it also became clear that American radio manufacturing would accelerate. Wireless telegraphy had become crucial to military strategy on land as well as at sea: orders could be relayed quickly to the front by wireless; reports on troop movements could be relayed to command posts many miles distant by wireless. The American Expeditionary Force (as it was called) was going to need wireless equipment: millions of vacuum tubes, thousands of mobile spark-gap transmitters, huge numbers of receivers and earphones. Dozens of companies vied for wartime contracts: General Electric and Westinghouse, for example, converted some of their lightbulb factories to vacuum tube factories; Western Electric, the manufacturing subsidiary of AT&T, began to manufacture receivers. Thousands of young men and women who had been radio amateurs responded to the government's call: the women staffed training schools for new recruits; the men went right into military service. In January 1917, there were 979 Navy radiomen; by Armistice Day, November 11, 1918, there were 6,700.

Even when the war was over, it was clear that the manufacture of wireless equipment for military purposes would continue to be a very profitable enterprise. Several very powerful men were determined to see to it that neither transmission nor manufacture would return to the prewar status quo in which the airwaves could be more or less freely used by anyone and in which both point-to-point transmission and the manufacture of equipment that made it possible would be dominated by companies owned by foreigners.

A few weeks after the Treaty of Versailles was signed, hearings were held in Congress on a bill sponsored by the Navy Department that would give the federal government a monopoly over radio transmission. Proponents argued that the successes of the war years would not have been possible if the government had not brought order to what had previously been a competitive free-for-all. Opponents were convinced, as one congressman put it, that "[h]aving just won a fight against autocracy, we would start an autocratic movement with this bill."[1] All over Europe, governments were taking precisely such autocratic control over the airwaves, but in the United States, the bill was tabled. The Navy Department was not, however, easily deflected; if it could not have its own monopoly, it could at least try to guarantee that American radio transmission and manufacturing capabilities would be wholly controlled by Americans.

Something had to be done with the vast Marconi facilities for transmitting and receiving that had been commandeered by the government during the war. Would they be returned to American Marconi, a subsidiary of a British company? Secretary of the Navy Josephus Daniels, Commander S. C. Hooper, the Navy officer in charge of radio operations, and Owen D. Young, president of General Electric, quickly came up with a different plan: a new company would be chartered, which could only have United States citizens as its directors and officers; only 20 percent of its stock would belong to foreigners. A representative of the government would sit on the board of directors. American Marconi would transfer all of its assets to this new company. Individual investors in American Marconi would receive shares of the new company's stock. General Electric would purchase the shares owned by British Marconi.

The officers of American Marconi had been made an offer they couldn't refuse. In February 1919, with the tacit approval of the president of the United States, the Radio Corporation of America (RCA) was born—holding almost complete control of all international and ship-to-shore wireless telegraphic facilities in the United States.

But that was not all. Within months, RCA had entered into cross-licensing agreements (in exchange for shares of its stock) with almost all the large companies that held crucial patents for wireless telegraphy and telephony. DeForest and Armstrong would be, henceforth, very rich men, and although their personal enmity led each to sue the other multiple times (two of their suits were carried as far as the Supreme Court), none of their suits interfered with the manufacture of the components and circuits that they had patented. Telefunken's American patent portfolio had been seized by the Alien Property Custodian in 1917 and subsequently sold to General Electric for $1,500, which meant that it became part of the portfolio that RCA could control. Each party to the RCA agreement agreed to respect the other parties' defined spheres of influence: RCA would not compete with AT&T on long-distance telephony; AT&T would not step on RCA's toes with regard to international wireless transmission; Western Electric (a subsidiary of AT&T) and GE would each specialize, without competition from the others, in the manufacture of specified radio components. If not a governmental monopoly, Daniels, Hooper, and Young assumed that American wireless would at least be a set of interlocking private monopolies—much easier for the government to control if control should, for whatever reason, become necessary.

Wireless Broadcasting: Radio

As clever as they were, the men who created RCA did not anticipate the character of the mass market that radio broadcasting would very shortly create. On the night that the *Titanic* sank, one of the wireless telegraphic operators who worked round the clock to relay messages for American Marconi was a young Russian-Jewish immigrant, David Sarnoff. By 1917, Sarnoff had worked his way up through the ranks to become the commercial manager of American Marconi, responsible for the maintenance and expansion of its services to businesses. Two years later, he was the commercial manager of RCA, and two years after that, he was promoted to general manager.

Like Fessenden, DeForest, Armstrong, and several dozen others, Sarnoff had imagined the possibilities of a mass market for wireless telephonic broadcasting—what we now call radio. In the fall of 1916, he had prepared a memorandum for the president of American Marconi in which he envisioned "a plan of development which would make radio a 'household utility' in the same sense as the piano or phonograph." Recent improvements in radio equipment, Sarnoff thought, would make such a scheme entirely feasible:

> [A] radio telephone transmitter having a range of, say, 25 to 50 miles can be installed at a fixed point where the instrumental or vocal music or both are produced. . . . The receiver can be designed in the form of a simple "Radio Music Box" and arranged for several different wave lengths, which should be changeable with the throwing of a single switch or pressing of a single button.

Sarnoff imagined that most of the profits from developing radio telephony as a household utility would come from manufacturing and selling the radio music boxes. But, he guessed, if the broadcasts carried something other than music, the potential market would be even greater:

> Events of national importance can be simultaneously announced and received. Baseball scores can be transmitted in the air by the use of one set installed at the Polo Grounds [where the New York Giants baseball team played]. The same

would be true in other cities. This proposition would be especially interesting to farmers and others living in outlying districts removed from cities. By the purchase of a "radio Music Box" they could enjoy concerts, lectures, music, recitals, etc.[2]

Sarnoff hoped that when the war was over American Marconi would begin to manufacture simple, durable receivers adapted for home use. Unfortunately, after 1918, the complications of setting RCA up in business left him little time to devote to the project. Others, elsewhere, however, were quicker to the punch.

In the last months of the war, Edwin Armstrong had invented a superheterodyne circuit, eight vacuum tubes that, when suitably modified and wired together in precisely the right way, could function as receiver, tuner, and amplifier all in one: a radio. The Westinghouse Corporation purchased the manufacturing rights under Armstrong's patent in 1920. Westinghouse, one of the country's major electrical manufacturers and innovators, had not been included in the RCA consortium.

In the immediate postwar years, radio amateurs all over the country were using vacuum tubes to build transmission facilities in sheds, garages, and attics. In Madison, Wisconsin, a professor and his students were broadcasting weather bulletins and phonograph music from a laboratory at the university; in Hollywood, an electrical engineer built a five-watt transmitter in his bedroom; in Charlotte, North Carolina, an electrical contractor built a transmitter using parts that he had acquired while doing war work for General Electric; in Detroit, a newspaper publisher who had set up a transmitter in his office used it to broadcast primary election returns. All the people assumed that other amateur radio buffs would receive their transmission.

Initially that was also the assumption made by Frank Conrad, who worked for Westinghouse at its headquarters in Pittsburgh, Pennsylvania. Conrad had been an amateur operator before the war; during the war, he had been in charge of Westinghouse's radio production facilities. In 1919, using vacuum tubes, he had built a transmitter over his garage and had used it to broadcast phonograph concerts and conversation; because he had access to the best vacuum tubes, his transmissions could be heard clearly for miles around. Sometime in 1920, with the help of his sons, he began broadcasting concerts from his home.

A department store in Pittsburgh, the Joseph Horne Company, had installed a receiving station on its premises so as to promote the sale of what it called amateur wireless sets. On September 29, 1920, the store ran a newspaper advertisement:

> Victrola music, played into the air over a wireless telephone, was "picked up" by listeners on the wireless receiving station which was recently installed here for patrons interested in wireless experiments. . . . The music was from a Victrola [a record player] pulled up close to the transmitter of a wireless telephone in the home of Frank Conrad. . . . Mr. Conrad is a wireless enthusiast and "puts on"

the wireless concerts periodically for the entertainment of the many people in this district who have wireless sets.

The ad went on to suggest that consumers purchase ready-made sets, "$10.00 and up."[3]

On September 30, the day after the Horne advertisement appeared, a Westinghouse executive, Harry Davis, asked Conrad if he could build a transmitter at corporate headquarters that could be finished in time to broadcast election results on November 2. The department store advertisement had given Davis an idea: if there were voice transmissions that *everyone* might be interested in receiving, then the potential market for receivers that Westinghouse could manufacture under the Armstrong patents would be not just amateur operators but, potentially, *every* American. If it could conquer that mass market, Westinghouse's star might just eclipse RCA's.

Construction began on a shack and a 100-watt transmitter on the roof of one of Westinghouse's factory buildings in East Pittsburgh. An application was filed with the U.S. Department of Commerce (as required by the Radio Act of 1912), and on October 27, the call letters KDKA were assigned. A local newspaper agreed to relay election returns from its teletype machine to the shack by telephone. On November 2, Warren G. Harding's defeat of James M. Cox became the very first radio newsflash, and radio broadcasting was born—with the Westinghouse Corporation as its midwife.

The idea was contagious. Within weeks after the election, Westinghouse built another transmitter atop its plant in Newark, New Jersey, and began broadcasting play-by-play descriptions of baseball games. A few months later, Westinghouse was also broadcasting in Chicago. In other cities, amateurs began building transmitting stations for local businesses: department stores were building stations because the performers who came to broadcast attracted crowds; newspapers were building stations because printing the news and entertainment schedule helped to sell more newspapers; colleges and universities were building stations to provide home-based education; some entrepreneurs were building stations so as to broadcast information about their products between concerts and lectures.

In the first half of 1921, the Department of Commerce issued five new licenses for radio transmission; in the second half, twenty-three; by February 1922, it was issuing twenty-four licenses *a month*; by July, that figure had trebled. And all over the country, amateurs were buying parts for radio receivers and putting them together for sale to friends and family; some of those amateurs were even going into the business of assembling and selling radios to the public. Sales of radio sets and parts for radio sets mushroomed: $60 million in 1922; $136 million in 1923. RCA and its allies had been trumped. One prominent GE executive, writing his memoirs a few years later, remarked that he was "amazed at our blindness . . . We had everything except the idea."[4]

The men in this picture were members of the first generation of broadcast radio engineers. They are preparing the equipment to broadcast a public event, possibly the Dempsey-Carpentier prize fight (1921), which was received by over 300,000 people, the largest radio audience up to that time and a landmark in the popular acceptance of radio broadcasting. (Courtesy David Sarnoff Research Center.)

After frantic negotiations, Westinghouse was brought into the RCA consortium in the spring of 1921; in exchange for 20 percent of the stock in RCA, Westinghouse agreed to put its patents—most especially the Armstrong superheterodyne patent—into the common pool. But even that move did not stop the proliferation across the country of new transmitting stations and new radio manufacturers; the country's appetite for radio seemed, at least for the moment, insatiable. Under the RCA agreements, the sole manufacturer of transmission equipment was supposed to be Western Electric, a subsidiary of AT&T, but the amateurs who were building stations all across the country did not know that—had they known it, they probably would not have cared. Controlling the radio market—now that broadcasting had become considerably more popular than point-to-point communications—was going to be a good deal more difficult than the founders of RCA had predicted just three years earlier.

As the cost of broadcasting began to escalate, the parties to the RCA agreements began to step on each other's toes. In the first, halcyon years of radio, performers had flocked to radio studios; for singers and pianists and opera stars and politicians, the free publicity was sufficient inducement.

But after a singer had been asked to fill up ten or twenty hours of airtime, she might start to feel that she was being exploited. After hearing one of its records played two thousand times—and realizing that its retail market was being undercut—a record company might begin to press for royalties. Composers and lyricists began to think that they, too, were entitled to some payment when their works were performed repeatedly. And when volunteer labor would no longer suffice, station managers began to have to pay salaries to the people who read the news and the weather or reported on baseball games.

Who was going to pay all these expenses? Westinghouse had originally assumed that it would underwrite the expenses of KDKA and its other stations in order to induce householders to buy its radio sets, but that assumption began to look worrisome as the costs of broadcasting began to increase. At RCA, Sarnoff proposed that perhaps it would be a good idea to add a small "tax" to the price of every vacuum tube, putting the money that RCA would collect into a pool from which stations could pay performers. But early in 1922, someone at AT&T came up with a different idea, which would develop into what we have come to understand as commercial broadcasting: why not sell airtime? Why not create, by analogy with the telephones, tolls for broadcasting?

In early August 1922, AT&T's station, WEAF, began broadcasting from atop one of its buildings in New York City. On August 28 at 5 P.M., it broadcast its first income-producing program—a ten-minute "message" describing the advantages of cooperative apartments for sale in Queens—and the radio commercial was born, with the phone company as its midwife. By 1924, WEAF was starting to become profitable, and AT&T began to think about creating other toll stations in other cities, linking them together with its phone lines, thereby creating an opportunity to sell airtime, not in just one locale, but all over the country.

When the officers of AT&T began toting up the potential profits from such a network, they also began to wonder why they were letting other stations use their phone lines for long-distance transmission of signals (for example, when a radio station broadcast a speech at a political convention, it was using phone lines to transmit the speech to its studio; live performances of operas or play-by-play descriptions of sporting events were all being transmitted the same way). Why should it make its facilities available to its competitors? AT&T wondered. And since some of its competitors were its partners in the RCA agreements, why should it remain a party to those agreements? And if it ceased to be a member of the RCA consortium, then couldn't its subsidiary, Western Electric, begin competing in the lucrative market for the manufacture of household radios?

At the very same time that AT&T was beginning to renegotiate its agreements, the Federal Trade Commission, which had received several complaints from small radio manufacturers, began an antitrust investigation of all the members of the consortium. RCA, just four years after its founding, was under siege from two sides: one of the crucial partners, AT&T, wanted to pull out and the government, which had initially welcomed the creation of RCA, now—under a different administration and for different reasons—wanted to break it up.

Young and Sarnoff were, however, brilliant negotiators. Their solution to one of these problems was to set the pattern of social control of radio (and subsequently television) broadcasting for at least the next fifty years. A new company would be created; ownership would be shared by RCA, GE, and Westinghouse. It would own all the broadcasting stations of all the consortium participants, as well as any others that it chose to buy subsequently. AT&T would lease phone lines to this company as needed, whether to permit live, off-site transmissions or to link the stations in one, or several, networks. The stations would have to generate revenue by selling airtime to advertisers. In January 1926, the first media network, the National Broadcasting Company (NBC), was born.

A year later, Congress, which had been again under considerable pressure to create some order in the airwaves, passed the Radio Act of 1927, reaffirming some of the principles of the Radio Act of 1912, specifying some of the powers of the Secretary of Commerce with regard to broadcasting (*broad*casting of voice transmissions had not even been considered as a possibility fifteen years earlier), and creating a regulatory body, the Federal Radio Commission, the FRC (later the Federal Communications Commission, the FCC) to enforce the law. The fundamental arrangements of 1912 remained intact: frequency channels could be used but not owned; licenses to use the channels for limited periods would be granted by the FRC. But more than just public safety was to guide the hand of the licensing commission; "the public interest, convenience or necessity" had to be taken into account. Finally, in an effort to protect the country against concentrations of communications power, no organization was to be given a license if it had been found guilty of violating the antitrust laws. RCA had no trouble understanding the content of that message; it proceeded to drop its infringement suits and began licensing some of its competitors in return for a royalty based on sales; its effort to control the manufacture of radios had failed.

That same year, Arthur Judson, who made his living as a manager of concert artists, George A. Coats, a business promoter, and J. Andrew White, a sports broadcaster, decided to go into business as a performance network, creating and producing radio shows that would be transmitted only to those stations that had contracted to be part of the network. They called themselves United Independent Broadcasters—a direct antimonopoly attack on NBC. Within a year, short of cash, they sold their company to a family that had become rich in the cigar business; the company then acquired both a new name, the Columbia Broadcasting System (CBS), and a new president, William Paley. Sarnoff may not have immediately realized it, but he had met his match. The government had seen to it that RCA was not going to control the manufacture of radio sets; now CBS was going to see to it that NBC, wholly owned by RCA, was not going to control the content of broadcasting either.

Television

The entire history of electronic communications follows the patterns established in the early decades of the twentieth century in the history of radio: there are many surprises for individuals and organizations that think they can remain in control of the business of communications. Just when an inventor, after years of painstaking work, would begin to count his royalties in the millions of dollars, along came another inventor with a patent on a new, slightly better device, stealing his thunder and his plans for future wealth. Or just when a company, after years of complex maneuvering, would begin to exert almost complete control over some marketplace, along came another company with a new idea to compete for market share. Control of electronic industries—by individuals, by companies, even by governments—has been from the very beginning evanescent, transitory, fleeting. Individuals and companies have had their moment in the sun, but eventually they have all lost monopolistic control of the market that had once proven so profitable.

In the early day of broadcast radio, many people thought that it would be the ultimate mode of human communication and that it would revolutionize relations between individuals and between nations. "Messenger of sympathy and love . . . consoler of the lonely . . . promoter of mutual acquaintance, of peace and good will among men and nations," one radio engineer wrote; ". . . the whisper that leaps the hemisphere . . . the wisdom of the ages revived in a single breath," remarked another."[5] Yet even as they wrote, several ingenious men were already attempting to create a communications medium that they knew would be, to put the matter ungrammatically, even *more* ultimate.

Visible light is composed of electromagnetic waves that are shorter in length and higher in frequency than radio waves or electric current. By 1920, electrical experts in several countries were experimenting with various ways to translate *images* into electromagnetic signals that could be carried through wires or projected into the atmosphere. One of these experimenters was Ernst Alexanderson, the engineer who had designed and built the mammoth alternators for General Electric. Alexanderson's experimental device was a revolving disk punched with a spiral pattern of small holes, through which a light beam was directed; he hoped that this device would be able to digitize images (turn them into a pattern of tiny dark and light spots) so that they could be translated into electric signals by a another device called a photoelectric cell. Another experimenter, who worked for General Electric's chief competitor, Westinghouse, was Vladymir Zworykin, who, as a former Russian Army officer, had specialized in wireless telegraphy. In 1920, Westinghouse gave Zworykin permission to begin developing an all-electronic scanner. The idea for such a scanner had also occurred to Philo T. Farnsworth, a devoted radio amateur who, while he was still a high school student in Idaho in the early 1920s, had begun to experiment with cathode ray tubes. Until then, these tubes had been used principally in scientific laboratories to demonstrate the behavior of electrons.

A cathode-ray tube is a special kind of lightbulb (an evacuated glass chamber) that has a filament (an emitter of electrons) at one end; the other end is painted with a fluorescent chemical so that when current is applied to the filament the pattern made by the stream of electrons can be discerned by looking at the surface of the bulb. Both Farnsworth and Zworykin believed that if that pattern could be controlled, could be made to correspond to a generating image, then these tubes would work better than mechanical scanners (such as Alexanderson's revolving disc) to translate photographs into electricity. Farnsworth and Zworykin turned out to be right.

The developmental work on both television systems (mechanical and electronic) was difficult, tedious, and expensive, but by the latter part of the 1920s, both Farnsworth and Zworykin had acquired patents and Alexanderson was conducting daily test broadcasts. *Television,* the very first periodical devoted to analysis of the new medium, began publication in 1927; Alexanderson broadcast the first televised melodrama on September 11, 1928.

Thus, when broadcast radio was still in its adolescence, the handwriting was already on the wall—or rather the images were already in the air. The Radio Act of 1927 defined "radio" as "any intelligence, message, signal, power, *picture* or communication of any nature transferred by electrical energy from one point to another without the aid of any wire connecting the points." The word "picture" was included in this definition because knowledgeable people understood that what had already come to be called television was soon going to be feasible. The medium that would supplant radio at the forefront of communications had already been born, and like the first radio network, it had been born with a silver spoon of powerful corporate backing already in its mouth.

The commercialization of television was delayed first by the Depression and then by World War II, but what ingenious men had suspected in the 1920s was clear by the 1950s: television had supplanted radio as the nation's prime means of communication, just as radio had earlier supplanted the newspapers, the telegraph, and wireless telegraphy. Television had extraordinary popular appeal, and it spread rapidly throughout the country, encouraged by the general expansion of the American economy after World War II: 8,000 American households had television sets in 1946; 45.7 million had them by 1960—just fourteen years later. Paeans to the new medium were cast virtually in the same terms with which some people had, just a few decades earlier, extolled the virtues of radio: "the first truly universal medium or extension of man" according to one commentator.[6]

Yet by 1970 broadcast television, which had reigned virtually supreme for the previous two decades, was being challenged by yet another new medium, potentially both more flexible and more global: cable television. In the beginning the technology of cable was not revolutionary; the initial cable systems (one was already in place in 1949) used wires (bound together into cables, hence the name) to transmit television signals into the homes that, for one reason or another (intervening mountains or very tall buildings or very great distances), could not receive broadcast signals clearly. Broadcast programs were received at the studios of the cable company (sometimes over telephone lines, sometimes through antennae) and were then retranslated into signals that went out along the cables to subscribers. Once a sufficient number of customers had become regular subscribers, the cable companies invested funds in producing some of their own local programs in their own local studios, but in general most of the material that came over the cables in the early years was produced by one of the television networks.

In 1965, however, the first communications satellite was put in orbit (see Chapter 11), and that changed both the technology and the potential of cable television. Communications satellites contain electronic devices (called transponders) that amplify and transmit electromagnetic signals. Because the satellites are located high above the earth's surface and because their orbits are synchronous with the earth's rotation, communications satellites can receive and transmit signals from and to virtually anyplace on the surface of the globe: connecting a production studio in Texas, let us say, with a receiving antenna in Cairo, or vice versa. The first several satellites were constructed and rocketed into orbit by Intelstat (the International Space Communications Consortium); 60 percent of the shares in Intelstat belonged to COMSAT, an American corporation which—like RCA—had been created by Congress as a profit-making, investor-owned business, belonging equally to four American corporations (AT&T, ITT, RCA, and Western Union).

Communications satellites made it possible for cable companies to stop paying AT&T for the use of its phone lines, and it also made it possible for some of the cable companies to begin thinking of creating national, even international, networks, beaming programs from one location to receiving antennae all over the country, if not all over the world. Satellites could transmit signals from dozens, even hundreds, of cable channels: cable companies were not limited to the twelve very high frequency (VHF) or the seventy ultra high frequency (UHF) channels that the FCC had assigned to television because, unlike broadcast transmissions, cable-satellite transmissions from the earth were directed upward, rather than outward, lessening the likelihood of interference. By radically cutting the cost of transmission, satellites made special interest channels economically feasible, channels that would carry programs that appealed to only a segment the nation's audience: cartoon channels for children; instructional channels for schools; religious channels for fundamentalists; movie, music, and shopping channels; even, by 1979, a channel that would broadcast nothing but congressional sessions and hearings. As a result, cable television exercised a powerful, decentralizing force, creating new means by which independent points of view could be heard.

Federal regulations, created in the era of radio and television network broadcasting, at first stood in the way of the expansion of cable programming, but a series of executive and judicial rulings (beginning with President Nixon's decision in 1972 that organizations other than COMSAT and Intelsat could orbit satellites) gradually lifted these restrictions. In 1977, there were just 100 cable receiving stations; two years later there were 1,500; in 1961 one million households had cable service; in 1990 the figure had reached 55 million—and the social control that the networks and their advertisers had once exercised over the content of broadcasts had ended.

Electronic Components: The Vacuum Tube and the Transistor

As it was with modes of communication, so it was with the components of electronic circuits. As a technological system, the vacuum tube had a fairly long life—it was a critical component of virtually all electronic circuits between 1915 and 1950—but RCA did not manage to control it for very long, partly because the DeForest and Armstrong patents expired and partly because the threat of antitrust action led the company to license competitors. In any event, the vacuum tube was eventually replaced, first by the transistor and subsequently by the integrated circuit.

The transistor, invented in 1947 by a team of physicists (William Shockley, John Bardeen, and Walter Brattain) working at Bell Laboratories (the research arm of AT&T), was a solid-state substitute for a vacuum tube. Instead of an evacuated glass chamber, it used a semiconductor crystal (semiconductors, like germanium and silicon, are fairly good, but not excellent, conductors of electrical currents). When wires are embedded in semiconductor crystals, the significant atomic changes occurred at the junctions, or transitions, between the wires and the crystal (in later versions of the transistor, which have different semiconductor crystals sandwiched together, the changes occur at the surfaces of the layers), hence the name transistor.

Considerably smaller than the vacuum tube, transistors did not need to be heated up (which meant that they began working as soon as current was applied to them), required much less power to operate, and were much more reliable (since they were unlikely to burn out, the way vacuum tubes regularly did). Transistors thus made miniaturization and battery operation of many electronic devices possible. For this reason, they were used extensively in airborne cold war weaponry and space exploration equipment throughout the 1950s. Late in the decade (when mass-production techniques had brought down their price), transistors also began to appear in consumer electronics: in radios, television sets, hearing aids, and calculators. Nonetheless, a little more than a decade after the transistor went into full-scale production it, too, was replaced, this time by a device that was at one and the same time potentially much more flexible and much more powerful: the integrated circuit.

An integrated circuit replaced the various components out of which electronic circuits were composed with lines of chemicals that were laid down on, etched into, or photographically reproduced onto sheets of various semiconductor materials. These sheets could then be laminated onto each other, forming even more complex circuits out of the three dimensions of one thick block: an electronic chip. Slightly different forms of the integrated circuit were patented in 1957 by two engineers, Jack Kilby and Robert Noyce. Shockley, Bardeen, and Brattain, the inventors of the transistor, had worked for Bell Labs, one of the world's premier industrial research laboratories, a subsidiary of one of the world's largest corporations. Kilby, by way of contrast, was an engineer employed by Texas Instruments, then a moderately sized company; Noyce, also an engineer, was one of the founders of Fairchild Semiconductor, a start-up firm that had just recently begun to develop and manufacture new kinds of transistors.

Bell Labs had been willing to license the transistor to anyone who wanted to manufacture it, but the royalties, which were substantial, quickly evaporated once integrated circuits began to appear on the market. The integrated circuit was not only smaller than the transistor (one chip the size of a transistor could hold three, five, even a dozen transistor equivalents, as well as other electronic components and all the connecting wires), it was also potentially more efficient (because many different kinds of circuits performing several different electronic tasks could be built onto the same chip). Texas Instruments and Fairchild Semiconductor were also willing to license the integrated circuit in return for a royalty, and within a very short time, other inventors and other companies had patented and were marketing new variations on the original idea: for example, a memory chip, which contained special kinds of circuits that could store electronic information; a logic chip, which contained circuits that could manipulate that stored information in certain defined ways; or a microprocessor chip, which was programmable so that it could manipulate stored electronic information in a variety of ways.

Thus the history of electronic components mimics the history of electronic communications: when new industrial and technical frontiers are opened, older companies suddenly discover that they cannot control areas they had once expected to dominate. Indeed, in the field of integrated circuits, neither the original inventors nor the original companies nor even the country in which the original innovations had been made succeeded in dominating for very long. Both Texas Instruments and Fairchild Semiconductor flourished, and both inventors were well rewarded for their creativity, but competition in their marketplace quickly became very intense, both with regard to innovative chips and with regard to innovative methods of producing chips. And in the end, because American producers were focusing on high-profit military markets, several Japanese manufacturers were able to capture the higher-volume consumer markets. By the 1980s, American firms had ceased to be in control of the market for electronic components, the very market that they had originally created.

Computers

The electronic computer was originally intended to do the same thing that mechanical and electromechanical calculating machines did with gears and levers, except to do it much, much faster. During World War II, vast numbers of guns were being manufactured for use on battleships, tanks, and fighter airplanes. To use those guns properly, gunners had to aim the gun in the right direction and raise the barrel of the gun to the right angle, taking into account not only the location of the target, but also the temperature of the air and the character of the wind. This meant that every gun had to have its own firing table, giving all the relevant variables. These tables were exceedingly difficult to compute; each required the solution of several thousand differential equations. Several hundred women with college degrees in mathematics were hired by the Ballistics Research Laboratory of the Ordnance Department to compute the tables, but even when those women were supplied with the most advanced electromechanical calculators, each table took roughly three months to calculate.

In 1942, an engineer on the faculty of the Moore School of Engineering at the University of Pennsylvania, John Mauchly, suggested, in a memo to the Ordnance Department, that he could build an electronic calculator based on high-speed vacuum tube devices which would solve the trajectory equations in a matter of seconds, rather than hours. In 1943, a contract was awarded to Mauchly and to one of his colleagues, J. Prosper Eckert.

Unfortunately for the war effort, ENIAC, which stood for electronic numerical integrator and computer, took three years to build. By May 1944, the machine could do simple calculations, but it was not fully operational until the winter of 1945–46, by which time the war had been over for several months (the Ballistics Research Laboratory had had to solve its backlog problem by hiring more women). When it was done, ENIAC contained more than 17,000 vacuum tubes; it weighed thirty tons, had cost roughly $450,000 to build, and filled a room the size of a squash court. But it worked; it could calculate very fast, multiplying 333 ten-digit numbers a second, finishing one of the trajectory equations, which had previously taken hours to complete, in twenty seconds.

And there was important work for it to do; the hot war was over, but the cold war was just beginning. In the winter of 1945–46, anxious to find out whether ENIAC could really be helpful, officials from Los Alamos National Laboratory asked that the machine be reprogrammed to work on some of the complex calculations needed to determine whether a hydrogen bomb could be built. When those calculations, which might have taken physicists

ENIAC, the first digital computer, pictured here in 1946, was built with vacuum tubes. Early programmers, such as the people in this picture, set up problems by plugging in cables and setting switches. (Courtesy Charles Babbage Institute.)

years to finish, were completed in a matter of weeks (someone had to punch and ENIAC had to process more than a million cards full of data), it was clear that the cold war was going to expand the military appetite for complex, high-speed calculations, which meant that it was also clear that the military would want electronic calculators to be developed as fast as possible.

In the summer of 1946, the Ordnance Department and the Office of Naval Research sponsored a conference at the Moore School so that the ENIAC team could explain how the machine worked to organizations that might be able to develop computers further. Many organizations were interested. Some of the country's biggest manufacturers of office equipment were already exploring development of large-scale, very fast electromechanical calculators; in 1944, International Business Machines Company (IBM), for example, had cooperated with Professor Howard Aiken in building the Harvard Mark I computer, which was not electronic but had pioneered the use of automatic operational sequences. Also in attendance at the Moore School meeting were representatives of some of the country's biggest manufacturers of electronic equipment, some universities that were famous for their graduate programs in electrical engineering, some research institutes that were devoted to high-quality work in physics and mathematics, and an official from the British military corps that had been responsible for the machine that had broken, and unscrambled the German secret military code.

That summer course lifted the cloak of secrecy in which ENIAC had been shrouded; if Eckert and Mauchly had harbored any hopes that they would be able to develop their ideas after the war without fear of competition, those hopes were dashed by the enthusiasm with which their course was greeted. In the fall of 1946, after a nasty dispute with the University of Pennsylvania about whether they were obligated to assign all of the rights to their patents to their employer, they resigned their academic positions and formed a company of their own. Unfortunately, at the very same time, several organizations—some of them with much easier access to money and research facilities—had already started to build their own computers.

By the end of 1947, at least nine computers were being constructed in various laboratories in the United States and Britain. At the Institute for Advanced Study in Princeton, New Jersey, the world-renowned mathematician John von Neumann was leading a team that would build, with funds contributed partly by the U.S. government and partly by the RCA Corporation, the first digital computer capable of parallel processing, able to perform calculations simultaneously rather than serially. At the University of Manchester in Britain, the mathematician Alan Turing was part of a team that was building what would turn out to be the first computer with a stored program. At MIT, the Servomechanisms Laboratory, which had been created during the war to develop flight training devices for pilots and stability analyzers for new aerodynamic designs, had put a young engineer, Jay Forrester, in charge of a contract to develop a new device for the Office of Naval Research, and Forrester had decided to turn that contract into an effort to develop a computer that could operate in real time, instantly analyzing changes in its own environment. IBM, the largest American manufacturer of such devices as adding machines and electric typewriters, had developed a fairly simple electronic multiplier in its own research laboratories, and this machine was already on the market; the first hundred produced had been snapped up in a matter of weeks. The Raytheon Corporation (a manufacturer of electronic components) had a contract with the Bureau of National Standards to supply an electronic calculator to the Census Bureau to assist in the tabulation of the 1950 census.

Clearly no one was going to have the kind of control over the development of the computer that the Marconi companies had once been able to exercise over the development of wireless telegraphy. Eckert and Mauchly, however, had what they regarded as an excellent new idea: their new company would build not a computer alone, but a computing system, a series of related machines that could be combined and adapted to the individual, particular needs of potential customers. The universal automatic computer (UNIVAC) would have a central calculating machine, to which could be added such peripheral machines as high-speed printers (for recording the output), magnetic-tape drives (for holding data), card punchers (for entering data), and converters (to translate the data into electronic signals on tape). Eckert and Mauchly also hoped to include in UNIVAC an idea that they had developed while still working on ENIAC: a stored program memory, a set of electronic circuits that could reprogram the calculating circuits without the necessity for pulling plugs.

UNIVAC was the first commercially successful computing system. One was purchased by the Census Bureau; a few went to research laboratories for scientific computing; a few others were purchased by industry, one was borrowed by CBS to provide an early prediction, correctly as it turned out, of the outcome of the 1952 presidential election. But the rewards of being first were very short-lived. By the end of 1950, short of capital, Eckert and Mauchly sold their company and their patent rights to Remington Rand, a manufacturer of business machines; henceforth, they would be employees, not entrepreneurs. A few years later, they even lost their legal claim to patent priority when company lawyers failed to defend them adequately against a claim by Eugene Atanasoff, a former teaching colleague, that he had the original idea on which ENIAC was constructed.

The computer industry was competitive at the beginning and remains competitive to this day; dominance has been fleeting and change has been rapid. Because the first electronic computer was built under government contract, many skilled people were able to use it and learn how it worked. Because the government, fighting the protracted cold war with the Soviet Union, believed that it would need better and better computational facilities, dozens of organizations—universities, research institutes, corporations—were awarded contracts to develop computers. With vast resources at its command, and fired by what appeared to be a pressing need for national defense, the government was able to subsidize the enormous cost of computer development. The government was the customer best able to afford the high prices of the first models of each new generation of computers, and the customer most eager to have the new capabilities of each of those generations. Much of the tax money spent on computer development and purchase was spent by military and quasi-military agencies (the armed services, the Atomic Energy Commission, or NASA), but some of it was spent by civilian agencies, such as the Census Bureau.

In part because so many young people received training in computer design under various governmental contracts, the pace of innovation in computers was probably faster than in any other previous technological domain. Each generation of computers was better—faster, more powerful, more flexible, and less expensive—than the generation that came before, and generations were measured in years, not decades. For inputting data, punchcards were replaced by magnetic tape, then both intermediate media became outmoded when data could be entered directly. Programming by plug pulling was replaced by programming by binary instruction, then programming by binary instruction was replaced by programming by computer languages, then programming by computer languages was replaced in many instances by prepackaged programming. Programs that could manipulate numbers were supplemented by programs that could manipulate words or play games or make judgments or draw or interpret human speech patterns. Machine memories became ever larger; machine speeds became ever faster—while the machines themselves, thanks to transistors and to integrated circuitry, became ever smaller and less expensive. Each of those innovations, and each of those generations, brought new, competitive corporate players into the computer marketplace. Every year, new companies were being formed; every year, some of those companies would start down the road to dominance; some would reach the goal, but most only managed to stay on top for very short periods of time. Even the giant eventually tumbled: IBM, which had dominated the mainframe computer business for close to a decade, fell on its face in the 1980s, when personal computing became the industry frontier. In 1992, the company—whose stock had been considered the safest possible investment for the previous fifty years, the bluest of all the bluechips—lost money for the first time since the end of World War II. The computer marketplace, like the communications and component marketplaces, turned out, despite many strenuous efforts, to be uncontrollable.

Conclusion: The Ultimate Failure of Efforts to Control Electronic Communication

From the invention of wireless telegraphy in the late nineteenth century to the creation of an electronic superhighway in the late twentieth, no one individual, and no one institution, has succeeded in controlling for very long the American market for electronic components or electronic devices or even the content of what is communicated by those devices. The history of electronics suggests that there have been, roughly speaking, three reasons for this. First, throughout the twentieth century, there have been many devoted electronics amateurs—ham operators in years past, hackers more recently—young people drawn by the mystery and magic of communicating over long distances, of calculating faster than the eye can blink, of imitating the functions of the human brain. Because there have been many amateurs, there have also been, regularly, a fair number of brilliant innovators: Armstrong got his start in amateur radio, so did Conrad; years later, the men who built the company that first succeeded in marketing a personal computer met each other at meetings of something called the Homebrew Computer Club.

The federal government has also played a multifaceted role in the decentralization of electronic technology and electronic communication. Sometimes, as when the Navy Department wished to create RCA as a virtual monopoly, there have been countervening tendencies, but on the whole, the government, whether through antitrust proceedings or the regulations of the Federal Communications Commission or the granting of military procurement contracts, has acted to weaken the hand of those who have wanted to monopolize the media.

Finally, the competitive possibilities of a relatively free market economy have also played a role. From the moment in 1899 when Guglielmo Marconi first made his spectacular publicity debut on American shores, clever men and women have understood that electronic circuits could be a route to riches, in fact, a fast route to riches. Even when one person or one company has managed somehow to build a gate around those riches, other people and companies have been willing to try just about anything to knock that gate down—and a fair number have succeeded.

As a consequence of these three factors, the pace of change in electronics has been fairly rapid, and no single individual, company, component, or medium has managed to dominate for very long. In the world of electronic media, centralization of power—and all the negative consequences that can follow from it—has always been a possibility, and many people—from presidents of multibillion dollar corporations to high school graduates fiddling with dials and switches in their garages—have tried to accomplish it. Nonetheless, no individual, no matter how creative or how powerful, has so far succeeded for very long; neither has any company, no matter how well capitalized, no matter how well supplied with highly trained and highly paid lawyers, no matter how many strings it was able to pull, no matter how many very high places to which those strings were connected. Partly because of the nature of the American economy, and partly because of the nature of electronic technology itself, the twentieth-century electronic media seem to have outwitted the critics who believed that those media had inescapably totalitarian tendencies.

Notes

1. Remarks by Congressman William S. Greene, *Government Control of Radio Communication: Hearings before the Committee on Merchant Marine and Fisheries* (1918), as quoted in Erik Barnouw, *A Tower in Babel: A History of Broadcasting in the United States* (New York: Oxford University Press, 1966), p. 55.

2. John Tebbel, *David Sarnoff: Putting Electronics to Work* (Chicago: Encyclopedia Britannica Press, 1963), pp. 99–106.

3. Pittsburgh *Sun*, September 19, 1920, as quoted in Barnouw, *Tower in Babel*, p. 63.

4. William C. White, from an unpublished memoir, as quoted in Barnouw, *Tower in Babel*, pp. 73–74.

5. The first phrases come from a poem by Alfred N. Goldsmith in his book *Radio Telephony* (New York: Wireless Press, 1918), p. 242; the second group comes from a poem by Robert Davis, in Alfred N. Goldsmith and Austin C. Lescarboura, *This Thing Called Broadcasting* (New York: Henry Holt, 1930), p. 344.

6. Frank J. Coppa, "The New Age of Television: The Communications Revolution in the Living Room," in Frank J. Coppa and Richard Harmond, eds., *Technology in the Twentieth Century* (Dubuque, IA: Kendall-Hunt, 1983), p. 131.

Suggestions for Further Reading

Hugh G. J. Aitken, *The Continuous Wave: Technology and American Radio, 1900–1932* (Princeton, 1985).

Hugh G. J. Aitken, *Syntony and Spark: The Origins of Radio* (Princeton, 1976).

William Aspray, *John von Neumann and the Origins of Modern Computing* (Cambridge, MA, 1990).

Erik Barnouw, *The Golden Web: A History of Broadcasting in the United States 1933–1953* (New York, 1968).

Erik Barnouw, *A Tower in Babel: A History of Broadcasting in the United States, to 1933* (New York, 1966).

Erik Barnouw, *Tube of Plenty: The Evolution of American Television*, 2nd ed., rev. (New York, 1990).

Ernest Braun and Stuart Macdonald, *Revolution in Miniature: The History and Impact of Semiconductor Electronics* (Cambridge, 1978).

James Brittain, *Alexanderson: Pioneer in American Electrical Engineering* (Baltimore, 1992).

Susan J. Douglas, *Inventing American Broadcasting, 1899–1922* (Baltimore, 1987).

Kenneth Flamm, *Creating the Computer: Government, Industry, and High Technology* (Washington, DC, 1988).

Herman H. Goldstine, *The Computer: From Pascal to von Neumann* (Princeton, 1972).

James A. Hijaya, *Lee De Forest and the Fatherhood of Radio* (Bethlehem, PA, 1992).

Tom Lewis, *Empire of the Air: The Men Who Made Radio* (New York, 1991).

Steven Lubar, *InfoCulture: The Smithsonian Institution Book of Information Age Inventions* (Boston, 1993).

David E. Lundstrom, *A Few Good Men from Univac* (Cambridge, MA, 1987).

Rene Moreau, *The Computer Comes of Age: The People, the Hardware, and the Software* (Cambridge, MA, 1984).

T. R. Reid, *The Chip: How Two Americans Invented the Microchip and Launched a Revolution* (New York, 1984).

Joel Shurkin, *Engines of the Mind: A History of the Computer* (New York, 1984).

Susan Smulyan, *Selling Radio: The Commercialization of American Broadcasting, 1920–1934* (Washington, DC, 1994).

Nancy Stern, *From ENIAC to UNIVAC: An Appraisal of the Eckert-Mauchly Computers* (Bedford, MA, 1981).

13

Biotechnology

FEW TECHNOLOGICAL INNOVATIONS worry people quite as much as the ones that we eat or the ones that we use as medications or the ones that we employ in our sexual and reproductive lives. At the same time, few of us ever think of foodstuffs and drugs and medical devices—such things as seedless grapes, antibiotics, and birth control pills—as technologies at all. We need to think again, for if we define technology as something created by human artifice in order to alter the environment so as to achieve human goals, then, of course, these artifacts qualify. They don't exist in nature; they certainly alter the environment, both biological and social. Just as certainly, they were created by some human beings to satisfy their own and other people's needs.

Indeed, one of the reasons why so many people worry about these kinds of technologies is that they seem to be altering our biological and social environments in disconcerting ways. New agricultural products—such as seedless grapes, hybrid corn, genetically engineered tomatoes—tend to drive old ones off the market and out of production. This means that genetic diversity is reduced and at the same time the uniform crops that remain are ever more susceptible to ever more devastating infestations. Our social environments change as well, sometimes dramatically. Antibiotics lower death rates, and this means that they increase the size of populations, leading, some say, to serious overcrowding and other ills of overpopulation. The various reproductive technologies, especially contraceptives and abortifacients, satisfy needs of some people that other people think ought to be left unsatisfied.

301

The history of the various twentieth-century biotechnologies thus raises all kinds of very knotty questions about the social and ethical impact of
technology. It also raises fundamental questions, as we shall see, about
whether we would be able to resolve such social and ethical paraodxes if
we understood that history better.

Science, Technology, and Technoscience

The history of biotechnology also raises questions about the differences between science and technology. If such things as seedless grapes, antibiotics,
and birth control pills are technologies, then surely the activities that created them—agricultural and medical research—must also be technological. Yet most of us, if asked, would say that the people who do agricultural and medical research are scientists. Scientists, we think, are the people
who discover new truths by using the experimental method, while somewhat different people—what shall we call them? engineers? technologists?—using very different methods eventually apply those truths in order
to create new products. We like to think of scientific research as the pursuit of something we call pure knowledge, and the pursuit of applications
as somehow different, certainly a "less pure" activity. But do such distinctions really make any sense in the domain of agricultural and medical research, where the ultimate goal has always been some new product that will
taste better, cure better, nourish better, flourish better, or work better?

Early in the nineteenth century, science and technology were pursued by
markedly different groups of people, and this meant that the differences between the two pursuits were fairly obvious. Science was taught and practiced by educated people who had jobs in colleges, universities, and museums; technologies were used and produced by artisans who had received
their training and practised their crafts in workshops and factories. Before
1850, if you had wanted a better mousetrap (or cast-iron stove or survey-
or's sextant), you probably wouldn't have thought to ask a college teacher
(or even a person educated at a college) for help. In this period of time, the
distinction between science and technology was essentially a social distinction: one group of people, with one set of traditions, did technical work;
another group of people, with a somewhat different culture, were scientists.

As the nineteenth century wore on, however, increasingly large numbers
of people began to believe that if artisans learned science, technological
progress would be accelerated, that if physicians and farmers learned the
rudiments of biology, more diseases could be cured and more food would
be produced. Many of the engineering colleges created in the nineteenth
century (see Chapter 6) were founded on this set of new assumptions; the
same assumptions guided the creation of agricultural colleges and new
medical schools. The faculty had to be trained in the sciences; education
was to be grounded in the sciences. No longer would a civil engineer get
his training on a canal site or a railroad line, or a farmer at his father's knee,
or a physician by apprenticeship. Instead, he (or occasionally she) had to

begin with the fundamentals of dynamics and mechanics, physiology and botany, anatomy and chemistry.

A social and intellectual hierarchy was created by this new set of assumptions: science is "better" or "more fundamental" than technology; research scientists, therefore, are "better" or "smarter" than engineers, or practicing physicians or working farmers. The person who works with his head is better than the person who works with his hands; the library and the laboratory are more honorable places of employment than the shop floor, the barnyard or the examining room. By the turn of the twentieth century, this distinction between science and technology—science discovers; technology applies; discovering is better than applying—had become a commonplace, which is why so many of us still believe it today.

Yet it is a problematic commonplace. Not surprisingly, engineers, farmers, and practicing physicians tend to resent the subsidiary position to which they are relegated in this hierarchical model, feeling—with some heat at times—that there is something especially challenging about the practical and clinical arts that the rather colorless word "applied" simply cannot convey. In addition, when scientists are applying for money to support their research, they tend to ignore the flattering distinctions at the heart of the model since philanthropists, industrialists, and governments usually want to know what technological payoff—not what truth—they are going to get in return for their investment.

Thus, after 1945, as scientific research became increasingly expensive and the scientists on the faculties of colleges and universities became increasingly dependent on outside funding for their research, the clear line that had once been drawn between science and technology began to look, to some people at least, a little fuzzy. In addition, as research began increasingly to depend on complex pieces of machinery (for example, linear accelerators and electron microscopes) and machine-based processes (for example, gel electrophoresis and computer modeling), the old distinction between those who work with their hands and those who work with their heads began to seem more than a little dubious. Accordingly, some scholars have coined a new term, *technoscience,* to describe such enterprises as agricultural, medical, and military research, enterprises in which science and technology, investigation and application, resemble two sides of the same coin.

The case studies in the history of biotechnology that are related in this chapter were thus chosen partly to illustrate the character of the enterprise of technoscience and partly to illuminate some of the social and ethical quandaries that the biologically based technosciences have created in our time.

Hybrid Corn

Human beings have been practicing genetics for several millennia; all domesticated animals and plants—dogs, horses, grains, fruits—were developed over the course of many centuries by selective breeding, deliberate attempts to take advantage of the natural variability of organisms: mating this ram with that ewe in the hope of getting lambs with thicker coats of wool; choosing seed from that row of wheat or this patch of rye in the hope of getting a higher yield next season. Some of this selective breeding was done informally; a farmer tried to choose seed carefully for one or two seasons; a herder managed to isolate a few pairs out of the flock every spring.

Corn is an easy crop on which to practice selective breeding. The tassel on the top of the stalk carries the pollen; the silks on the branches receive the pollen; the seeds, which develop after the pollen lands on the silk, are the kernels attached to the ears. Many nineteenth-century American farmers were selective breeders. They let the corn in their fields pollinate freely, the wind scattering the pollen among the silks of many plants; after the harvest, they retained some ears to use as seed, deliberately choosing those ears that had very regular rows or exceptionally sweet kernels or that had come from stalks with a large number of ears.

In the waning decades of the nineteenth century, some faculty members in the agricultural colleges began experiments in which they *artificially* pollinated corn to achieve selective breeding; they would detassel a row of corn or put paper bags over the silks and tassels; then researchers armed with little paintbrushes would control the transfer of pollen. Also by the end of the nineteenth century, a few seed companies had come into existence, often started by farmers who had found it more profitable to raise produce for seed than to sell it as food. Corn did not then account for a large part of seed company business since most farmers used their own, homegrown corn for seed. As a result, the seed companies—with one exception—had virtually no financial incentive to invest in experiments with artificial pollination.

In any event, none of the corn-breeding experiments was particularly successful. Experimental seeds could not be guaranteed to produce a new generation with the improvements the experimenters were aiming for; the corn that resulted was of variable quality, as variable as corn grown from openly pollinated plants.

But the experimenters kept trying, and in the first decade of the twentieth century, their efforts were galvanized by new discoveries and new theories in the study of heredity. In 1889, an Englishman, Francis Galton, had concluded, on the basis of his studies of the heights of parents and children, that the offspring of abnormal parents have a tendency to "regress toward the population mean." A decade or so later, a Dutch biologist, Hugo DeVries, discovered "mutations," major variations in a trait, which could arise spontaneously and be inherited. Both discoveries—that some unusual plants could breed true and that others would, after a generation or two of breeding, revert right back to their parental types—had important implications for breeders.

So did the work of a Czech priest, Gregor Mendel, who had published some papers about experiments that he had done on the artificial pollination of ornamental sweet peas. Mendel had concluded that some of the characteristics of a plant could be thought of as units, that some units (for example, the green color for seeds) dominated over others (for example, the yellow color for seeds), and that in inheritance these units operated independently and according to the mathematical laws of probability. In 1905, an American, W. S. Sutton, demonstrated that the probabilistic behavior of Mendel's units would make physiological sense if those units were located on the chromosomes in the nucleus of cells. A few years later, in 1910, another American, Thomas Hunt Morgan, was able to demonstrate this phenomenon experimentally when he determined that a particular unit character, eye color, was carried on the same chromosome that determined the sex of the common fruit fly, *Drosophila melanogaster*. The Mendel-Morgan hypothesis (later to be called the gene theory) also had important implications for breeders because it suggested that stable new species, or at least stable new types, could be created by artificially and carefully crossing one type with another over the course of several generations.

In addition, early in the twentieth century, two Scandinavian biologists, Hjalmar Nilsson and Wilhelm Johannsen, had isolated "pure lines" (Johannsen later called them genotypes) by inbreeding, or self-fertilizing, plants for several generations. Nilsson and Johannsen had been studying plants like wheat and oats, which normally self-fertilize, and they had discovered that, by carefully selecting the plants for several seasons, they could create seed stocks that would consistently breed true to type. Their stocks would not, to put the matter another way, vary with regard to one or two crucial characteristics. It remained to be seen whether careful crossing and subsequent inbreeding would have the same results in corn, which normally cross-fertilizes.

Two Americans, E. M. East and George Shull, created the techniques that would eventually solve the problem. They belonged to the first generation of American scientists who acquired a doctorate as part of their professional training. East had a doctorate in plant physiology from the University of Illinois. He became the director of the Connecticut Agricultural Experiment Station in 1905 and began experiments intended to create (and cross) inbred (or pure) lines of corn—varieties of corn created, as Johannsen had done, by self-pollinating plants for two or three seasons. George Shull had received a doctorate from the University of Chicago. He began studying the inheritance of row patterns in corn (the number of rows of kernels per ear, a significant factor in yield calculations) in 1906, when he became an employee of the Station for Experimental Evolution in Cold Spring Harbor, New York.

Looked at in one way, both East and Shull were scientists, and their experiments were purely scientific. East wanted to find out which corn characteristics were dominant and which recessive; Shull was trying to find out what were the unit characters in corn that would correspond to the unit characters (such as seed color and seed shape) that Mendel had identified in sweet peas. Both researchers were using corn in much the same way that

Morgan had used fruit flies: as suitable medium through which to do experiments. Looked at another way, however, both East and Shull were technologists, pursuing very practical goals. Both were trying to ascertain how the rules of Mendelism could be applied to a plant that had enormous economic importance; they wanted to find out about the patterns of heredity in corn in order to breed better corn plants. This is one of the characteristics of technoscience: the pursuit of abstract knowledge and practical goals can be carried out simultaneously.

Unfortunately, East found, season after season, as Galton had predicted, that his hybrid plants lost vigor after the first year; the seed produced by plants grown from hybrid seed did not have very high yields. When East left the agricultural station in 1909 to accept a professorial position at Harvard, his inbreeding experiments were continued by several of his students. One of these men, Donald F. Jones, made a crucial guess about the reason why hybrids between two inbred lines lost vigor. If, as Morgan's experiments with *Drosophila* had suggested, several different unit characters were linked on each chromosome, then the hybrids of two pure lines might not be vigorous because they were too pure—contained too few variations of each genetic unit. Why not try creating a hybrid of four lines, a cross between two hybrids, a double cross?

Jones tested his double crosses in the growing seasons of 1916 and 1917 and was delighted to discover that they did not lose their vigor. He wanted to try crossing different pure lines and to test his hybrid seed under normal growing conditions in other parts of the country. One of the farmers who volunteered to do this for him was Henry Wallace. Wallace was the scion of an eminent Iowa farming family who had shown considerable academic promise (and interest in scientific breeding) when a student at Iowa State University. In the early 1920s, he was managing his own farm and writing for an agricultural magazine, *Wallaces' Farmer*, which his grandfather had founded. *Wallaces' Farmer* encouraged what it called progressive farming—experimentation with new products and processes—and Henry Wallace practiced what he preached. Wallace planted some test plots of Jones's seed, but he also crossed the Connecticut seed with some of his own inbred lines until he had developed a strain of corn that was well adapted to the particular growing conditions in Iowa.

In 1926, the Hi-Bred Company, founded by Wallace, began advertising and marketing the seed for this new strain, making it the first company to commercialize hybrid corn. The Hi-Bred Company was immensely successful. Wallace went on to be Secretary of Agriculture for two of Franklin Delano Roosevelt's terms and eventually a presidential candidate himself. Although Wallace was the first to commercialize hybrid seed, other companies soon followed; at least one of those companies, Funk Brothers, had actually begun experimenting with double-crossed seed before either Jones had published about it or Wallace had tried planting it. The Funk Brothers Company had been founded in 1901 by Eugene Duncan Funk, who had attended the Sheffield Scientific School at Yale and had, even in the

1880s, developed an enthusiasm for scientific agriculture. In 1893, he had developed a new strain of corn (by crossing two varieties that had been developed by selection, not by inbreeding), and it was the commercial potential of this strain (called Funk's 90-day, to signify the length of time from planting to harvest) that led him and the members of his family to combine their resources so as to be able to sponsor more research.

By 1910, Funk Brothers had a reputation as a research-oriented organization; it had as many people employed in, and as much land devoted to, research work as most agricultural experiment stations funded by the USDA, and it had endowed a chemical laboratory at Wesleyan University, where the chemical analyses of its new strains was carried out. In 1915, the company hired J. R. Holbert as its chief agronomist. Holbert had just graduated from Purdue with a bachelor of science degree in plant pathology; he was immediately put in charge of the corn-breeding experiments. Between 1916 and 1918, Holbert developed several pure lines based on Funk's seeds; between 1918 and 1920 he developed several single crosses of these lines—and a year later, a double cross. Small quantities of Pure Line Double Cross No. 250 were being sold by Funk Brothers by 1922, but they did not actually advertise it in their catalogue until 1928, after Wallace had already demonstrated that there were a significant number of farmers who could be induced to try—and retry—this new type of seed.

The path from Mendel's sweet peas to Wallace and Holbert's hybrid corn is pretty much a straight line, but determining where on that straight line science stopped and technology began is not easy. Some people would say that Mendel and Galton and Johannsen were scientists because they discovered general principles, but that Jones and Holbert and Wallace were agricultural inventors because they were trying to create something that hadn't been there before: a hybrid, something unnatural, the result of human interference with the natural process of pollination.

Unfortunately, Mendel was "doing technology" also. Mendel's hobby was gardening; like many serious gardeners of his day, he had been trying to create new varieties of garden plants by artificial crossing, that is, by hybridizing. So his goal was, at least in part, precisely the same as Wallace's and Holbert's—that is, technological. So was his procedure, for when he crossed the plants that resulted from planting yellow seeds with those that resulted from planting green seeds, he was creating an entity that had never existed before: a sweet pea hybrid.

Thus one could say, without stretching the truth at all, that Mendel discovered the scientific laws of dominance and independent segregation while in pursuit of a technological development program. The same kind of statement could be made about Francis Galton, who discovered the law of regression to the population mean while trying to figure out how to improve the human race, and about E. M. East, who discovered that corn has unit characters while trying to figure out how to help farmers increase their yields. In which case, we end up arguing that scientific discoveries were made by people who were, despite their advanced training in the sciences, fundamentally technologists. And if we end up saying that, then what we have ended up saying is that from its inception genetics has been not so much a *science* as a *technoscience*. The story of hybrid corn thus teaches us that technosciences are enterprises in which people seek the truth about nature in order to transform it; conversely technosciences are also enterprises in which some truths about nature have been uncovered in the practice of trying to transform it.

Virtually all of the corn grown in the United States today is hybrid corn, the product of artificial pollination of inbred strains. There are hundreds of varieties: some bred for high productivity; some for early, or late, maturation; some that are appealing because they are sweeter and their rows of kernels more uniform. American farmers started making the transition to hybrid corn during the Depression, and completed the transition during World War II when the loss of so much farm manpower to the military made consistently high yields imperative for individual farmers. By the early 1930s, the great commercial potential of hybrid corn had become very clear, and several seed companies began developing and marketing their own inbred strains from which double-cross seed could be derived.

Both commercial experimenters and experimenters employed at universities and agricultural research stations had been in the habit of exchanging the seed of their inbred lines, but in the 1930s, as the profits from the use of that seed began to mount, the exchanges stopped. Some growers—both commercial and noncommercial—tried to patent specific crosses, and Jones tried to patent the idea of double crossing, but such claims were repeatedly rejected by the patent office and by the courts. As a result, all the commercial parties have had to protect their financial stake in their lines by preventing the sale or theft of the seed from their inbred varieties.

Hybrid corn looks and tastes much different from traditional varieties, but it is also very different socially and economically. Because it results from artificial, controlled pollination, farmers cannot produce the seed for hybrid corn themselves. Production of hybrid seed requires access to the seed of four different lines of inbred corn. That seed has to be planted on special plots and the silks have to be artificially pollinated. This has to be done afresh every growing season in order to produce hybrid seed for the next growing season. The seed from hybrid plants, openly pollinated, is degenerate seed; the next crop cannot be guaranteed to have anything like the same characteristics as the crop from which the seed had been drawn.

Hybrid corn is thus implicated in the process by which small-scale farms, owned and managed by farming families, have given way to very large farms, owned by large corporations, managed by corporate employees. To grow hybrid corn, a farmer must be dependent on commercial seed companies—and must go to the added expense of purchasing seed. Thus, like petroleum-based fertilizers, petroleum-fueled tractors, electrically heated incubators, and on-line meteorological databases, hybrid corn is a technological system that has increased the capital expenses required for successful farm operations. The more capital that is required, the larger the plot

Hybrid corn probably did improve yields and profits for many farmers as this advertisement suggests, but it also increased their costs since they became dependent on seed companies such as this one. (Courtesy DeKalb Genetics Corporation.)

of land that needs to be farmed to make a profit—and the fewer people there are who can afford to continue farming independently.

The end of family farming is an outcome that the people who experimented with hybrid corn never intended; indeed, at least one of those people, Henry Wallace, would have abhorred it if he had lived to see it. Even something as apparently natural as a succulent ear of corn-on-the-cob can be a technology, and like all technologies, it can have unexpected and unintended social and economic consequences. These unexpected and unintended outcomes are common features of virtually all technological changes, ones that we all must learn to appreciate. Because our social and economic system is complex—some would say infinitely complex—even the most sophisticated, best informed, thoroughly objective experts cannot predict all the consequences of a significant technological change.

Penicillin

By the beginning of the twentieth century, most physicians in the United States and Europe had come to accept the germ theory of disease, the notion that some diseases are caused by microscopically small organisms. By the turn of the century, in fact, a new science had been born: bacteriology—the effort to classify these tiny organisms and to understand both their metabolism and their modes of reproduction.

Alexander Fleming was a British bacteriologist; he was trained as a physician, but in the 1920s, he was employed as a medical researcher and teacher. Fleming was interested in developing therapies for bacterial infections by studying the ways in which white blood cells, leukocytes, destroy bacteria. Fleming's experiments were done on colonies of bacteria grown in laboratory dishes; the antiseptic effectiveness of the materials he was studying was gauged by the speed with which bacteria died. One day, Fleming noticed that one of the plates in his laboratory, which must have been accidentally exposed to the air, had some mold growing on it and—this is the significant part—that in the area near the mold the bacterial colony was dying. Fleming was not particularly interested in the mold—he was experimenting with leukocytes—but he was interested enough to send a sample off to a mycologist (someone who specializes in the study of molds) for identification. Fleming also grew some of the mold in test tubes and injected some of the broth that developed in those test tubes into some animals in his laboratory to see whether it was toxic. When the animals appeared unaffected by the injections, he tried dropping a very weak solution of the broth into the eyes of an animal that had an eye infection—and the infection was gone in a few days. Satisfied that the fluid produced by the mold, *Penicillium notatum*, was both nontoxic and powerfully antibacterial, he published a short paper on his researches in the *British Journal of Experimental Pathology* in 1929.

Neither Fleming nor anyone else seems to have paid much attention to the antibacterial capacity of *Penicillium* for the next decade. One or two biochemists tried to isolate the active chemical in the broth by distilling the broth or by evaporating it or by mixing it with various kinds of solvents, but all such efforts failed, and most knowledgeable people assumed that if the active chemical couldn't be isolated, then the antibiotic couldn't be manufactured in quantities large enough to do anyone any good as a medication.

A decade later, however, two men renewed interest in the *Penicillium* mold. Ernest Chain, a German-Jewish biochemist had fled his homeland and found a job in the medical research laboratory headed by Howard Florey, an Australian-born professor of pathology at Oxford University. Florey and Chain, aware that Europe was headed toward war, feared that existing antibiotics would be inadequate. Florey's lab was also running a deficit; research on a promising new antibiotic was, he thought, a sure road to new government funding. On August 27, 1939—as Hitler's army was attacking Poland—Florey applied to the Medical Research Council (a British government agency that supported medical research) for a small grant to study *Penicillium*. Three days later, Britain and France declared war on Germany.

The Medical Research Council was, however, also strapped for funds; most of the money that Florey and Chain needed to start their research eventually came from the Rockefeller Foundation in New York, which was then one of the major nongovernmental underwriters of medical research.

Penicillin (as the active ingredient in the broth produced by the mold was eventually called) turned out to be an exceptionally fickle substance. The mold had to be grown under absolutely sterile conditions since, ironically, it was susceptible to being killed by airborne bacteria. Extracting the active ingredient in the broth proved to be a major stumbling block as well since heat, acids, and alkalis—the usual extraction media—all seemed to destroy the bacteriolytic action of the chemical. By March 1940, after months of work and the cooperation of many people in the laboratory—biochemists, physicians, and technicians—Chain had only been able to produce one tenth of a gram of a brown powder from the fluid that had accumulated in hundreds of incubated dishes. Nonetheless, when this impure powder was tested in petri dishes, it proved to be immensely powerful at killing a wide spectrum of bacteria; a little might go a very long way. When injected at various dilutions into laboratory mice, it proved to be, as expected, nontoxic. Then mice were deliberately infected with staphylococcus. Those mice that also received injections of penicillin survived; untreated mice were dead within a day. The entire laboratory staff was encouraged by the results.

But the war in Europe was going very badly. In the first days of April 1940, in rapid succession, Norway, Denmark, Holland, and Belgium were attacked and overrun; over the course of the next month, British and French troops were pushed back to the sea, then forced to flee across the English Channel, leaving much of western Europe occupied by the Germans. As the mouse experiments were being finished and the results written up for a paper in the prestigious medical journal *Lancet*, men, women, and children were being killed and wounded by German bombs which were falling on London, just an hour's train ride from the Oxford laboratory.

Penicillium continued, however, to be recalcitrant. The researchers calculated that they would need thirty grams of penicillin powder to treat a human being, but using all the dishes and staff and extraction apparatus at its command, the Oxford laboratory could only produce three grams a week. Florey tried to enlist the aid of British pharmaceutical companies, but they were all operating at full capacity to meet the demands of the by now very stressed British population. By turning virtually his entire laboratory into a penicillin manufactory, Florey was finally able, by February 1941, to accumulate enough of the powder to treat six patients—five of whom were cured of massive, life-threatening infections. Florey understood that penicillin had the potential to be a powerful weapon not only in the fight against disease but also in the fight against the Nazis; he also understood that there was no possibility of having it produced in sufficient quantities under wartime conditions in Britain.

In June 1941, Florey and a chemist, Norman G. Heatley, departed for the United States, taking samples of the mold with them. Heatley had succeeded in developing an effective system for extracting penicillin from the culture broth—a crucial step in the process of turning a mold into a medicine. Heatley's method was called back separation; it required the use of two solvents, the first to get the penicillin separated from the broth and the second to separate the penicillin from the first. Heatley had also devised a system for measuring the potency of any batch of penicillin powder, a unit measurement (later called the Oxford Unit), which created a standard of measurement and a way of determining safe doses for individual patients. Florey hoped that he could convince some American organization—either the government or a pharmaceutical company—to begin growing large quantities of *Penicillium notatum* and to build a fairly large extraction apparatus.

The question of who, if anyone, would be the first to patent penicillin or some part of the system for obtaining it from its natural condition had already been raised by Chain before Florey departed—and would be raised again, numerous times in subsequent years. Chain's father had been a chemical manufacturer in Germany; it seemed natural to him that the drug should be protected by patents, so as to raise money for the Oxford laboratory. Florey was dubious, and the physicians in charge of the Medical Research Council were adamantly opposed, believing that it was unethical for medical researchers to benefit in any way from the commercial exploitation of their discoveries. And there the matter rested, temporarily, as Florey and Heatley prepared to depart.

Florey, Heatley, and *Penicillium* received a very friendly reception in the United States; in short order, they had met with two people whose cooperation would turn out to be crucial. The first was Dr. Robert Coghill, an agricultural chemist, who was the director of the Fermentation Division of the Northern Regional Research Laboratory in Peoria, Illinois. Located in the middle of the farm belt, this laboratory, which was part of the United States Department of Agriculture, was devoted to the task of finding industrial applications for American farm products, particularly corn. Heatley and Florey had been told to see Coghill because the process by which *Penicillium* produces penicillin out of the chemicals in its growing medium is a fermentative process, like the processes involved in making beer and wine, a living organism changing one substance into another. Coghill's staff was expert both at growing a variety of fermentative organisms and at devising means for extracting the chemicals they produced. That summer, Coghill and his staff were working with an abundant industrial by-product for which they were trying to find some use. Corn steep liquor was a sticky substance, a bit like molasses—the residue that remains when starch is extracted from corn. Within a month, the laboratory staff had discovered that if corn steep liquor was used as a growing medium, penicillin production could be increased tenfold.

The other crucial person whom Florey and Heatley met that summer was A. N. Richards. Like Florey, Richards had been trained as a physician, and also like Florey, he had made his reputation at the laboratory bench rather than the bedside. In 1941, he was a professor of pharmaceutical sciences at the University of Pennsylvania Medical School and one of the country's leading experts on the development and testing of new medications; he was also a scientific advisor to a number of pharmaceutical companies. That summer, Richards had just accepted a new appointment, as head of the medical branch of the Office of Scientific Research and Development (OSRD; see Chapter 11), a federal agency that was to mobilize technical and scientific talent in the event of war.

Richards was immediately convinced that penicillin would be crucial to the war effort, but there was little he could do to assist Florey and Heatley because the United States was not then at war and OSRD had no power to overrule, or even to influence, the decisions made by pharmaceutical companies. Several American pharmaceutical companies were already investigating *Penicillium* broth. Research scientists working for these firms had read Florey's report about his success with mice. As a result, there were several small research projects in the works in which efforts were being made to isolate and characterize the active ingredient in the broth, precisely what Chain was working on at Oxford. Florey and Heatley had called on some of these industrial research laboratories, but, wishing to protect their investments, several of the companies had been reluctant to talk with them or to reveal very much about their research programs.

Florey returned to Britain in October; Heatley remained in Peoria. On December 7, the Japanese attacked Pearl Harbor. The next day, Congress declared war on Japan, Germany, and Italy. Five days later, Richards called a meeting in Washington to initiate a crash program for the production of penicillin.

Coghill was present; he told the committee about the benefits of using

corn steep liquor as a growing medium. Several representatives of pharmaceutical companies were present. Richards told them that under new
wartime regulations they would not be prosecuted for violation of the antitrust laws if they agreed voluntarily to pool all the information that they
were acquiring about penicillin. He also assured them that no one would
be allowed to take out a patent on the production process, or any part of
it, for the duration of the war. Under new tax regulations, the companies
were going to be allowed to plow back 85 percent of their profits into war-
related research. A scientific advisory board was going to be formed to suggest ways to increase penicillin production. The companies and the Department of Agriculture agreed tentatively to cooperate. The American war
to conquer the secrets of the penicillin molecule had begun; the British researchers would join the effort, but the initiative, and most of the funding,
was now American.

The penicillin war proceeded on several fronts. The chemical engineers
working for pharmaceutical companies designed systems for scaling up the
incubation of the mold and the extraction of the powder. Within a year,
several companies had opened massive "bottle plants," capable of handling
100,000 one-liter bottles at a time, each containing 200 cubic centimeters
of corn steep liquor. Eighteen months earlier Florey's group at Oxford was
obtaining one to two units of penicillin for every cubic centimeter of *Penicillium* it cultured; because of the improved medium being used in the
United States and because of the improved extraction systems that chemists
and chemical engineers had developed, the bottle plants were obtaining
1,500 units for every cubic centimeter of culture.

In March 1942, a woman in New Haven, Connecticut who was close to
death from a staphylococcus infection became the first American to receive
and be cured by *commercially produced* penicillin. A year later, the annual
rate of American production was 2 billion units, and 100 patients had been
treated. In the summer of 1943, the commanders of the Allied forces began making plans for a massive invasion of Europe, an invasion that was
going to be both difficult and bloody. Among other things that would be
needed were vastly increased supplies of penicillin. In the late summer of
1943, OSRD, the War Production Board, and the Medical Research Council decided to appoint a penicillin "czar" who would be given unusual powers to command materials and unusual resources with which to contract for
research. Production soared: 684 billion units in the first half of 1944,
2.489 trillion units in the second, 7.5 trillion units by the time peace was
declared in August 1945. The price also fell, from $200 to $35 per million
units. (A few years after the war, with continued improvements in the production process, the price fell as low as $.50 per million units.) The OSRD
had succeeded, both according to plan and beyond its wildest dreams;
penicillin had saved more lives than any medication previously devised.

All of the penicillin produced during the war was derived from a natural organism, but in the summer of 1943, OSRD had also begun an effort
to synthesize the antibiotic ingredient. As a first step, biochemists had to

Pharmaceutical companies had to build entirely new facilities for manufacturing penicillin very quickly in the closing years of World War II. (Courtesy Library of Congress.)

crystallize a pure sample of the active ingredient in the *Penicillium* broth so that they could determine what elements the penicillin molecule contained and how they were arranged. In the summer of 1943, a group of chemists working under Oskar Wintersteiner at the G. Squibb Company succeeded in crystallizing pure penicillin. Shortly thereafter, Roger Adams, a biochemist who was head of a very large research group at the University of Illinois and editor of a very prestigious biochemistry journal (some of his colleagues called him the pope of biochemistry), was put in charge of the synthesis program. Ten companies, with the largest and best organized research laboratories, were given government contracts to pursue research, and four academic chemists, who had previously had appointments at universities and medical schools, were recruited as consultants.

Once again, the penicillin molecule proved recalcitrant; the synthesis of penicillin was not accomplished for another fourteen years. During the war years, the constituent elements of the molecule were identified and its structure was determined, but before anything more could be done, the war ended and the penicillin crash program drew to a close. Academic chemists left government service and returned to their laboratories and to the problems they had dropped when the war broke out. Most of the pharmaceutical companies lost interest in penicillin synthesis, partly because their fermentation plants were in place and functioning well and partly because it was clear that no developments made during the war under OSRD contracts were going to receive patent protection.

During the war, John Sheehan had worked on penicillin synthesis as an employee in the research laboratories of the Merck company. After the war, he returned to an academic position at MIT but continued to struggle with the synthesis of penicillin because he believed that, once synthesized, the penicillin molecule could be modified so as to be targeted against specific bacteria. Sheehan's research was funded partly by MIT and partly by small grants from a pharmaceutical company. He discovered that it was relatively easy to create a linear molecule out of the constituent elements of penicillin, but that it was exceedingly difficult to get that molecule to form into a ring. He didn't solve the problem until 1957 and the solution, by his account, depended on happening across an article in a professional journal that suggested to him that one group of chemicals—the carbodiimides—had just the characteristics that were needed to get the reaction to proceed. Sheehan's synthesis opened a whole new era in penicillin research and manufacturing because, as he had suspected, the synthesis of the basic compound was just the first step in the creation of variations that would adapt an all-purpose medication to several specific situations.

The history of penicillin reveals, just as the history of hybrid corn did, that there is a continuum—not a neat distinction—between science and technology. Florey, Chain, Fleming, Coghill, Heatley, and Sheehan all succeeded in discovering something, and each discovery had a specific practical meaning for human health. They were able to make these discoveries partly because they were skilled at experimental procedure and partly because, due to their training in the sciences, they managed to make connections between various phenomena. Yet at each step of the way, each of those scientists was also exploring applications. Fleming injected some of the mold broth into mice to see if it would kill them; Florey injected it into humans to make sure that it was safe; Chain searched the literature to discover what other substances could be used in the same way Florey hoped to use penicillin; Heatley knew he had to find a technique for purifying penicillin that could, potentially, be scaled up; Sheehan took out a patent because he knew that more than one company was going to be interested in scaling up his synthesis. The ultimate goal (indeed, in the case of Florey, Chain and Heatley, the *immediate* goal) was to alleviate the pain and suffering that some bacteria can inflict on human beings.

In addition, the history of penicillin reveals that, more often than not, successful technoscience requires a great deal of money, plus the cooperation of many different kinds of people and many different social institutions; technoscience, to put the matter another way, is very often "big science." Commercial production of penicillin and its artificial variants involved research physicians, biochemists, chemical engineers, and agricultural scientists. It also involved government agencies, private companies, philanthropic foundations, and universities. In addition, it was an international effort, which may have culminated in the United States but depended nonetheless on discoveries made by people trained in Britain, Australia, and Germany. Finally, technoscience depends on skillful managers—the people who make connections between all these individuals and institutions—as much as, or perhaps more than it depends on the individual genius of researchers.

Some people object to big science precisely because of that latter feature, because it is a team effort rather than a solitary pursuit. Others people object to it because frequently that team effort has been directed toward military or destructive goals; the atomic and hydrogen bombs, for example, and the intercontinental ballistic missile were all products of big science. The history of penicillin reminds us, however, that while there were many lives destroyed and put in peril by big science, there were also many that were saved and eased by the same kind of enterprise.

Like hybrid corn, some of the impact of penicillin, and the hundreds of other antibiotics based on it, has been unexpected and ironic. Everyone who labored so mightily to create penicillin hoped that it would lower death rates, but none of the researchers anticipated that inexpensive antibiotics, distributed all over the globe, would so profoundly lower death rates that, in many places, severe overpopulation, overcrowding, and environmental degradation would result. Similarly, few of the researchers could have guessed that their work would profoundly affect familial relationships. Yet when physicians were able to use antibiotics to reduce the severity of such common and widespread diseases as pneumonia and tuberculosis, mothers and wives began to realize that their home-nursing duties were no longer as onerous as they had once been. The end result was that married women began to take full-time jobs outside their homes without fearing that they were threatening the lives of their children and husbands. Antibiotics were not the only causes of such complex social phenomena as overpopulation and married women joining the workforce, but they weren't insignificant causes either, and they were certainly unexpected.

Finally, no one working in antibiotic research in the 1940s and 1950s envisioned what we now know grievously to be the case. Bacteria mutate. An antibiotic, by killing off old forms of a bacterium, makes it easier for newer, antibiotic-resistant forms to survive. Each new generation of bacteria can therefore require a new generation of antibiotics—and in the lag time between the natural creation of the one and the artificial creation of the other, people can become dangerously, even fatally, ill. The bacterial environment, to put the matter another way, is profoundly altered by antibiotics, altered in ways that threaten human life. This means that in the antibiotic wars there are only temporary victories: hard-won to be sure, and probably worthwhile, but never permanent.

The Birth Control Pill

Because it affected the most intimate and personal aspects of people's lives, the birth control pill (often referred to simply as *the* Pill) was, unquestionably one of the most controversial technologies of its day. Indeed, one indicator of its revolutionary impact is the fact that people who were born after the controversy had died down have a hard time imagining what all the fuss was about.

Like hybrid corn, the birth control pill was the product of what might be called little technoscience. No governmental agencies of any kind were committed to its development, which meant that the number of people involved was smaller and the amount of money spent was smaller. Aside from that, however, the development of the Pill was as much a complex, managed technoscientific enterprise as the development of penicillin.

In the latter years of the nineteenth century, several European medical researchers had become interested in certain chemicals that circulated in the blood. Apparently, these chemicals, released by some of the body's internal organs, were able to affect the behavior of other organs; by 1905, the whole class of such chemicals had been given the name "hormone," from a Greek word meaning "to incite to activity."

A series of experiments performed on laboratory and domestic animals in the early decades of the twentieth century convinced medical researchers that some of these hormones affected both the development of secondary sex characteristics and sexual behavior: these substances came to be called sex hormones. By 1928, two American physicians, George Corner and Willard M. Allen, having carefully studied all the changes that occur in mammalian ovaries during ovulation, discovered that there was such a substance in the corpus luteum (the vessel formed out of the follicle after an egg has been released). This substance seemed to prevent the ripening of additional eggs; it also seemed to increase and alter the lining of the uterus to prepare for implantation of eggs and the commencement of pregnancy. They named this substance "progesterone," from the Greek words meaning "in favor of" and "bearing offspring."

Subsequently, researchers began to understand that the complex feedback relations between progesterone and three other hormones (estrogen; follicle-stimulating hormone, or FSH; and luteinizing hormone, or LH) governed the human menstrual cycle. They also reasoned that progesterone, because it prepared the uterus for pregnancy, might be used as a therapy for women whose pregnancies regularly miscarried or for women who struggled with other menstrual disorders. Not surprisingly, some medical scientists also realized that, since progesterone prevented ovulation, it had potential as a contraceptive. Since women do not generally ovulate when they are pregnant and since maintenance of a pregnancy seemed to depend on maintenance of a certain level of progesterone, artificially establishing that same level would actually prevent a pregnancy by preventing ovulation, by "fooling" the body into "thinking" it was already pregnant. Fuller Albright, an endocrinologist at Harvard Medical School, referred to this in 1945 as "birth control by hormone therapy."[1]

Betwixt cup and lip there were, however, two enormous problems, one technical, the other social. Supplies of either human or animal progesterone were extremely limited. There was no known way either to extract large quantities of it from living tissue or to synthesize it, which meant, among other things, that the limited amount available was being used only to treat those patients who could afford the very high price. Furthermore, all published discussions of birth control—including professional research reports about contraceptives—had to be very carefully circumscribed because in many states dissemination of information about birth control was illegal. Albright hid his suggestion about birth control by hormone therapy in a technical article about treating menstrual disorders. If someone unfriendly to birth control had noticed what he had written and had brought the matter to the attention of the authorities, Albright could have been fined and jailed.

Fortunately or unfortunately (depending on how one feels about the Pill), there were two women in the United States who had already done battle with the authorities over birth control and who had no reservations about continuing to do so. One was Margaret Sanger, a trained nurse who was devoted to the cause of birth control. Sanger and her sister had opened the country's first birth control clinic in Brooklyn in 1916; both women had spent time in jail as punishment. By the late 1940s, Margaret Sanger was internationally famous for her efforts to spare American women from unwanted pregnancies. The Birth Control League, which she and several others had founded in 1915, had been transformed into the Planned Parenthood Federation of America, which coordinated the activities of more than two hundred local clinics. And the Supreme Court case that she had initiated (by having someone mail her a diaphragm from overseas) had been won in her favor, thereby voiding the federal laws that had made it illegal to use federal facilities such as the postal service to disseminate birth control devices or information. Sanger regarded herself as a rebel with a cause; she could afford to continue to be rebellious because she had been left a considerable fortune by her second husband.

As wealthy as she was, Sanger did not have enough money to sponsor a research and development project; however, by 1951 there was someone else who did. Katherine Dexter McCormick was the heir to one large fortune (her father's) and one enormous one (her late husband's). As a young woman, she had earned a bachelor of science degree in biology from MIT (at a time when this was a very unusual thing for a young woman to do) and had become an ardent suffragette. During the 1920s and 1930s, she cooperated with Margaret Sanger by arranging to smuggle European birth control devices into this country so that they could be dispensed in clinics. As a young married woman, she had struggled tragically with her husband's collapse into schizophrenia—and had vowed not to have children by him. While he was still alive, she had gained control of his financial affairs and donated large sums of money to neuropsychiatric research. After he died in 1947, she sold several of their large estates and decided to spend her money on her own political causes. Birth control was one of them.

In 1951, McCormick asked Sanger what, at that point in time, the birth control movement most needed. Sanger responded that it needed a safe, inexpensive contraceptive that would combat over-population and that women, rather than men, could control. Sanger then introduced McCormick to Gregory Pincus, a physiologist, who was one of the world's leading experts on the mammalian egg. Pincus was the head of a private, not-for-profit scientific research institute, the Worcester Foundation for Experimental Biology, which he had founded with a colleague in 1944 when he was denied tenure at Harvard. The Worcester Foundation had struggled along, doing biomedical research under contract to a variety of organizations (including Planned Parenthood), but in 1951, it was close to bankruptcy. McCormick made Pincus an offer: she would foot the bill if he would turn his foundation to the task of developing birth control by hormonal therapy. Between 1951 and 1959, Pincus acted, in fact if not in title, as the manager of the Pill project.

Unbeknownst to Pincus, in the early 1940s, a maverick American chemist, Russell Marker, had developed a process for converting a steroid found abundantly in some plants, sapogenin, into progesterone. Marker was a professor at Penn State at the time, and his research costs were being underwritten by a pharmaceutical company. When he discovered that one of the world's richest sources of sapogenin was a plant that grew only in the deserts of Mexico, Marker proposed to executives of the pharmaceutical company that they create a laboratory for extracting sapogenin and making progesterone in Mexico. When they turned him down, Marker, aware of the great therapeutic demand for progesterone, promptly quit his job and went into business with a Mexican entrepreneur.

Eventually Marker was forced out of that company—and another that he helped found—but his process (which he had refused to patent, making it available to anyone) created the foundation for what soon became a flourishing Mexican industry: the synthesis of various steroids, including progesterone, from desert plants. In 1949, needing an expert chemist, one of those companies, Syntex, hired Carl Djerassi, a young American who had recently completed his doctoral dissertation. Within a few months, Djerassi had figured out a way to make cortisone, which was being used extensively as a treatment for arthritis, from the chemicals in one of those plants. A few months later, he took on another interesting project: making a version of progesterone that would work better than progesterone.

The problem with synthetic progesterone was that it had to be administered by injection and in very large doses because it was degraded and rendered ineffective when it passed through a patient's liver. Djerassi thought up several different ways to try modifying progesterone, and after several months, he found a molecule that proved to be highly active when *fed* to laboratory animals. After a physician tested the compound successfully on three women volunteers who were suffering from excessive menstrual bleeding, Djerassi applied for a patent on November 22, 1951, on *norethindone*, an orally active variant of progesterone. In those very same months, Frank B. Colton, chief research chemist for the G. D. Searle Company, was also trying to alter the progesterone molecule so that it could be taken orally; Colton's successful version, to which Searle gave the trade name Enovid was not patented for another year and a half.

In April 1951, when Pincus turned his laboratory to contraceptive research, he knew nothing about these developments; the first task he set his research associates, Min-Chueh Chang and Anne Merrill, was to find out whether regular injections of progesterone would actually *prevent* female laboratory animals from becoming pregnant. Not long after starting these trials, Pincus met a Boston gynecologist, John Rock, who was testing out a new therapy for infertility. Rock had been regulating his patients' menstrual cycles by administering doses of estrogen and progesterone every day for several months—deliberately fooling their bodies into thinking they were pregnant—and then taking his patients off the medication and letting their natural menstrual cycles return. In effect, Rock was already testing a chemical contraceptive on human beings and had discovered—in pursuit of a cure for infertility—that it worked.

Rock's work meant that Pincus could stop fussing with laboratory animals, but he couldn't know whether a chemical contraceptive would be safe for long-term use until he conducted a clinical trial. He also knew that he couldn't proceed to a clinical trial (and also satisfy McCormick and Sanger's requirement that the new contraceptive be cheap) until he located, or developed, a form of the hormone that didn't have to be given by daily injection. A series of inquiries to pharmaceutical companies yielded the information that two of them—Searle and Syntex—would be happy to sell him samples of the orally administered progestins that Djerassi and Colton had just recently developed.

Clinical trials of progestin could now proceed. Rock agreed that he would very quietly test oral progestin on fifty women volunteers—Massachusetts state law still made it illegal to discuss birth control publicly. Pincus announced the successful results (not one of the fifty women had ovulated) at a meeting of the International Planned Parenthood League in October 1955. A few months earlier, he had begun to try to locate a place where an even larger, longer-term clinical trial could be organized involving poor women (who were, after all, McCormick and Sanger's target population). Pincus and Rock also wanted women volunteers who could be counted on both to want a contraceptive and eventually to want to go off it since the two needed to be assured that the effect would be reversible and that subsequent babies would not be harmed by their mothers' chemical regimen. Not surprisingly, they also wanted a place in which there was no law banning discussion of birth control.

The places they chose were Puerto Rico, Haiti, and Mexico City. The clinical trials were paid for by McCormick, but organized with the help of local physicians, nurses, and social workers who were already birth control advocates. Before the trials began, Pincus and Rock decided to use Searle's progestin, Enovid, because it appeared to have fewer side effects. While the Puerto Rican trials were in progress, they discovered that the reason for this was that several Enovid batches were contaminated with small amounts of estrogen; thereafter, all formulations of the Pill contained trace amounts of estrogen. By 1957, the team had carefully monitored 25,421 monthly cycles of pill administration and had calculated a contraceptive failure rate of 1.7 percent (by comparison, the failure rates for diaphragms in those years were 33.6 percent and for condoms, 28.3 percent). McCormick and Sanger were, of course, delighted.

But the pharmaceutical companies, even Searle and Syntex, were hesitant. In 1957, Searle and applied for and received FDA approval to market Enovid as therapy for menstrual disorders and repeated miscarriages. Two years later, 500,000 American women were taking the Pill every day, far more than the number that were thought to suffer from such disorders. Clearly some physicians knew, and had told their patients, that Enovid could be used as a contraceptive. Despite this unexpected phenomenon, the companies perceived that the political risks involved in marketing oral progestins as contraceptives were very great. The Catholic Church had clearly indicated that it would use the laws that were then on the books, as well as its considerable political clout, against any effort to popularize birth control.

Eventually Searle capitulated, both to Pincus's remonstrations and to its own potential profits. In 1959, it applied to the FDA for permission to advertise Enovid as a contraceptive. The FDA took a long time examining the reports of all the clinical trials; this would be, after all, the first medication intended to be taken by healthy people, who would be taking it for very long periods of time, possibly for decades. Finally, on May 11, 1960, the Pill was formally declared safe for prolonged and regular use as a contraceptive.

American married women were enthusiastic; so were many of their physicians. By 1966, researchers estimated that 56 percent of all married American women under the age of twenty and 25 percent of those under the age of forty-five were using the Pill, a total of at least 5 million women. The number of users continued to rise until the middle 1970s when some feminist groups became concerned about the relationship between the Pill and the incidence of other diseases, and an alternative contraceptive, the IUD, became widely available.

By that time, however, as far as the opponents of birth control were concerned, the Pill had already done far too much social damage. In the decade between 1965 and 1975, a multifaceted revolution in American sexual behavior occurred. Scholars today debate the extent to which that revolution can be attributed to the oral contraceptive, but even those who doubt that it was the sole causative factor admit that its role was far from insignificant.

Ironically, the revolution turned out not to be quite the one McCormick and Sanger had anticipated. The most likely users of oral contraceptives turned out not to be poor women in undeveloped countries but fairly well-educated, middle-class women in relatively wealthy countries. The problems of overpopulation were not solved by the Pill. Even where governments were eager to try contraception, the difficulty of distributing the requisite daily doses and educating illiterate women about how to take them proved insurmountable.

For reasons no one can quite explain, the oral contraceptive made it possible for Americans to discuss birth control, which had previously been a taboo subject in public and in polite conversation. Married women learned about it from their friends and began demanding it from their doctors. As more and more women discovered that there was a simple and reasonably safe way to control their fertility themselves, more were encouraged, and found the time, to assert themselves in other fields of endeavor as well. And as they discovered that they could control their fertility without constraining their sexual desires, more and more women began admitting that they had such desires and that they wanted to satisfy them.

In short order, the rumor spread to unmarried women as well. Some traveled to distant cities and pretended to be married in order to get a physician to write a prescription; others soon learned which nearby physicians were willing to look the other way, ignoring the absence of a wedding ring. At one college after another, infirmary physicians began distributing the Pill to students who asked for it; at one Planned Parenthood clinic after another, physicians and nurses stopped insisting that patients demonstrate their marital status before being examined. In 1961, two leaders of Planned Parenthood in Connecticut informed the state police that they were going to publicly open a birth control clinic. When they were duly arrested and fined, they went to court to challenge Connecticut's restrictive law. Four years later, when the Supreme Court decided (in *Griswold v. Connecticut*) that the law was unconstitutional, it thereby invalidated all the restrictive state laws about dispensing birth control devices and information to married women. In 1972, when it resolved a related case (*Eisenstadt v. Baird*) the Supreme Court finally extended the right to obtain birth control to unmarried women as well.

The rebellious generation that came of age in the late 1960s therefore reached sexual maturity at a time when birth control could be publicly discussed. In addition, all kinds of contraceptives were fairly easy to obtain, including one—the oral contraceptive—that did not interfere with sexual pleasure and did not require a man's cooperation to be effective. The two strongest disincentives to premarital sexual behavior, the fear of being disgraced and the fear of getting pregnant, had evaporated. Millions of young people, not surprisingly, took advantage of the opportunities thus presented. In the process, they totally transformed the nation's tolerance for explicit sexual discussion and for premarital sexual activity. This was an outcome that only a very few birth control advocates had ever expected, as completely unintended as some of the consequences of penicillin and hybrid corn.

Even as venerable an institution as the Roman Catholic Church was rocked to its foundations by the oral contraceptive. When the 1960s dawned, the only birth control technique that the Church approved—and only lukewarmly—was the rhythm method. American Catholics were not, however, unanimous in their adherence to this teaching and the advent of the Pill made them even more disgruntled. By 1964, some 60 percent of the Catholics surveyed in a national poll said that they were dissatisfied with their Church's position; and two-thirds of the women were willing to admit that they were using some contraceptive method other than rhythm; a third of those were using the Pill.

All over the country, parishioners were putting pressure on priests to deviate from the Church's rules and in fairly short order the American bishops began putting pressure on the Roman hierarchy to pay attention to what was happening among their congregants. In the spring of 1965, Pope Paul VI convened an extraordinary meeting in Rome; twenty-one clergymen and thirty-four lay people (including three married couples) were asked to give the pope advice on whether the Church ought to change its position on birth control. After weighty deliberations, a majority of the commission concluded that it should. The pope took several years to consider those recommendations, but in 1968, in the encyclical *Humanae Vitae*, he announced that he was rejecting it completely. The pope's decision was a shattering blow from which, many Catholics argue, the American Church has still not recovered. The number of Catholics (both married and unmarried) using birth control continued to rise and the number attending mass began to plummet; 1967 was the last year in which the majority of American Catholics said that they believed that the pope's authority was paramount.

By 1970, both birth control and abortion had become discussable—and protestable—subjects. These demonstrators are gathered across the street from St. Patrick's Cathedral (Roman Catholic) in Manhattan in 1970. (Courtesy Temma Kaplan.)

Such was the power of sexual desire and of the little pill that finally made it safe for large numbers of women to consider succumbing. Like the research and development project that produced hybrid corn, it had been a relatively small undertaking, engaging the efforts of two wealthy and determined women, plus a handful of university scientists, a few pharmaceutical companies, a not-for-profit research institute, a few thousand volunteer patients, and one federal regulatory agency. Like the project that produced commercial penicillin, it was managed by a few clever and persistent individuals who were able to manipulate several social institutions so as to achieve their objectives in a relatively short period of time—eight years from start to finish. And whatever its faults and its unintended consequences may be, most people whose lives were altered by the product of that technoscientific undertaking continued to approve of it.

Conclusion

In recent decades, the power of biomedical technoscience has increased, partly because so many people believe that its past achievements were positive and partly because new frontiers were opened by the technoscience that developed after James Watson and Francis Crick discovered the double helical structure of DNA in 1952. The same techniques of selective breeding that produced hybrid corn have since yielded new forms of many traditional grains and vegetables and garden flowers. New breeding techniques involving the micromanipulation and splicing together of chromosomes (called recombinant DNA techniques) have given us such new life-forms as bacteria that can digest oil spills and fully ripe tomatoes that can be harvested mechanically. Manipulation of the same hormones that are in the oral contraceptive now make it possible for physicians to harvest eggs from women's ovaries, the fundamental starting point for in vitro fertilization. And the accumulated knowledge about how to manage technoscientific projects has made it possible for molecular geneticists to create the Human Genome Project—an international effort, coordinating the activities of universities, medical schools, government agencies, and businesses—to identify the location and chemical structure of every active gene in the nuclei of human cells. Some people hope that when the genome project is finished it will be possible to cure every disease that afflicts human beings; other people fear that when the project is finished technoscientists will be able to manipulate other people as easily as they can now manipulate bacteria and grasses.

The history of hybrid corn, commercial penicillin, and the birth control pill has several things to teach us as we contemplate the future that biomedical technoscience may or may not create. First, no new technology has ever been the unalloyed blessing that its advocates say it is—or the unalloyed curse that its opponents insist it is. All technological changes have unintended and unexpected social and ethical outcomes, few of which have been predicted by even the best of experts. Second, since at least 1945 (and possibly earlier) even the best of these experts have not been disinterested. Because of the nature of technoscience, virtually all researchers earn their livings, in one way or another, on some goal-directed project, which means that they all have some vested interest in the politics and policies that will affect those projects. Finally, some people will be affected positively and others negatively by the outcome of any technological change; indeed, the same person can be affected negatively or positively depending on which of several possible social roles that person happens to be playing. Thus the most important lessons that the history of technology can teach us are these: technology is complicated; so is life; so is the history that we create as we live our lives together. Those us who care about the future need to pay very close attention to these complexities. Every technological change has profound social and ethical consequences, and we cannot rely on experts to make wise decisions about those consequences for us.

Note

1. Fuller Albright, "Disorders of the Female Gonads," in John H. Musser, ed., *Internal Medicine: Its Theory and Practice* (Philadelphia, 1945), p. 966, as quoted in Bernard Asbell, *The Pill: A Biography of the Drug that Changed the World* (New York: Random House, 1995), p. 18.

Suggestions for Further Reading

Bernard Asbell, *The Pill: A Biography of the Drug that Changed the World* (New York, 1995).

Stuart S. Blume, *Insight and Industry: On the Dynamics of Technological Change in Medicine* (Cambridge, MA, 1991).

Robert Bud, *The Uses of Life: A History of Biotechnology* (Cambridge, 1993).

Lawrence Busch, William B. Lacey, Jeffrey Burckhardt, and Laura R. Lacy, *Plants, Power and Profit: Social, Economic and Ethical Consequences of the New Biotechnologies* (Cambridge, MA, 1991).

Albert L. Elder, ed., *The History of Penicillin Production* (New York, 1970).

Deborah Fitzgerald, *The Business of Breeding: Hybrid Corn in Illinois, 1890–1940* (Ithaca, NY, 1990).

Stephen S. Hall, *Invisible Frontiers: The Race to Synthesize a Human Gene* (New York, 1987).

Jim Hightower, *Hard Tomatoes, Hard Times* (Cambridge, MA, 1978).

David Hounshell and John Kenly Smith, Jr., *Science and Corporate Strategy: Du Pont R&D, 1902–1980* (Cambridge, MA, 1988).

Joel Howell, *Technology in the Hospital: Transforming Patient Care in the Early Twentieth Century* (Baltimore, 1995).

Sheldon Krimsky, *Genetic Alchemy: The Social History of the Recombinant DNA Controversy* (Cambridge, MA, 1982).

Jonathon Liebenau, *Medical Science and Medical Industry: The Formation of the American Pharmaceutical Industry* (Baltimore, 1987).

Alan Marcus, *Agricultural Science and the Quest for Legitimacy* (Ames, IA, 1985).

Jean L. Marx, *A Revolution in Biotechnology* (New York, 1989).

James Reed, *From Private Vice to Public Virtue: The Birth Control Movement and American Society since 1830* (New York, 1978).

John C. Sheehan, *The Enchanted Ring: The Untold Story of Penicillin* (Cambridge, MA, 1982).

John Swann, *Academic Scientists and the Pharmaceutical Industry: Cooperative Research in Twentieth-Century America* (Baltimore, 1988).

Arnold Thackray, Ed., *Private Science: The Biotechnology Industry and the Rise of Contemporary Molecular Biology* (Philadelphia, 1994).

Trevor I. Williams, *Howard Florey, Penicillin and After* (Oxford, 1984).

Edward Yoxen, *The Gene Business: Who Should Control Biotechnology?* (New York, 1983).

Index

Note: Many objects, companies, and places are mentioned just once in this text; to keep the index from being overly long, such entries have been consolidated. To locate information about an object which is not indexed, consult the generic category for the object (e.g., Tools, Electronic components, etc.). To locate information about a company not indexed, consult entries for the appropriate products or services (e.g., Railroad, companies; Automobile industry, etc.). Many rivers are indexed as subentries in the entry, Rivers; places can usually be found as subentries under the state in which they are now located (e.g., for Schuykill River look in Rivers; for Waltham, MA look in Massachusetts).